The
Carnivorous
Dinosaurs

LIFE OF THE PAST
James O. Farlow, Editor

The Carnivorous Dinosaurs

Edited by Kenneth Carpenter

Indiana University Press
Bloomington & Indianapolis

This book is a publication of

Indiana University Press

601 North Morton Street

Bloomington, IN 47404-3797 USA

http://iupress.indiana.edu

Telephone orders   800-842-6796

Fax orders   812-855-7931

Orders by e-mail   iuporder@indiana.edu

© 2005 by Indiana University Press

The paper used in this publication meets
the minimum requirements of American
National Standard for Information
Sciences—Permanence of Paper for
Printed Library Materials, ANSI
Z39.48-1984.

Manufactured in the United States of
America

**Library of Congress Cataloging-in-
Publication Data**

The carnivorous dinosaurs / edited by
    Kenneth Carpenter.
        p. cm. — (Life of the past)
    Includes bibliographical references
    (p.) and index.
    ISBN 0-253-34539-1 (cloth : alk.
    paper)
    1. Saurischia—Juvenile literature.
        I. Carpenter, Kenneth, date-
        II. Series.
    QE862.S3C27 2005
    567.912—dc22
                        2004023225

1 2 3 4 5 10 09 08 07 06 05

This volume on theropods is dedicated to

William Buckland, 1784–1856,

first Professor of Geology at Oxford University, England. Buckland's careful analysis of large bones from the Jurassic near Oxford led him to describe and name the first theropod in 1824, making him the first individual to give a scientifically valid name to a taxon of dinosaur. He led the way. . . .

Buckland, W. 1824. Notice on the Megalosaurus or great fossil lizard of Stonesfield. *Transactions of the Geological Society of London*, ser. 2, 1: 390–396.

CONTENTS

Contributors

Ronan Allain, Laboratoire de Paléontologie, UMR 8569 du CNRS, Muséum National d'Histoire Naturelle, 8 rue Buffon, 75005 Paris, France.

Rinchen Barsbold, Geological Institute, Mongolian Academy of Sciences, Ulaan Baatar 11, Mongolia.

Kenneth Carpenter, Department of Earth Sciences, Denver Museum of Natural History, 2001 Colorado Blvd., Denver, CO 80205 USA.

Karen Cloward, Western Paleontological Laboratories, 2929 Thanksgiving Way, Lehi, UT 84043 USA.

Rodolfo A. Coria, CONICET—Dirección Provincial de Cultura—Museo Carmen Funes, Av. Córdoba 55 (8318) Plaza Huincul, Neuquén, Argentina.

Philip J. Currie, Royal Tyrrell Museum of Palaeontology, Box 7500, Drumheller, Alberta T0J 0Y0, Canada.

Peter M. Galton, College of Naturopathic Medicine, University of Bridgeport, Bridgeport, CT 06601 USA.

Robert Gay, Department of Geology, Northern Arizona University, 400 E. McConnell Dr. #11, Flagstaff, AZ 86001 USA.

Donald M. Henderson, Department of Biological Sciences, University of Calgary, Calgary, Alberta T2N 1N4, Canada.

Dong Huang, Heyuan Museum of Guangdong Province, Heyuan 517000, China.

James I. Kirkland, Utah Geological Survey P.O. Box 146100, Salt Lake City, UT 84114 USA.

Yoshitsugu Kobayashi, Fukui Prefectural Dinosaur Museum, 51–11 Muroko, Terao, Katsuyama, Fukui 911-8601, Japan.

Eva B. Koppelhus, Royal Tyrrell Museum of Palaeontology, Box 7500, Drumheller, Alberta T0J 0Y0, Canada.

Peter Larson, Black Hills Institute of Geological Research, Inc., P.O. Box 643, Hill City, SD 57745 USA.

Junchang Lü, Department of Geological Sciences, Southern Methodist University, Dallas, TX 75275.

Lorrie A. McWhinney, Department of Earth Sciences, Denver Museum of Natural History, 2001 Colorado Blvd., Denver, CO 80205 USA.

Clifford Miles, Western Paleontological Laboratories, 2929 Thanksgiving Way, Lehi, UT 84043 USA.

Ralph E. Molnar, Museum of Northern Arizona, 3101 North Fort Valley Road, Flagstaff, AZ 86001 USA.

Nate Murphy, Philips County Museum, Highway 2 East, Malta, MT 59538 USA.

John H. Ostrom, Vertebrate Paleontology, Peabody Museum of Natural History, New Haven, CT 06520 USA.

Gregory S. Paul, 3109 N. Calvert St., Baltimore, MD 21218 USA.

Licheng Qiu, Institute of Cultural Relics and Archaeology of Guandong Province, Guangzhou 510075, China.

J. Keith Rigby Jr., Department of Civil Engineering and Geological Sciences, University of Notre Dame, Notre Dame, IN 46556 USA.

Bruce Rothschild, Arthritis Center of Northeast Ohio, 5500 Market, Youngstown, OH 44512.

Christopher B. Ruff, Center for Functional Anatomy and Evolution, Johns Hopkins University, School of Medicine, Baltimore, MD 21205 USA.

Leonardo Salgado, CONICET—Museo de la Universidad Nacional del Comahue, Buenos Aires 1400, (8300) Neuquén, Argentina.

Frank Sanders, Department of Earth Sciences, Denver Museum of Natural History, 2001 Colorado Blvd., Denver, CO 80205 USA.

Julia T. Sankey, Department of Physics, Physical Sciences, and Geology, California State University, Stanislaus 801 West Monte Vista Ave. Turlock, CA 95382 USA.

Judith A. Schiebout, Museum of Natural Science, Louisiana State University, Baton Rouge, LA 70803 USA.

David K. Smith, Earth Science Museum, Brigham Young University Provo, Utah 84602 USA.

Barbara R. Standhardt, Museum of Natural Science, Louisiana State University, Baton Rouge, LA 70803 USA.

Kathy Stokosa, Museum of the Rockies, Montana State University, 600 W. Kagy Blvd. Bozeman, MT 59717 USA.

Darren H. Tanke, Royal Tyrrell Museum of Palaeontology, Box 7500, Drumheller, Alberta, T0J 0Y0 Canada.

François Therrien, Center for Functional Anatomy and Evolution, Johns Hopkins University, School of Medicine, Baltimore, MD 21205 USA.

David Trexler, Timescale Adventures Research and Interpretive Center, Box 786, Bynum, MT 59419 USA.

Kelly Wicks, Timescale Adventures Research and Interpretive Center, Box 786, Bynum, MT 59419 USA.

Douglas G. Wolfe, Mesa Southwest Museum, 53 N. MacDonald St. Mesa, AZ 85012 USA.

Lowell Wood, University of California Lawrence Livermore National Laboratory, Livermore, CA 94551 USA.

Acknowledgments

This volume was made possible by the support of Jim Farlow and Bob Sloan, Indiana University Press. Thanks also to Jane Lyle, Managing Editor, and Carlotta Shearson, copyeditor. Finally, thanks to all the authors for their contributions; I hope they are pleased.

Theropod dinosaurs remain one of the most prolific areas of dinosaur research in part because of the theropod-bird link. There is a growing body of evidence that "non-avian theropods" share many features that have traditionally been thought to be restricted to "avian theropods." This topic has been explored in several recent works (e.g., Dingus and Rowe 1997; Gauthier and Gall 2001; Paul 2002); this book is not one of them. Instead, this volume brings together eighteen papers on a diversity of topics organized into three sections. Part 1, "Theropods Old and New," presents morphological details for understanding theropod systematics. The individual chapters are arranged in geochronological order. One new theropod is named, and new information is presented for several others previously described. Part 2, "Theropod Working Parts," focuses on specific regions of theropod anatomy and biomechanics. Part 3, "Theropods as Living Animals," examines various lines of evidence that reveal information about theropods as once living creatures.

References Cited

Dingus, L., and Rowe, T. 1997. *The Mistaken Extinction: Dinosaur Evolution and the Origin of Birds*. New York: W. H. Freeman.

Gauthier, J., and L. Gall (eds.). 2001. *New Perspectives on the Origin and Early Evolution of Birds*. New Haven, Conn.: Yale Peabody Museum Special Publications.

Paul, G. S. 2002. *Dinosaurs of the Air*. Baltimore: Johns Hopkins University Press.

I. Theropods
Old and New

1. Tibiae of Small Theropod Dinosaurs from Southern England

From the Middle Jurassic of Stonesfield near Oxford and the Lower Cretaceous of the Isle of Wight

PETER M. GALTON AND RALPH E. MOLNAR

Abstract

The distal part of a small theropod tibia from the Stonesfield Slate (Taunton Limestone Formation, middle Bathonian, Middle Jurassic) of Oxfordshire is described. This tibia probably represents a plesiomorphic tetanuran. Another tibia from the Lower Cretaceous of the Isle of Wight is of normal theropod form, representing a coelurosaur, rather than a left tibia with a form totally unique for any dinosaur as suggested recently.

Introduction

The distal end of a small tibia from the Stonesfield Slate was located by one of us (PMG) while examining drawers of dinosaur bones in the Berlin Museum für Naturkunde. There is little information about the specimen, MB R2352 from the v. Dechen Collection (Quenstedt-Katalog, p. 136, translation: "bone fragment from Stonesfield, perhaps from the limb of a small animal"). A hand-written label by W. Janensch reads "Theropode, Coelurosaurier?, tibia, distalende" (D.-W. Heinrich, pers. comm.). Given the rarity of small dinosaur bones from these beds, or any other beds of Middle Jurassic age, this specimen is described.

Lydekker (1891) described a small tibia from the Wealden (Wessex Formation, Lower Cretaceous) of the southwestern coast of the Isle of Wight. He regarded it as a right tibia and provisionally referred it to the coelurosaur *Calamosaurus foxi*. However, Naish et al. (2001) contend that it is a left tibia, and because Lydekker (1891) misidentified the cnemial crest as the inner (medial) condyle, he inadvertently identified the posterior face as anterior and vice versa. Because the re-identification of Naish et al. (2001) results in a tibia with a form that is totally unique for any dinosaur, the problem of identifying which side of the body this tibia comes from is discussed.

Institutional Abbreviations. BMNH, The Natural History Museum [formerly British Museum (Natural History)], London; MB, Museum für Naturkunde, Humboldt Universität, Berlin; OUMNH, Oxford Museum of Natural History, Oxford; YPM, Yale Peabody Museum of Natural History, New Haven, Conn.

Stonesfield Theropod Tibia

The tibia (MB R2352) derives from the Stonesfield Slate (or "tilestones") that was mined between the seventeenth century and the early 1900s (Aston 1974). The slate is a recurrent lithological facies of laminated calcareous limestone that occurs at three levels within the Taunton Limestone Formation (Boneham and Wyatt 1993). The workable stone was restricted to an area within about a kilometer of the village of Stonesfield, 15 km northwest of Oxford in Oxfordshire, southern England (UK National Grid Ref. SP 394 173, map of quarries and section in Boneham and Wyatt 1993, figs. 1, 2; Benton and Spencer 1995, fig. 6.6; see also Arkell 1947, p. 139; Aston 1974; Cope et al. 1980; Unwin 1996).

The Stonesfield Slate was deposited during the Middle Jurassic (earliest middle Bathonian, *Procerites progracilis* Zone; Cope et al. 1980; Boneham and Wyatt 1993). This unit was probably deposited in a shallow, nearshore sea. The terrestrial fauna, which presumably lived on the nearby landmass, includes pterosaurs (Unwin 1996) and dinosaurs: a small ornithischian (probably an ornithopod; Galton 1975, 1980), a sauropod, and theropods. These theropods include the small *Iliosuchus incognitus* (Galton 1976) and the much more common large *Megalosaurus bucklandi,* the first dinosaur to be described (Lluyd 1699) and named (Buckland 1824). Small tetrapods, such as lepidosaurs, are not represented, possibly because of the winnowing action of currents. Additional information about the unit, its fauna, and the conditions of deposition is given by Benton and Spencer (1995).

MB R2351 is the distal portion of a left tibia (Fig. 1.1A–D), probably comprising no more than 20 percent of the total length; the maximum preserved length and distal width are 45.8 and 32.1 mm, respectively. The distal extremity is somewhat abraded, particularly the articular surface (Fig. 1.1C), the tibial buttress (Fig. 1.1A), and the distal part of the posterior ridge (Fig. 1.1D). This abrasion is consistent

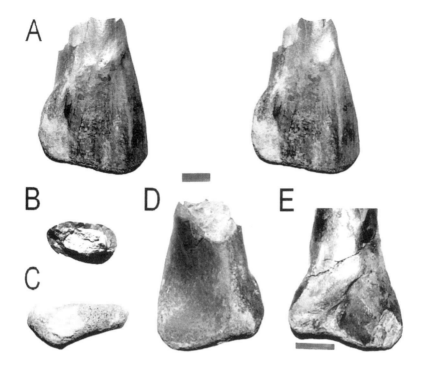

Figure 1.1. The Stonesfield distal left tibia (MB R2351) in (A) anterior, (B) proximal, (C) distal, and (D) posterior views and (E) Megalosaurus bucklandi (OUMNH J13562 from Stonesfield) distal left tibia (right in reverse) in anterior view. Scale bars = 1 cm (A–D) and 5 cm (E).

with the action of currents on the bottom of a shallow, nearshore sea. The remainder of the surface is little worn.

Description. In general form, the distal portion of the tibia (Figs. 1.1A, 1.2E) is similar to that of *Poekilopleuron bucklandii* (Fig. 1.2B; Allain 2001; Allain and Chure 2002; Eudes-Deslongchamps 1838) and those attributed to *Megalosaurus bucklandi* (Figs. 1.1E, 1.2A; Owen 1857; Huxley 1870, fig. 1). Seen anteriorly (Figs. 1.1A, 1.2E), the shaft is slightly expanded laterally to form an angular postfibular flange that is now slightly rounded by abrasion, and medially into a truncated medial epicondyle. A broad, rounded prominence, the medial buttress of Molnar et al. (1996), bounds the facet for the ascending process of the astragalus. Proximally, this buttress narrows and tapers into the anterior face of the shaft. This proximal restriction of the buttress is a feature unique to this tibia, to our knowledge found in no other theropod. The part of the anterior surface bordering this ridge is slightly impressed into the body of the bone, so that the buttress projects only very slightly (less than 1 mm) anterior to the general surface of the facet.

The posterior face of the element (Fig. 1.1D) is almost flat, but with a slight longitudinal concavity. This concavity is set off from the medial (or, more accurately, medioposterior) face by a bluntly angulate ridge, which extends proximally.

The distal surface is slightly inclined 10° from perpendicular to the axis of the shaft (Fig. 1.1A,D). This surface is very slightly concave in anterior view, unlike the tibiae of *Poekilopleuron* and those attributed to *Megalosaurus*. In these tibiae, the distal margin comprises three

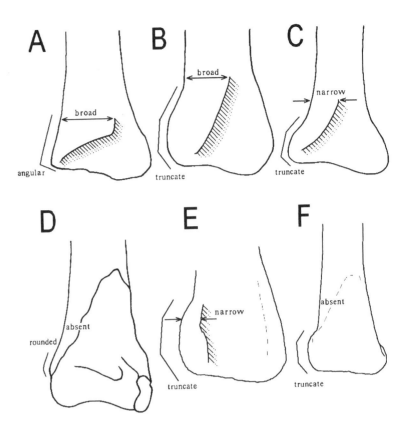

Figure 1.2. Distal left tibiae in anterior view, showing the character states used here. (A) Megalosaurus bucklandi (modified after Huxley 1870), (B) Poekilopleuron bucklandii (modified after Eudes-Deslongchamps 1838), (C) Allosaurus fragilis (modified after Madsen 1976a), (D) Gorgosaurus libratus (modified after Lambe 1917; depicted with astragalus and calcaneum in place), (E) tibia MB R2351 (see Fig. 1.1A), and (F) tibia BMNH R186 (in reverse, see Fig. 1.7A). The medial buttress is on the left of A, B, C, and E and is marked by the hatching along the margin adjacent to the astragalus. The form of the medial epicondyle is schematically depicted by a line adjacent to that condyle. Not to scale.

straight-lined segments, the medial and lateral perpendicular to the long axis of the shaft, and the central segment slightly inclined (Figs. 1.1E, 1.2A,B). The distal end is triangular in outline (Fig. 1.1C), with the anterior surface forming the longest edge and the inclined medio-posterior edge the shortest.

The anterior surface (Fig. 1.1A) suggests that the ascending process of the astragalus was moderately low and broad, extending across most of the tibial shaft. Thus, the form of the ascending process would have been unlike that of later Cretaceous groups such as ornithomimosaurs, in which the process extends completely across the shaft. However, the impression for the ascending process is less distinct than in later forms, rendering uncertain any conclusions regarding the form of this process. Laterally, a low angulation (shown dashed in Fig. 1.2E) parallels the lateral margin: this angulation may laterally delimit the impression for the ascending process or may indicate a close approach of the distal part of the fibula.

Phylogenetic Position. Molnar et al. (1996) briefly analyzed the phylogenetic distribution of certain characters of the distal end of the tibiae of theropods (Molnar et al. 1996, fig. 3; for larger figures of tibiae, see Molnar and Pledge 1980, fig. 3; Long and Molnar 1998, fig. 5). The characters used were the form of the medial epicondyle, the presence of a medial buttress, and the degree of lateral and medial expansion of the distal end of the tibia (i.e., the epicondyles, referred to by

Molnar et al. 1996 as "malleoli," but this mammalian term may be inappropriate). Additionally, the breadth and height of the astragalar ascending process were also included (Molnar et al. 1996). However, an examination of Molnar et al. 1996 (table 1) strongly suggests that these two features, at least as used there, are not independent. All ascending processes considered to be "high" are also considered to be "broad," and all processes considered to be "low" are also regarded as either "narrow" or "moderate" (in width). "Broad" as used by Molnar et al. (1996) indicates an ascending process extending across, or almost across, the entire anterior face of the distal tibia. Thus, this character is not independent of the presence or absence of a medial buttress, because this buttress restricts the ascending process from extending across the entire anterior face of the tibia. In view of these considerations, the character states of the astragalar ascending process are not used here.

Molnar et al. (1996) mapped the tibial character states onto the cladogram of Holtz (1994), modified by inclusion of the conclusions of Sereno et al. (1994). However, theropod phylogenies have been further refined since then, so the cladogram of Rauhut (2003a) is used here. Of the four tibial characters, the distal expansion of the medial and lateral margins of the Stonesfield tibia does not, of itself, contribute any information concerning its phylogenetic placement. Consequently, three of these characters are used here to estimate the phylogenetic relationships of the Stonesfield tibia (Table 1.1): the form of the medial epicondyle; the presence of a medial buttress; and the form of the medial buttress. These states are shown diagrammatically for several theropod taxa (Fig. 1.3). In addition, distal tibial development (Rauhut 2003a) is used as a fourth character.

Form of Medial Epicondyle. Although the distribution of truncate and angular epicondyles is not as simple as one might wish, this character still gives a reasonably straightforward distribution of states (Fig. 1.3). Small to moderate-size plesiomorphic theropods show slight development, ceratosaurs and some plesiomorphic tetanurans have an angular medial epicondyle, carnosaurs (*sensu* Holtz 1998a) (Figs. 1.2C, 1.4A) usually have a truncate epicondyle, and maniraptoriforms (and possibly coelurosaurs) have a slightly developed or rounded epicondyle. The distribution of this character among maniraptoriforms suggests that a slightly developed epicondyle occurs in the smaller or more gracile taxa, and a rounded epicondyle in the larger or more robust taxa. This distribution may be true for theropods in general because some large ceratosaurs (*Ceratosaurus dentisulcatus,* abelisauroids) also show rounded medial epicondyles.

The presence of a truncate medial epicondyle suggests that the Stonesfield tibia (Figs. 1.1A, 1.2E) derives from a theropod more plesiomorphic than coelurosaurs but more derived than taxa such as spinosaurs, *Torvosaurus,* and *Magnosaurus.* Thus this theropod might be related to creatures such as *Poekilopleuron, Allosaurus,* or carcharodontosaurids.

Presence of Medial Buttress. The distribution of this character is more straightforward (Fig. 1.3). The buttress is absent in small plesiomorphic theropods (coelophysids) and maniraptoriforms. It may also

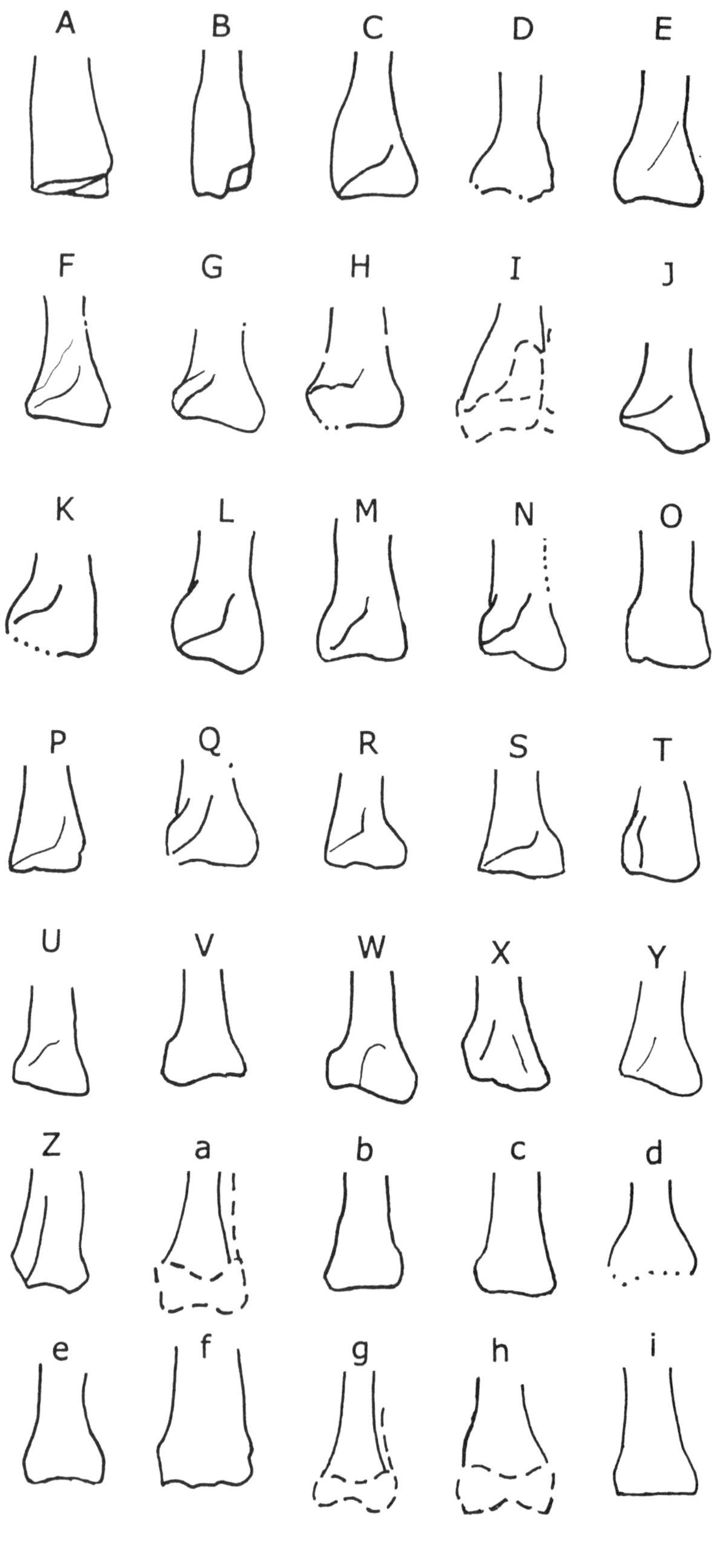

Figure 1.3. Diagrams of the distal left tibiae in anterior view of most of the taxa represented in Table 1.1. (A) Herrerasaurus ischigualastensis, (B) Coelophysidae, (C) Dilophosaurus wetherilli (the distal tibia of Magnosaurus nethercombensis is quite similar to this: Huene 1926b, fig. 2), (D) Elaphrosaurus bambergi, (E) Xenotarsosaurus bonapartei, (F) Ceratosaurus nasicornis, (G) Ceratosaurus magnicornis, (H) Ceratosaurus dentisulcatus, (I) Suchomimus tenerensis, (J) Torvosaurus tanneri, (K) Streptospondylus altdorfensis, (L) Megalosaurus bucklandi, (M) Eustreptospondylus oxoniensis, (N) Poekilopleuron bucklandii, (O) Erectopus superbus, (P) Szechuanosaurus zigongensis, (Q) the Hidden Lake Formation tibia, (R) Piatnitzkysaurus floresi, (S) Sinraptor dongi, (T) the Stonesfield tibia, (U) Allosaurus fragilis, (V) Neoventor saleri, (W) Acrocanthosaurus atokensis, (X) Carcharodontosaurus saharicus, (Y) Dryptosaurus aquilunguis, (Z) Coelurus fragilis, (a) Bagaraatan ostromi, (b) BMNH R186, if a left tibia, (c) BMNH R186, if a right tibia (shown reversed), (d) Gallimimus bullatus, (e) Tyrannosauridae, (f) Chirostenotes pergracilis, (g) Avimimus portentosus, (h) Troodon formosus, and (i) Deinonychus antirrhopus. Dotted lines indicate missing or obscured features, dashed lines indicate obscuring elements, thin lines indicate subdued features. Not to scale. Redrawn from references in Table 1 and Molnar et al. 1996, except Q, T, and Z, which are original.

8 • Peter M. Galton and Ralph E. Molnar

TABLE 1.1.

Characters of the Theropod Distal Tibia

Taxon	Distal tibial development	Form of medial epicondyle	Presence of medial buttress	Form of medial buttress	Source
Herrerasauridae	slight	slight	absent	—	Molnar et al. 1996
Coelophysidae	slight	slight	absent	—	Molnar et al. 1996
Dilophosaurus wetherilli	lateral	slight	present	Broad	Welles 1984
Elaphrosaurus bambergi	both	angular	?	?	Janensch 1925
Xenotarsosaurus bonapartei	both	rounded	?	?	Martínez et al. 1987
Ceratosaurus nasicornis	both	angular	present	broad	Gilmore 1920
Ceratosaurus magnicornis	both	angular	present	broad	Madsen and Welles 2000
Ceratosaurus dentisulcatus	both	rounded	present	broad	Madsen and Welles 2000
Suchomimus tenerensis	both	angular	?	?	Holtz 1998b
Magnosaurus nethercombensis	slight	slight	present	broad	Huene 1926b
Streptospondylus altdorfensis	both	angular	present	broad	Allain 2001
Torvosaurus tanneri	lateral	angular	present	broad	Britt 1991
Megalosaurus bucklandi	lateral	truncate	present	broad	Huene 1926a
Poekilopleuron bucklandii	lateral	truncate	present	broad	Eudes-Deslongchamps 1837
Eustreptospondylus oxoniensis	both	truncate	present	broad	Huene 1926a
Erectopus sauvagei	lateral	truncate	?	?	Sauvage 1882
Hidden Lake tibia	lateral	truncate	present	broad	Molnar et al. 1996
Chilantaisaurus tashuikouensis	lateral	angular	present	narrow	Hu 1964
Szechuanosaurus zigongensis	slight	slight	present	broad	Gao 1993
Piatnitzkysaurus floresi	lateral	slight	present	probably broad	Bonaparte 1986
Yangchuanosaurus shangyuensis	both	angular?	present	broad	Dong et al. 1983
Sinraptor dongi	lateral	slight	present	broad	Currie and Zhao 1993
Neovenator salerii	both	truncate	?	?	Naish et al. 2001

TABLE 1.1. *(cont.)*

Characters of the Theropod Distal Tibia

Taxon	Distal tibial development	Form of medial epicondyle	Presence of medial buttress	Form of medial buttress	Source
Allosaurus fragilis	lateral	truncate	present	narrow	Gilmore 1920; Madsen 1976a
Carcharodontosaurus saharicus	lateral	angular	present	narrow	Stromer 1934; Sereno et al. 1994
Acrocanthosaurus atokensis	both	truncate	present	narrow	Stovall and Langston 1950; Currie and Carpenter 2000
Stonesfield tibia	both	truncate	present	narrow	Original
Gasosaurus constructus	lateral	angular	absent?	?	Dong and Tang 1985
Dryptosaurus aquilunguis	lateral	angular	absent?	—	Carpenter et al. 1997
Coelurus fragilis	lateral	angular	present	narrow	Original
BMNH R186	both	angular	absent	—	Lydekker 1891; original
Bagaraatan ostromi	slight	flattened	absent	—	Osmólska 1996
Ornithomimidae	both	slight	absent	—	Molnar et al. 1996
Tyrannosauridae	both	rounded	absent	—	Molnar et al. 1996
Oviraptoridae	both	slight	low or absent	—	Molnar et al. 1996
Avimimus portentosus	both	slight	absent	—	Kurzanov 1981
Chirostenotes pergracilis	medial	slight	absent	—	Currie and Russell 1988
Troodontidae	both	slight	absent	—	Molnar et al. 1996
Deinonychus antirrhopus	both	rounded	absent	—	Ostrom 1969

be absent, or poorly developed, in *Dryptosaurus,* although it is present in *Coelurus* (Fig. 1.4D), suggesting that the buttress may have been in the process of becoming reduced in plesiomorphic coelurosaurs or, perhaps, that *Coelurus* is more plesiomorphic than these forms as regards this character. The presence of a medial buttress refines the conclusion drawn from the previous character, although a medial buttress is also found in neoceratosaurs.

Form of Medial Buttress. In Table 1.1, the buttress is described as consisting of two forms: narrow and broad. A narrow buttress does not expand substantially at any point and is approximately no more than one-half the width of the tibial shaft just above the epicondyles (Figs. 1.2C, 1.4A). A broad buttress widens proximally more (Fig. 1.2A) or less (Fig. 1.2B) abruptly, to become more than one-half the width of the

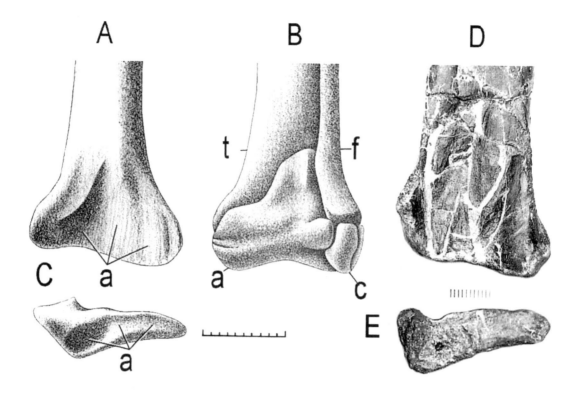

Figure 1.4. Distal left tibiae from the Upper Jurassic of Wyoming in (A, B, D) anterior and (C, E) distal views (B with astragalus, calcaneum, and distal fibula). (A–C) Allosaurus fragilis from Madsen (1976a) and (D, E) Coelurus fragilis, YPM 2010. Abbreviations: a—astragalus or sutural surface for it; c—calcaneum; f—fibula; t—tibia. Scale bars = 10 cm (A–C) and 10 mm (D, E).

tibial shaft just above the epicondyles. There may well be a continuous spectrum of forms from broad to narrow, as suggested by Figure 1.2A–C. Broad, triangular buttresses are found in ceratosaurs (in those with buttresses) and plesiomorphic tetanurans (except *Chilantaisaurus, Allosaurus*, and carcharodontosaurids; Figs. 1.2C, 1.4A). A narrow buttress also occurs in *Coelurus* (Fig. 1.4D). Presumably at some stage among coelurosaurs the buttress was lost, and it is absent in maniraptoriforms (perhaps indicating descent from small forms).

The narrow form of the buttress seen in the Stonesfield tibia (Figs. 1.1A, 1.2E) serves to distinguish this Middle Jurassic tibia from those attributed to *Megalosaurus bucklandi* and those of *Poekilopleuron* (Figs. 1.1E, 1.2A,B).

Distal Articular Surface. The outline of the distal articular surface forms a narrow triangle that is strongly expanded mediolaterally (Fig. 1.1C; 2.24 times as wide transversely as anteroposteriorly), as in *Megalosaurus, Poekilopleuron, Allosaurus* (Fig. 1.4C), and others. However, it is not subrectangular in outline as in *Coelurus* (Fig. 1.4E), *Avimimus*, dromaeosaurids, and troodontids (Rauhut 2003a: character 208, modified, see below).

The first three character states indicate that the Stonesfield tibia derives from a plesiomorphic tetanuran, an animal more derived than abelisaurs but less derived than *Coelurus* (Figs. 1,2C, 1.4A,B,D).

Discussion. Working with such incomplete material always raises the problem of reliability, but with the advent of phylogenetic methods, this should be less an issue. If the distribution of character states,

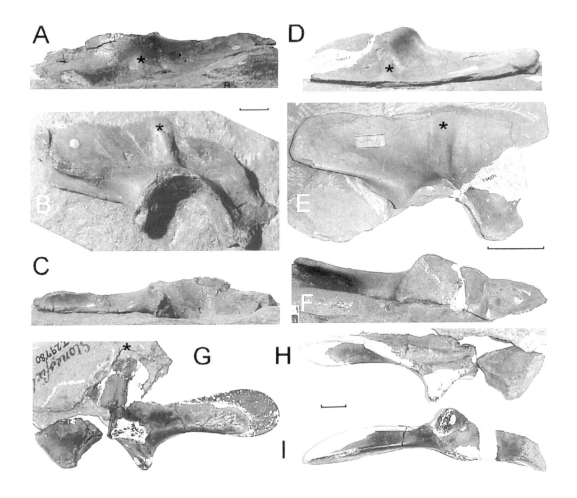

A

D

B

E

C

F

G

H

I

Figure 1.5. Ilia of Iliosuchus incognitus Huene 1932 from the Middle Jurassic of Stonesfield. (A–C) Holotype right ilium, BMNH R83, in (A) dorsal, (B) lateral, and (C) ventral views (figured in Huene 1932; Galton 1976; Foster and Chure 2000). (D–I) Ilia in W. Buckland Collection: (D–G) right ilium, OUMNH J28971, in (D) dorsal, (E) lateral, and (F) ventral views (mentioned in Phillips 1871 [p. 213]; figured in Foster and Chure 2000); (G–I) left ilium, OUMNH J29780, in (G) lateral, (H) medial, and (I) ventral views (figured in Phillips 1871 [p. 213, figs. LXVII.1, 2]; Galton 1976). * = broad sub-vertical ridge. Scale bars = 1 cm (A–C, G–I) and 5 cm (D–F).

especially apomorphies, is understood, then any material, regardless of how incomplete, can be correctly identified taxonomically if it possesses synapomorphies or autapomorphies. Thus, if the phylogeny of Rauhut (2003a) and the character state distribution presented here are accurate, then the conclusions are reliable.

The obvious taxonomic identification for this Stonesfield tibia is the small theropod Iliosuchus incognitus Huene 1932, also known from the Stonesfield Slate. Unfortunately, Iliosuchus is known only from three isolated ilia (Fig. 1.5; Galton 1976; Foster and Chure 2000), so it cannot be directly compared with this tibia. Furthermore, for readily apparent reasons, Iliosuchus has not been included in published phylogenetic analyses. However, these ilia do have a median vertical ridge on the external surface, a derived character state used in the analyses of Holtz (1998a, character 295) and Rauhut (2003a, character 172). This vertical ridge separates either the insertion areas for the M. iliofemoralis externus anteriorly from the M. iliofibularis (Carrano and Hutchinson 2002) or the M. iliotrochantericus posterior (=M. pubo-ischio-femoralis internus pars I) anteriorly from the M. piriformis pars iliofemoralis (probably =M. ilio-femoralis of crocodiles) (Rauhut 2003a).

Unfortunately, this ridge has a rather sporadic distribution in thero-pods, occurring in the basal tetanuran *Piatnitzkysaurus;* the therizino-sauroid *Segnosaurus;* the tyrannosaurs *Albertosaurus, Siamotyrannus,* and *Tyrannosaurus;* and some specimens attributed to *Megalosaurus* (Foster and Chure 2000). However, in these cases the ridge is relatively narrow, whereas in *Iliosuchus* and *Stokesosaurus* (Upper Jurassic, Morrison Formation, western United States), it is broad and robust (Madsen 1974; Foster and Chure 2000). *Stokesosaurus* is also known from ilia, along with referred cranial material (Madsen 1974; Chure and Madsen 1998), and it is generically distinct from *Iliosuchus* (Galton and Jensen 1979 *contra* Galton 1976; Foster and Chure 2000; Rauhut 2003b). Rauhut (2000) reported the presence of a *Stokesosaurus* sp. from the Upper Jurassic of Portugal on the basis of an ilium with a subvertical ridge. Rauhut (2003b) makes this ilium the holotype of a new genus of tyrannosaurid, *Aviatyrannis,* the family to which he refers *Stokesosaurus* (as did Madsen 1974). He notes that *Iliosuchus* may represent an earlier occurrence of the same lineage, which, if correct, would indicate that the Stonesfield tibia does not pertain to *Iliosuchus.* This would not be surprising, because the fauna probably included several taxa of small theropods, as is the case in the better represented Morrison and Tendaguru faunas of the Upper Jurassic. The well-known Morrison theropod fauna of the western United States includes at least seven more or less contemporaneous small theropods (*Coelurus* Marsh 1879; *Koparion* Chure 1994; *Marshosaurus* Madsen 1976b; *Ornitholestes* Osborn 1903; *Stokesosaurus* Madsen 1974; and *Tanyco-lagreus* Carpenter et al. this volume; as well as *Elaphrosaurus* Janensch 1920; Galton 1982; Chure 2001). That this is not an unusual feature of the Morrison fauna is indicated by the less well known fauna of the Tendaguru Beds of Tanzania, East Africa, which includes the small coelurosaurian tibiae A, B, and C of Janensch (1925) and, in addition, *Elaphrosaurus* Janensch 1920 (tibia distal width 44 mm).

The Isle of Wight Small Theropod Tibia

Rev. William Fox (1813–1881; see Blows 1983) collected a small, gracile tibia (BMNH R186, Fig. 1.6A–J) from the Wessex Formation, Lower Cretaceous (Barremian; Insole and Hutt 1994) on the south-western coast of the Isle of Wight (for details on geology, localities, and dinosaurs, see Martill and Naish 2001). The specimen has a maximum length of 160 mm, a proximal width of 28.5 mm, and a distal width of 25.3 mm. Lydekker (1888) initially identified it as a left tibia and suggested that it might have come from the same individual of the ornithopod dinosaur *Hypsilophodon foxii* as the pair of femora BMNH R184 and R185, but this suggestion is incorrect. These femora are the holotype of the dryosaurid ornithopod *Valdosaurus caniculatus* Galton 1977. Lydekker (1891) subsequently identified the tibia as that of a small *Coelurus*-like theropod, following the suggestion made by Othniel Charles Marsh of Yale College on seeing the specimen in the

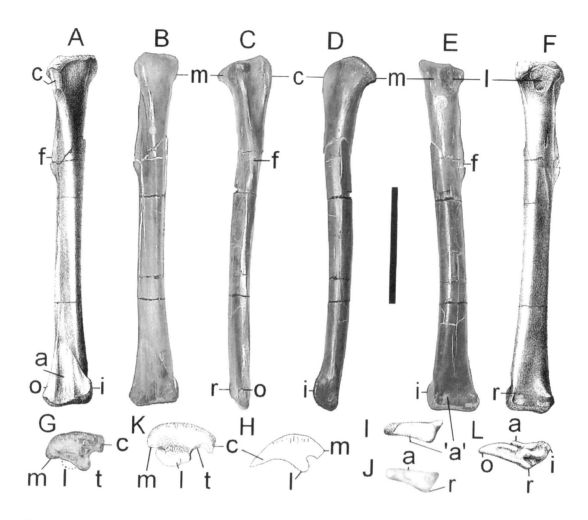

(continued on page 15)

Figure 1.6. (A–J) BMNH R186 from Lower Cretaceous of the Isle of Wight, a right theropod tibia in (A, B) anterior, (C) lateral, (D) medial, (E, F) posterior, (G, H) proximal, and (I, J) distal views (A, F, I from Lydekker 1891, H from Naish 1999 and labeled as a left tibia). (K) Right tibia of ornithomimid Gallimimus *(Upper Cretaceous, Mongolia) in proximal view, from Osmólska et al. 1972. (L) Right tibia of carnosaur* Sinraptor *(Upper Jurassic, China) in distal view, from Currie and Zhao 1993. Abbreviations: a—articular surface for astragalus anteriorly on distal part of tibia;*

BMNH. Lydekker (1891) provisionally referred it to the coelurosaur *Calamosaurus foxi* and described it as a right tibia of "normal" theropod form, an identification that has been accepted by subsequent workers (e.g., Welles and Long 1974; Norman 1990, fig. 13.26; Brookes 1997; Long and Molnar 1998; Kirkland et al. 1998).

Right or Left? Naish et al. (2001, p. 304, plate 36, figs. 4, 5) contend that, owing to the misidentification of the proximal inner (medial) condyle as the cnemial crest, Lydekker (1891) inadvertently identified the posterior face as anterior and vice versa. They note that this tibia is highly unusual in many features, but no other reasons for this new interpretation as a left tibia are given except for a reference to Naish 1999. Naish (1999) noted that the proximal "pitted area," interpreted as the broken lateral condyle by Lydekker (1891) (Fig. 1.6E,F), is a gentle convexity (Fig. 1.6H). If this area is interpreted as the broken lateral condyle of a right tibia, the usual interpretation, then it is situated nearly midway along the mediolateral axis. However, if the cnemial crest is actually the medial condyle of a left tibia, then the lateral condyle would be on the posterolateral corner, the normal position (Fig. 1.6H). In addition, both the distal malleoli (=epicondyles) of this left tibia flare outward moderately (Fig. 1.6A,B), being unlike those of other theropods. A faint triangular impression on the anterior face (as a left tibia) is interpreted as the scar made by the ascending astragalar process (Fig. 1.6E,F), not the more prominent flat triangular impression on the opposite surface (Fig. 1.6A,B; which would be anterior with tibia as a right and has been universally interpreted as being for the ascending process). Naish et al. (2001) noted that the lack of a tall scar for the ascending process of the astragalus (Fig. 1.6E,F) probably excludes such groups as tyrannosaurids, ornithomimosaurs, oviraptorosaurs, and troodontids as possible identifications; as a result, the tibia may have affinities with basal or stem-group coelurosaurs such as compsognathids.

It is possible that the tibia was inadvertently misaligned at one of the several breaks along the shaft (Fig. 1.6B–E). However, a careful examination of the specimen indicates that this is not the case, so the evidence for orientating the tibia needs to be examined. In lateral view (Fig. 1.6C), the proximal structures identified by Lydekker (1891) as the cnemial crest and the lateral (outer) and medial (inner) condyles of a right tibia (c, l, m, Fig. 1.6A–F) appear to be correctly identified when compared to the tibiae of the other theropods (e.g., *Allosaurus* Gilmore 1920, Madsen 1976a; coelurosaur *Nedcolbertia* Kirkland et al. 1998; possible compsognathid Naish et al. 2001; ornithomimid *Gallimimus* Osmólska et al. 1972; *Acrocanthosaurus* Stovall and Langston 1950, Currie and Carpenter 2000; *Bagaraatan* Osmólska 1996; *Ceratosaurus* Gilmore 1920, Madsen and Welles 2000; *Dilophosaurus* Welles 1984; *Elaphrosaurus* Janensch 1925; *Piatnitzkysaurus* Bonaparte 1986; *Avimimus* Kurzanov 1981; Vickers-Rich et al. 2002; *Chirostenotes* Currie and Russell 1988; *Deinonychus* Ostrom 1969; *Neovenator* Naish et al. 2001). In addition, the curvatures of the shaft and the form of the distal end are similar to those of other theropods as well. The "pitted area" (Fig. 1.6E,F) is slightly concave in proximal view (Fig. 1.6G). The area is not continuous with the medial condyle (as the cnemial crest) as shown (Fig. 1.6H), because there is a slight step (Fig. 1.6G). This area shows cancellous or spongy bone (Fig. 1.6E), as also illustrated by Lydekker (1891; Fig. 1.6F). Consequently, the "pitted area" is not the bone surface, but instead it is the broken base of the posterolateral part of the lateral condyle (Fig. 1.6E–G), as was interpreted by Lydekker (1891).

A proximal view of the right tibia of the ornithomimid *Gallimimus* is given for comparison with that of BMNH R186. In both tibiae (Fig. 1.6G,K), the cnemial crest (c) has a rounded-off square outline; the crest is bordered laterally by a prominent semicircular tibial incisure (t); the lateral condyle (l, broken in BMNH R186) for the fibula is large, wide anteroposteriorly, and laterally situated; and the posteriorly positioned medial condyle (m) tapers to a rounded-off pointed end. Indeed, a judicious break across the posterolateral part of the lateral condyle of the tibia of *Gallimimus* would give a proximal end that, apart from the proportionally more elongate form and more strongly curved medial edge of the cnemial crest, would be almost identical to that of BMNH R186.

(continued from page 14)
'a'—anterior articular surface for astragalus according to Naish (1999); c—cnemial crest; f—fibular crest; i—inner or medial epicondyle (malleolus); l—lateral or fibular condyle, in BMNH R186 its broken base, the "pitted area" of Naish 1999; m—medial condyle; o—outer or lateral epicondyle (malleolus); r—posteromedial ridge; t—tibial incisure. Scale bar = 5 cm for BMNH R186.

Distal views of the tibiae of the carnosaur *Sinraptor* and BMNH R186 show a subtriangular outline, with the longest edge anterior and the thickened angle or ridge posteromedially (r, Fig. 1.6I,J,L), as in the Stonesfield tibia (Fig. 1.1C) and *Allosaurus* (Fig. 1.4C). The former surface bears the distinct subtriangular flat impression against which the closely applied ascending process of the astragalus fitted (Fig. 1.6A,B cf. Fig. 1.4A,B), none of which is attached (*contra* Galton 1973). The "scar for ascending process of astragalus" of Naish (1999, fig. 22A,B) is the slight depression on the wider lateral part of the posterior surface of the distal end ('a,' Fig. 1.6E,F,I,J). This transversely very gently concave surface is not on the longest side of the triangular distal end, and it does not resemble the astragalar surface of other theropods. There are no striations indicative of its being a sutural area, and it resembles the posterolateral surface of many theropod tibiae, including *Allosaurus* (Fig. 1.4C), *Coelurus* (Fig. 1.4E), *Sinraptor* (Fig. 1.6L), and the one from Stonesfield (Fig. 1.1C,D).

In anterior or posterior view, the distal part of the tibiae of several theropods shows a lateral epicondyle that extends further distally than does the medial one (Figs. 1.2A–C, 1.4A,D). The proximal margin of the medial epicondyle diverges more abruptly from the shaft than does that of the lateral epicondyle (Fig. 1.3K–P,T,U). The distal ends of these tibiae are inclined so that the lateral corner is more distal than the medial (Fig. 1.3F,G,J,L,N,O,U). All of these features are shown by tibiae of *Ceratosaurus nasicornis, Torvosaurus tanneri, Megalosaurus bucklandi* (Figs. 1.1E, 1.2A), *Poekilopleuron bucklandii* (Fig. 1.2B), *Erectopus superbus, Sinraptor dongi, Acrocanthosaurus atokensis, Allosaurus fragilis* (Fig. 1.4A), *Coelurus agilis* (Fig. 1.4D), and *Carcharodontosaurus saharicus*, among others (Fig. 1.3F–H,J,L,N,O,R,U,V). These features are generally not shown by plesiomorphic forms, such as the coelophysids and *Dilophosaurus wetherilli* (Fig. 1.3B,C), or by advanced coelurosaurs, such as ornithomimosaurs, tyrannosaurs, and dromaeosaurs (Fig. 1.3D,c,d,h). If the tibia BMNH R186 is from the right, then the form of the distal end (Fig. 1.6A,B) matches those listed above in these three features; if a left tibia, then it matches no other known theropod tibia. In this case this tibia would be the only theropod tibia known with a truncated lateral epicondyle.

Re-identifying the medial condyle of BMNH R186 as the cnemial crest results in a tibia in which the curves of the bone in lateral view are reversed relative to those of other theropods, with all the structures on the anterior surface and lateral side having the form of those on the posterior surface and medial side of other theropods, and vice versa. Consequently, we consider that Lydekker (1891) correctly described BMNH R186 as a right tibia, the form of which is normal for a theropod dinosaur (Fig. 1.6A–G,I,J); whereas the incorrect re-identification as a left (Fig. 1.6H; Naish et al. 2001, p. 304, plate 36, figs. 4, 5) gives a tibia with a totally unique form for any theropod, or dinosaur for that matter. In addition, there is no explanation for the sutural area usually interpreted as the attachment area of the astragalus.

Nature of the Knee. The tibia, BMNH R186, is also unusual in that the longest axes of the proximal and distal ends are not perpendicular to each other, but instead are nearly parallel, which implies that the femur-tibia articulation was extremely unusual for a theropod or that this specimen exhibited a pathological left leg. However, lines drawn through the greatest widths of the ends intersect at an angle of about 75°, so, with the distal anterior surface oriented transversely, the anterior end of the proximal line is directed slightly laterally. More significantly, a line bisecting the apex of the cnemial crest is directed more or less perpendicular to the distal anterior surface. Consequently, the femur-tibia articulation was normal for a theropod, and there is no need to postulate that this tibia was part of a pathological leg.

Phylogenetic Position. Proximally, a fibular condyle that is strongly offset from the cnemial crest by a well-developed tibial incisure, as in BMNH R186 (Fig. 1.6G), is a possible apomorphy under DELTRAN for Ceratosauria + Tetanurae (Rauhut 2003a, p. 141, character 204). The distal end of the tibia BMNH R186 as a right tibia is expanded both medially and laterally, has an angular medial epicondyle, and lacks a medial buttress (Figs. 1.2F, 1.6A,B). Among tetanurans, an angular medial epicondyle is found (sporadically) among plesiomorphic tetanurans and in plesiomorphic coelurosaurs (Fig. 1.3). The medial buttress is lost among maniraptorans and, as far as we can tell from the literature, may be absent in *Dryptosaurus*. It is, however, present in *Coelurus* (Fig. 1.4D). Thus the presence of an angular epicondyle is consistent with the identification of BMNH R186 as either a plesiomorphic coelurosaur or a maniraptoran. The absence of a medial buttress also suggests this conclusion, if the buttress is in fact absent in *Dryptosaurus*. If not, the absence indicates that BMNH R186 is a plesiomorphic maniraptoran. Thus, this animal might have been closely related either to *Coelurus* (Fig. 1.4D), as traditionally thought, or to *Compsognathus,* as suggested by Naish et al. (2001). However, it should be noted that *Coelurus* and *Compsognathus* are themselves rather closely related in the cladograms of Holtz (1998a) and Rauhut (2003a).

Rauhut (2003a, p. 117, character 208) notes that the distal articular surface of the tibia in *Coelurus* (YPM 2010), *Avimimus*, birds (*Rahonavis*), dromaeosaurids, and troodontids is compressed so that the distal outline is broadly "rectangular and more than three times wider transversely than anteroposteriorly." However, although distally the tibia of *Coelurus agilis* (YPM 2010, Fig. 1.4E) is broadly rectangular, the transverse to anteroposterior width is 2.6, compared to 2.7 for *Allosaurus* (Fig. 1.4C), so this character state needs to be restricted to shape. This ratio is 3.0 in BMNH R186, and, because the distal end is subrectangular rather than triangular as in *Compsognathus* (Rauhut 2003a), it is regarded as being closer to *Coelurus*. It is definitely not an ornithomimoid as identified by Welles and Long (1974).

Acknowledgments. We wish to thank the following for all their assistance while we studied the specimens described in this paper: S. Chapman and A. Milner (BMNH), D.-W. Heinrich, J. Helms and the late H. Jaeger (MB), T. Holtz Jr. (University of Maryland), and H. P. Powell (OUMNH). We also thank T. R. Lipka (Baltimore) for a copy of the section on BMNH R186 from Naish 1999; O. W. M. Rauhut (MB) for a pre-publication copy of his paper on the ilium from Portugal, and J.-C. Lu (Southern Methodist University, Dallas) and K. Carpenter for their comments on the paper.

References Cited

Allain, R. 2001. Redescription de *Streptospondylus altdorfensis*, le dinosaur théropode de Cuvier, du Jurassique de Normandie. *Geodiversitas* 23: 349–367.

Allain, R., and D. J. Chure. 2002. *Poekilopleuron bucklandii*, the theropod dinosaur from the Middle Jurassic (Bathonian) of Normandy. *Palaeontology* 45: 1107–1121.

Arkell, W. J. 1947. *The Geology of Oxford*. Oxford: Clarendon Press.

Aston, M. 1974. *Stonesfield Slate*. Publication no. 5. Oxford: Oxfordshire County Council, Department of Museum Services.

Benton, M. J., and P. S. Spencer. 1995. *Fossil Reptiles of Great Britain*. London: Chapman and Hall.

Blows, W. T. 1983. William Fox (1813–1881), a neglected dinosaur collector of the Isle of Wight. *Archives of Natural History* 11: 299–313.

Bonaparte, J. F. 1986. Les dinosaures (carnosaures, allosauridés, sauropodes, cétiosauridés) du Jurassique moyen de Cerro Cóndor (Chubut, Argentina). *Annales de Paléontologie (Vertébrés-Invertébrés)* 72: 247–289.

Boneham, B. F. W., and R. J. Wyatt. 1993. The stratigraphical position of the Middle Jurassic (Bathonian) Stonesfield Slate of Stonesfield, Oxfordshire, UK. *Proceedings of the Geologists' Association of London* 104: 123–136.

Britt, B. B. 1991. Theropods of Dry Mesa Quarry (Morrison Formation, Late Jurassic), Colorado, with emphasis on the osteology of *Torvosaurus tanneri*. *Brigham Young University Geology Studies* 37: 1–72.

Brookes, A. 1997. A reassessment of the small theropod material from the Wealden of southern England. M.Sc. dissertation, University of Bristol.

Buckland, W. 1824. Notice on the *Megalosaurus* or great fossil lizard of Stonesfield. *Transactions of the Geological Society of London* 2: 390–396.

Carpenter, K., D. A. Russell, D. Baird, and R. Denton. 1997. Redescription of the holotype of *Dryptosaurus aquilunguis* (Dinosauria: Theropoda) from the Upper Cretaceous of New Jersey. *Journal of Vertebrate Paleontology* 17: 561–573.

Carrano, M. T., and J. R. Hutchinson. 2002. Pelvic and hindlimb musculature of *Tyrannosaurus rex* (Dinosauria: Theropoda). *Journal of Morphology* 253: 207–228.

Chure, D. J. 1994. *Koparion douglassi*, a new dinosaur from the Morrison Formation (Upper Jurassic) of Dinosaur National Monument: The oldest troodontid (Theropoda: Maniraptora). *Brigham Young University Geology Studies* 40: 11–15.

Chure, D. J. 2001. The second record of the African theropod *Elaphrosaurus* (Dinosauria, Ceratosauria) from the Western Hemisphere. *Neues Jahrbuch für Geologie und Paläontologie, Monatshefte* 2001: 565–576.

Chure, D. J., and J. H. Madsen. 1998. An unusual braincase (?*Stokesosaurus clevelandi*) from the Cleveland-Lloyd Quarry, Utah (Morrison Formation; Late Jurassic). *Journal of Vertebrate Paleontology* 18: 115–125.

Cope, J. C. W., K. L. Duff, C. F. Parsons, H. S. Torrens, W. A. Wimbledon, and J. K. Wright. 1980. A correlation of Jurassic rocks in the British Isles. Part 2, Middle and Upper Jurassic. *Geological Society of London, Special Report* 15: 1–109.

Currie, P. J., and K. Carpenter. 2000. A new specimen of *Acrocanthosaurus atokensis* (Theropoda, Dinosauria) from the Lower Cretaceous Antlers Formation (Lower Cretaceous, Aptian) of Oklahoma, USA. *Geodiversitas* 22: 207–246.

Currie, P. J., and D. A. Russell. 1988. Osteology and relationships of *Chirostenotes pergracilis* (Saurischia, Theropoda) from the Judith River (Oldman) Formation of Alberta, Canada. *Canadian Journal of Earth Sciences* 25: 972–986.

Currie, P. J., and Zhao X.-J. 1993. A new carnosaur (Dinosauria, Theropoda) from the Jurassic of Xinjiang, People's Republic of China. *Canadian Journal of Earth Sciences* 30: 2037–2081.

Dong Z. and Tang Z. 1985. A new mid-Jurassic theropod (*Gasosaurus constructus* gen. et sp. nov.) from Dashanpu, Zigong, Sichuan Province, China. *Vertebrata PalAsiatica* 23: 77–83.

Dong Z., Zhou S., and Zhang Y. 1983. The dinosaurian remains from Sichuan Basin, China. *Palaeontologia Sinica*, n.s. C, no. 23: 1–145.

Eudes-Deslongchamps, M. 1838. Mémoire sur le *Poekilopleuron bucklandii,* grand saurien fossile, intermédiare entre les crocodiles et les lézards. *Mémoires de la Société Linnéenne de Normandie* 6: 37–146.

Foster, J. R., and D. J. Chure. 2000. An ilium of a juvenile *Stokesosaurus* (Dinosauria, Theropoda) from the Morrison Formation (Upper Jurassic: Kimmeridgian), Meade County, South Dakota. *Brigham Young University Geological Studies* 45: 5–10.

Galton, P. M. 1973. A femur of a small theropod from the Lower Cretaceous of England. *Journal of Paleontology* 47: 996–997.

Galton, P. M. 1975. English hypsilophodontid dinosaurs (Reptilia: Ornithischia). *Palaeontology* 18: 741–752.

Galton, P. M. 1976. *Iliosuchus,* a Jurassic dinosaur from Oxfordshire and Utah. *Palaeontology* 19: 587–589.

Galton, P. M. 1977. The Upper Jurassic ornithopoid dinosaur *Dryosaurus* and a Laurasia-Gondwanaland connection in the Upper Jurassic. *Nature* 268: 230–232.

Galton, P. M. 1980. European Jurassic ornithopod dinosaurs of the families Hypsilophodontidae and Camptosauridae. *Neues Jahrbuch für Geologie und Paläontologie, Abhandlungen* 160: 73–95.

Galton, P. M. 1982. *Elaphrosaurus,* an ornithomimid dinosaur from the Upper Jurassic of North America and Africa. *Paläontologische Zeitschrift* 56: 265–275.

Galton, P. M., and J. Jensen. 1979. A new large theropod dinosaur from the Upper Jurassic of Colorado. *Brigham Young University Geological Studies* 26: 1–12.

Gao Y. 1993. A new species of *Szechuanosaurus* from the Middle Jurassic of Dashanpu, Zigong, Sichuan. *Vertebrata PalAsiatica* 31: 308–314.

Gilmore, C. W. 1920. *Osteology of the Carnivorous Dinosauria in the United States National Museum, with Special Reference to the Genera* Antrodemus (Allosaurus) *and* Ceratosaurus. U.S. National Museum Bulletin no. 110. Washington, D.C.: Government Printing Office.

Holtz, T. R., Jr. 1994. The phylogenetic position of the Tyrannosauridae: Implications for theropod systematics. *Journal of Paleontology* 68: 1100–1117.

Holtz, T. R., Jr. 1998a. A new phylogeny of the carnivorous dinosaurs. B.

P. Pérez-Moreno, T. Holtz Jr., J. L. Sanz, and J. Moratalla (eds.), *Gaia: Aspects of Theropod Paleobiology,* vol. 15, pp. 5–61. Lisbon: Museu Nacional de História Natural.

Holtz, T. R., Jr. 1998b. Spinosaurs as crocodile mimics. *Science* 1282: 1276–1277.

Hu S.-Y. 1964. Carnosaurian remains from Alashan, Inner Mongolia. *Vertebrata PalAsiatica* 8: 42–63.

Huene, F. 1926a. The carnivorous Saurischia in the Jura and Cretaceous formations, principally in Europe. *Revista del Museum de La Plata* 29: 35–167.

Huene, F. 1926b. On several known and unknown reptiles of the Order Saurischia from England and France. *Annals and Magazine of Natural History,* ser. 9, 17: 473–489.

Huene, F. 1932. Die fossile Reptil-Ordnung Saurischia, ihre Entwicklung und Geschichte. *Monographien zur Geologie und Palaeontologie,* ser. 1, 4: 1–361.

Huxley, T. H. 1870. Further evidences of the affinity between the dinosaurian reptiles and birds. *Quarterly Journal of the Geological Society of London* 26: 12–31.

Insole, A. N., and S. Hutt. 1994. Palaeoecology of the dinosaurs of the Wessex Formation (Wealden Group, Early Cretaceous), Isle of Wight, southern England. *Zoological Journal of the Linnean Society of London* 112: 135–150.

Janensch, W. 1920. Uber *Elaphrosaurus bambergi* und die Megalosaurier aus dem Tendaguru-Schichten Deutsch-Ostafrikas. *Sitzungsberichte der Gesellschaft naturforschender Freunde Berlin* 1920: 225–235.

Janensch, W. 1925. Die Coelurosaurier und Theropoden der Tendaguru-Schichten Deutsch-Ostafrikas. *Palaeontographica* 1 (suppl. 7): 1–100.

Kirkland, J. I., B. Britt, C. H. Whittle, S. K. Madsen, and D. L. Burge. 1998. A small coelurosaurian theropod from the Yellow Cat Member of the Cedar Mountain Formation (Lower Cretaceous, Barremian) of eastern Utah. In S. G. Lucas, J. I. Kirkland, and J. W. Estep (eds.), *Lower and Middle Cretaceous Terrestrial Ecosystems,* pp. 239–248. New Mexico Museum of Natural History and Science Bulletin, no. 14. Albuquerque: New Mexico Museum of Natural History and Science.

Kurzanov, S. M. 1981. O neobichnikh teropodakh iz verkhnego Mela MNR. *Trudy Sovmestnaia Sovetsko-Mongol'skaia Paleontologicheskaia Ekspeditsiia* 15: 39–50.

Lambe, L. M. 1917. The Cretaceous theropodous dinosaur *Gorgosaurus*. *Geological Survey of Canada Memoir* 100: 1–84.

Lhuyd, E. 1699. Lithophylacii Britannici Ichnographia, sive, lapidium aliorumque fossilium Britannicorum singulari figura insignium. London: Gleditsch & Weidmann.

Long, J. A., and R. E. Molnar. 1998. A new Jurassic theropod dinosaur from Western Australia. *Records of the Western Australian Museum* 19: 121–129.

Lydekker, R. 1888. *Catalogue of the Fossil Reptilia and Amphibia in the British Museum. Part I.* London: British Museum (Natural History).

Lydekker, R. 1891. On certain ornithosaurian and dinosaurian remains. *Quarterly Journal of the Geological Society of London* 47: 43–44.

Madsen, J. H., Jr. 1974. A new theropod dinosaur from the Upper Jurassic of Utah. *Journal of Paleontology* 48: 27–31.

Madsen, J. H., Jr. 1976a. Allosaurus fragilis: *A Revised Osteology.* Utah Geological and Mineral Survey Bulletin, no. 109. Salt Lake City: Utah

Geological and Mineral Survey, Utah Department of Natural Resources.

Madsen, J. H., Jr. 1976b. A second new theropod from the Late Jurassic of east-central Utah. *Utah Geology* 3: 51–60.

Madsen, J. H., Jr., and S. P. Welles. 2000. *Ceratosaurus* (Dinosauria, Theropoda), a revised osteology. *Utah Geological Survey Miscellaneous Publication* 00-2: 1–80.

Marsh, O. C. 1879. Notice of new Jurassic reptiles. *American Journal of Science,* ser. 3, 18: 501–505.

Martill, D. M., and D. Naish. 2001. The geology of the Isle of Wight. In D. M. Martill and D. Naish (eds.), *Dinosaurs of the Isle of Wight,* pp. 25–44. The Palaeontological Association Field Guides to Fossils, no. 10. London: Palaeontological Association.

Martínez, R., O. Giménez, J. Rodríguez, and G. Bochatey. 1987. *Xenotarsosaurus bonapartei* nov. gen. et sp. (Carnosauria, Abelisauridae), un nuevo Theropoda de la formacion Bajo Barreal Chubut, Argentina. *Congreso Argentino de Paleontología y Bioestratigrafía, Actas* 4: 23–31.

Molnar, R. E., and N. S. Pledge. 1980. A new theropod dinosaur from South Australia. *Alcheringa* 4: 281–287.

Molnar, R. E., A. Lopez Angriman, and Z. Gasparini. 1996. An Antarctic Cretaceous theropod. *Memoirs of the Queensland Museum* 39: 669–674.

Naish, D. S. 1999. Studies on Wealden Group theropods: An investigation into the historical taxonomy and phylogenetic affinities of new and previously neglected specimens. M.P. dissertation, University of Portsmouth.

Naish, D., S. Hutt, and D. M. Martill. 2001. Saurischian dinosaurs 2: Theropods. In D. M. Martill and D. Naish (eds.), *Dinosaurs of the Isle of Wight,* pp. 242–309. Palaeontological Association Field Guides to Fossils, no. 10: London: Palaeontological Association.

Norman, D. 1990. Problematic Theropoda: "Coelurosaurs." In D. B. Weishampel, P. Dodson, and H. Osmólska (eds.), *The Dinosauria,* pp. 280–305. Berkeley: University of California Press.

Osborn, H. F. 1903. *Ornitholestes hermanni,* a new compsognathoid dinosaur from the Upper Jurassic. *Bulletin of the American Museum of Natural History* 19: 459–464.

Osmólska, H. 1996. An unusual theropod dinosaur from the Late Cretaceous Nemegt Formation of Mongolia. *Acta Palaeontologica Polonica* 41: 1–38.

Osmólska, H., E. Roniewicz, and R. Barsbold. 1972. Results of the Polish-Mongolian palaeontological expeditions—Part IV. A new dinosaur, *Gallimimus bullatus* n. gen., n. sp. (Ornithomimidae) from the Upper Cretaceous of Mongolia. *Palaeontologica Polonica* 27: 103–143.

Ostrom, J. H. 1969. Osteology of *Deinonychus antirrhopus,* an unusual theropod from the Lower Cretaceous of Montana. *Bulletin of the Peabody Museum of Natural History* 30: 1–165.

Owen, R. 1856. *Monograph on the Fossil Reptilia of the Wealden Formations.* Part III: *Megalosaurus bucklandi,* pp. 1–26. Palaeontographical Society Monographs, no. 9. London: Palaeontographical Society.

Phillips, J. 1871. *Geology of Oxford and the Valley of the Thames.* Oxford: Clarendon Press.

Rauhut, O. W. M. 2000. The dinosaur fauna from the Guimarota mine. In T. Martin and B. Krebs (eds.), *Guimarota: A Jurassic Ecosystem,* pp. 75–82. Munich: Pfeil.

Rauhut, O. W. M. 2003a. *The Interrelationships and Evolution of Basal Theropod Dinosaurs.* Special Papers in Palaeontology, no. 69. London: Palaeontological Association.

Rauhut, O. W. M. 2003b. A tyrannosaurid dinosaur from the Late Jurassic of Portugal. *Palaeontology* 46: 903–910.

Sauvage, H. E. 1882. Recherches sur les reptiles trouvés dans le Gault de l'est du Bassin de Paris. *Mémoires de la Société géologique de France* 4: 1–42.

Sereno, P. C., J. A. Wilson, H. C. E. Larsson, D. B. Dutheil, and H.-D. Sues. 1994. Early Cretaceous dinosaurs from the Sahara. *Science* 266: 267–271.

Stovall, J. W., and W. Langston Jr. 1950. *Acrocanthosaurus atokensis*, a new genus and species of Lower Cretaceous Theropoda from Oklahoma. *American Midland Naturalist* 43: 696–728.

Stromer, E. 1934. Ergebnisse der Forschungsreisen Prof. E. Stromer in den Wüsten Ägyptens. II: Wirbeltierreste der Baharije-Stuf e unterstes Cenoman. 13. Dinosauria. *Abhandlungen der Bayerische Akademie der Wissenschaften Mathematische-Naturwissenschaftliche Abteilung (Neue Folge)* 22: 1–79.

Unwin, D. M. 1996. The fossil record of Middle Jurassic pterosaurs. *Museum of Northern Arizona Bulletin* 60: 291–304.

Vickers-Rich, P., L. M. Chiappe, and S. Kurzanov. 2002. The enigmatic birdlike dinosaur *Avimimus portentosus*. Comments with a pictorial atlas. In L. M. Chiappe and L. Witmer (eds.), *Mesozoic Birds: Above the Heads of Dinosaurs*, pp. 65–86. Berkeley: University of California Press.

Welles, S. P. 1984. *Dilophosaurus wetherilli* (Dinosauria, Theropoda): Osteology and comparisons. *Palaeontographica Abt. A* 185: 85–180.

Welles, S. P., and R. A. Long. 1974. The tarsus of theropod dinosaurs. *Annals of the South African Museum* 64: 191–218.

2. New Small Theropod from the Upper Jurassic Morrison Formation of Wyoming

KENNETH CARPENTER, CLIFFORD MILES, AND KAREN CLOWARD

Abstract

A partial skeleton from the Upper Jurassic Morrison Formation at Bone Cabin Quarry West, Wyoming, is named *Tanycolagreus topwilsoni*. The taxon is phyletically closest to *Coelurus fragilis* but differs in the absence of pleurocoels in the anterior dorsal centra and in the moderate-length prezygapophyses on the distal caudals, the straight humeral shaft, and the relative length of the metatarsals to the humerus. It differs from *Ornitholestes hermanni* from nearby Bone Cabin Quarry in the presence of a centrodiapophyseal lamina, more elongate dorsals, shorter prezygapophyses on the distal caudals, and a bowed radius. Small theropods during the Late Jurassic were clearly more diverse than previously realized.

Introduction

Small to medium-size theropods (< ~5 m) from the Upper Jurassic (Oxfordian-Tithonian) Morrison Formation of the western United States are considerably less common and diverse than they are in the Late Cretaceous (Campanian-Maastrichtian). The difference in the number of specimens is multiple orders of magnitude and appears to be real, but research in the area is only preliminary (e.g., Chure et al. 2000;

Foster et al. 2001). The known small to medium-size theropods from the Morrison Formation include *Coelurus fragilis* Marsh 1879, *Ornitholestes hermanni* Osborn 1903, *Elaphrosaurus?* sp., *Stokesosaurus clevelandi* Madsen 1974, *Marshosaurus bicentesimus* Madsen 1976, *Koparion douglassi* Chure 1994, and the new taxon named below as *Tanycolagreus topwilsoni*. Of these taxa, *Coelurus, Ornitholestes,* and *Tanycolagreus* are known from partial skeletons, the rest from isolated bones or teeth.

Coelurus fragilis was named by O. C. Marsh (1879) for material recovered from Reed's Quarry 13 at Como Bluff, Wyoming. The locality was an extensive bone bed known for its numerous specimens of *Stegosaurus* and *Camptosaurus* (Gilmore 1909, 1914). In describing the specimen, Marsh emphasized the extreme hollowness of the vertebrae and illustrated three of them in 1881 (repeated again in Marsh 1896). A second species of *Coelurus, C. agilis,* was named a few years later by Marsh (1884) for a partial skeleton, of which only a pair of pubes was described and illustrated (repeated in Marsh 1896). Ostrom (1980) has concluded that *C. fragilis* and *C. agilis* are different parts of the same individual, a conclusion we agree with as well. A third species of *Coelurus, C. gracilis,* was erected by Marsh in 1888 for metapodials from the Lower Cretaceous Potomac Formation of Maryland. However, the taxon is considered a nomen dubium (Ostrom 1980); it is certainly not a species of *Coelurus*.

Ornitholestes hermanni was named by H. F. Osborn (1903) for a partial skeleton from Bone Cabin Quarry, north of Como Bluff. The site is also a multi-taxa bone bed, although little of the material from this quarry has been described. Osborn (1903) figured some of the *Ornitholestes* material and offered a skeletal reconstruction. The skull was figured at a later date, along with a slightly modified rendition of the skeleton (Osborn 1916). Gilmore (1920) synonymized *Ornitholestes hermanni* with *Coelurus fragilis,* arguing that the differences in the vertebrae were minor. However, as was shown by Ostrom (1980), numerous characters separate the two taxa (Carpenter et al., Chapter 3). More recently, Western Paleontological Laboratories, Inc., of Lehi, Utah, renewed excavation a few tens of meters west of Bone Cabin Quarry, calling their site Bone Cabin Quarry West (BCQ West; Miles and Hamblin 1999). Among the numerous specimens found was a partial theropod skeleton collected in 1995 (Fig. 2.1), originally identified as *Coelurus fragilis* (Miles et al. 1998). However, a comparison with the holotype material of *C. fragilis* demonstrated that the BCQ West specimen was of a new taxon, described below as *Tanycolagreus topwilsoni*.

Other small theropods from the Morrison are based on a few isolated bones. *Stokesosaurus clevelandi* was named by Madsen (1974) for an ilium from the Cleveland-Lloyd Quarry. The ilium is characterized by a prominent vertical ridge on the lateral surface just dorsal to the acetabulum. Another ilium and a premaxilla were also referred to this taxon; however, as will be shown below, the premaxilla is referable to *Tanycolagreus*. *Marshosaurus bicentesimus* was also named by

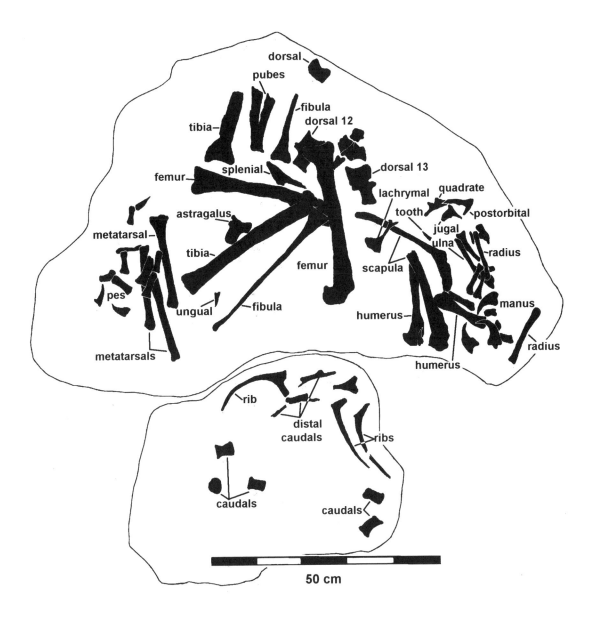

Figure 2.1. Distribution of the holotype skeleton of Tanycolagreus topwilsoni as uncovered in the field jackets (outline).

Madsen (1976) for an ilium from the Cleveland-Lloyd Quarry; a premaxilla, several maxillae, ilia, pubes, and ischia were also referred to this taxon. An *Elaphrosaurus* sp. was identified in the Morrison Formation by Peter Galton (1982) from a humerus collected at Felch Quarry 1 and from a tibia from the Small Quarry described by Dan Chure (2001). *Elaphrosaurus* was originally described from a partial skeleton from the Tendaguru Formation of Tanzania (Janensch 1925). Russell (1972) suggested that the pubes of *Coelurus agilis* were very similar to those of *Elaphrosaurus*, and later referred to the species as *Elaphrosaurus agilis* (Russell et al. 1980). Russell was apparently not aware that the holotype of *C. agilis* consisted of a partial skeleton, parts of which make the synonymy untenable (Galton 1982).

Chure (1994) named a tooth from Dinosaur National Monument *Koparion douglassi* and considered it the oldest troodontid (although its status as such has yet to be verified). Nothing of the rest of the skeleton is known. Another small theropod is known only from cervical vertebrae from Quarry 9; Ostrom (1980) listed it as *Coelurus*, but Makovicky (1997) demonstrated that it belongs to a different, unnamed taxon, a point with which we agree. Additional specimens of this unnamed taxon probably include the vertebrae from Quarry 9 listed by Gilmore (1920) as *Coelurus agilis*, as well as two vertebrae from the Small Quarry at Cañon City (Carpenter, unpublished).

Numerous isolated, small to medium theropod bones occur in most collections from the Morrison Formation (Carpenter, notes), suggesting that these theropods were more diverse than previously recognized. The main collections are held at the Earth Science Museum of Brigham Young University, the Denver Museum of Natural History, the Yale Peabody Museum of Natural History, and the National Museum of Natural History. Although some of the specimens are of juveniles of known large adults, others cannot be referred to any of the recognized small to medium-size theropods.

Institutional Abbreviations. AMNH, American Museum of Natural History, New York; TPII, Thanksgiving Point Institute, Inc. (North American Museum of Ancient Life), Lehi, Utah; UMNH, Utah Museum of Natural History, Salt Lake City, Utah; YPM, Yale Peabody Museum of Natural History, New Haven, Conn.

Systematic Paleontology
Order Theropoda
Maniraptora
Coeluridae
Tanycolagreus n. g.

Etymology. From Greek *tany-* "long, stretched out" + Greek *kolon* "limb" + Greek *agreus* "hunter."

Diagnosis. As for the species.

Holotype. TPII 2000-09-29, partial skeleton including left premaxilla, premaxillary tooth, left partial nasal, left lachrymal, left postorbital, left quadratojugal, right quadrate, left squamosal fragment, right splenial, left articular, two cheek teeth, two anterior dorsal centra, four complete posterior-most dorsals, first sacral centrum, two anterior caudal centra, two mid-caudal centra, three distal caudals, fourteen ribs, seven chevrons, gastralia fragments, left and right scapula and coracoid, left and right humeri, left and right ulnae, left and right radii, left capalia and ulnare, right and left mani, distal half of paired pubes, left and right femur, left and right tibiae, proximal half of left fibula, right fibula, right astragalus and calcaneum, right metatarsals I–V, complete right foot. Measurements of select bones are given in Figures 2.2 and 2.3. A cast of this specimen is also housed at the Denver Museum of Nature and Science.

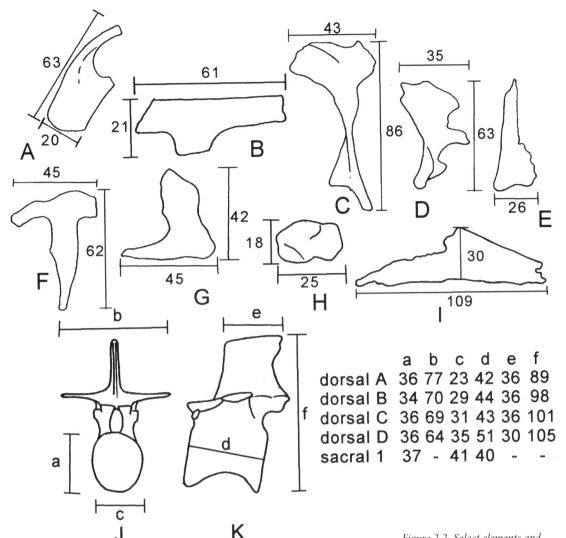

	a	b	c	d	e	f
dorsal A	36	77	23	42	36	89
dorsal B	34	70	29	44	36	98
dorsal C	36	69	31	43	36	101
dorsal D	36	64	35	51	30	105
sacral 1	37	-	41	40	-	-

Figure 2.2. Select elements and their measurements (in mm) for the holotype Tanycolagreus topwilsoni *(TPII 2000-09-29). Compare with Figures 2.4 and 2.7. (A) Premaxilla, (B) nasal, (C) lachrymal, (D, E) quadrate in lateral and posterior views, (F) postorbital, (G) quadratojugal, (H) articular in dorsal view, (I) splenial, and (J, K) dorsal in anterior and posterior views (measurements of best vertebrae given in the table). Not to scale.*

Tanycolagreus topwilsoni n. sp.
Ornitholestes hermanni (in part)
Stokesosaurus clevelandi (in part)

Etymology. Named for George "Top" Wilson, retired, United States Marine Corps.

Diagnosis. Medium-size tetanuran having short, deep-bodied premaxilla pierced by narial foramen at base of nasal process, orbital process on postorbital, T-shaped quadratojugal, centrodiapophyseal lamina on dorsals. Differs from *Coelurus* in the absence of pleurocoel on anterior dorsals; posterior caudal prezygapophyses elongated to one-third centrum length, rather than short; straight, rather than sigmoidal, humeral shaft; bowed, rather than straight, radius; flat-bottomed rather than arced pubic foot; straight rather than sigmoidal

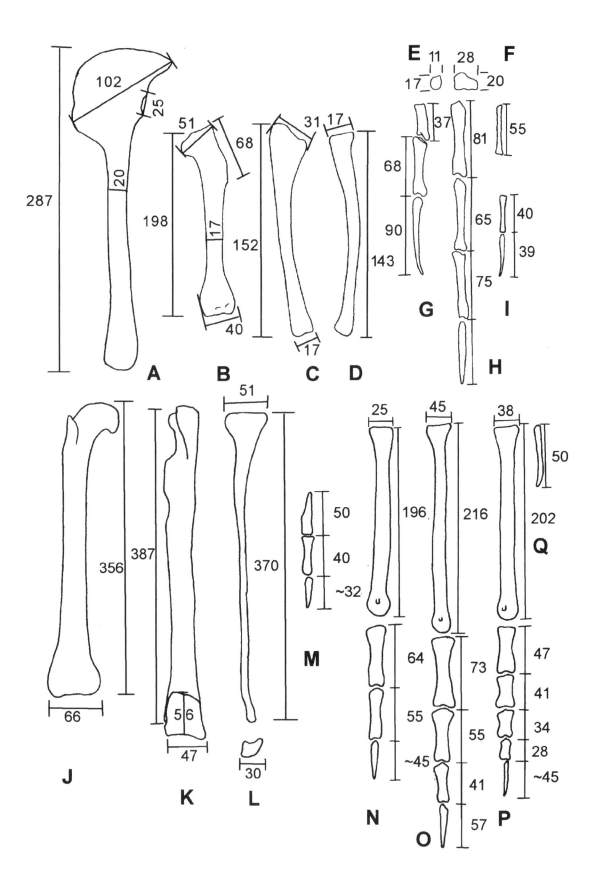

femoral shaft; metatarsal length subequal to humeral length, rather than 1.75 times humeral length. Differs from *Ornitholestes* in straight anterior margin of premaxilla, rather than rounded; T-shaped rather than L-shaped quadratojugal; elongate neural spine that overhangs centrum, rather than short neural spine; posterior caudal prezygapophyses only one-third centrum length, rather than one-half centrum length; bowed, slender radius, rather than straight, robust radius.

Type Locality. Morrison Formation, Bone Cabin Quarry West, Albany County, Wyoming.

Paratypes. AMNH 587, partial manus previously referred to *Ornitholestes hermanni* from Bone Cabin Quarry, Albany County, Wyoming. UMNH 7821 (formerly UUVP 2999), premaxilla previously referred to *Stokesosaurus clevelandi* from the Cleveland-Lloyd Quarry, Emery County, Utah.

Description

Cranial

The skull of *Tanycolagreus* is known only from a few elements: left premaxilla, premaxillary tooth, left nasal, left lachrymal, left postorbital, left quadratojugal, right quadrate, left squamosal fragment, left articular, right splenial, and two cheek teeth (Fig. 2.4). Unfortunately, nothing of the skull roof or braincase has been found. The tentative reconstruction of the skull presented in Figure 2.5 is long and low as is typical of small to medium-size theropods (i.e., non-"carnosaurian"). Comparisons to the skull of *Ornitholestes* are based on a cast of the skull (YPM 53262). A detailed description of the *Ornitholestes* skull is in development by Mark Norell (AMNH) following additional preparation of this specimen.

Premaxilla. The premaxilla is very deep and very short anteroposteriorly (Fig. 2.4A–C) and is similar to that referred to as "*Stokesosaurus clevelandi*" by Madsen (1974) (see Fig. 2.6A–C). Both are anteroposteriorly short and deep, although these need not imply an extremely short-snouted skull as envisioned by Madsen (1974) (see Fig. 2.5). There are four subrectangular alveoli, and the one damaged premaxillary tooth present (Fig. 2.4D) has the asymmetrical cross-section typical of theropods. On the external surface, the narial fossa extends from the external nares onto the lateral surface and is pierced by the narial foramen at the base of the nasal process (Figs. 2.4A, 2.6A,D). The placement of such a foramen is unusual in the theropods. Just dorsal to the dental margin is a series of shallow vertical grooves for the blood vessels and facial nerves to the now missing soft tissue. The narial process is thin and short, and it gently curves dorsoposteriorly. Along the posterior margin of the premaxilla, the maxillary suture extends as a shallow vertical groove that is pierced by a deep fossa (Fig. 2.6B,E). The sutural groove extends more dorsally than as described by Madsen in "*Stokesosaurus clevelandi*" because the groove is damaged in the latter specimen (Fig. 2.4B). On the medial side (Figs. 2.4C, 2.6C,F), the sutural surface for the right premaxilla is a smooth,

(Opposite page)
Figure 2.3. Select elements and their measurements (in mm) for the holotype Tanycolagreus topwilsoni *(TPII 2000-09-29). (A) Scapulocoracoid, (B) humerus, (C) ulna, (D) radius, (E) semilunate, (F) distal carpal, (G) manus digit I (H) manus digit II, (I) manus digit III, (J) femur, (K) tibia and astragalus, (L) fibula and calcaneum, (M) pes digit I, (N) pes digit II, (O) pes digit III, (P) pes digit IV, and (Q) metacarpal V. Not to scale.*

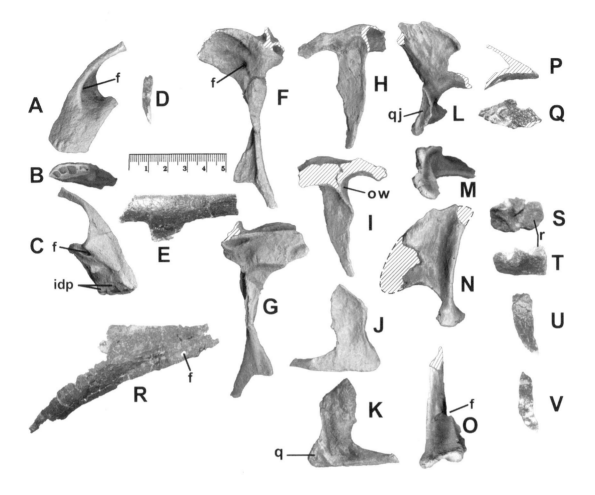

Figure 2.4. Cranial elements of the holotype Tanycolagreus topwilsoni (TPII 2000-09-29). Left premaxilla in lateral (A), occlussal, showing the alveoli (B), and medial (C) views. Premaxillary tooth in lateral view (D). Partial nasal in dorsal view (E). Left lacrimal in lateral (F) and medial (G) views. Left postorbital in lateral (H) and medial (I) views. Left quadratojugal in lateral (J) and medial (K) views. Right quadrate in lateral (L), ventral (M), medial (N), and posterior (O) views. Left squamosal fragment in dorsal (P) and lateral (Q) views. Right splenial (R). Left articular in dorsal (S) and lateral (T) views.

(continued on page 31)

inverted triangle, the widest part being at the level of the external nares. The sutural surface extends to the level of the tooth row, indicating that a subnarial gap was not present. The angle formed by the surface of the symphyseal and the lateral surface is very low, indicating that united, the two premaxillae would form a V shape in ventral view. A large foramen, which Madsen (1974) has referred to as the foramen for the facial nerve, pierces the body of the premaxilla medially below the external nares (Fig. 2.6C,F). Ventral to the foramen is the short, triangular maxillary process, or ala, to which the medial premaxillary process of the maxilla articulates. This ala is unlike the well-developed prong or strap seen in large theropods, such as *Allosaurus* or *Sinraptor* (Currie and Zhao 1993). The interdental plates of both specimens of *Tanycolagreus* are slightly damaged, but enough remains to show that they are fused to each other.

Nasal. Most of the left nasal is preserved in the holotype, although it is damaged at each end (Fig. 2.4E). The element is slightly curved along its length. The medial edge is flat for contact with the other nasal. The descending, or ventral, process, which forms the posterior margin

of the external nares, is incomplete; thus it is not known whether the process contacted the premaxilla ventral to the external nares.

Lachrymal. The left lachrymal is nearly complete, except for the tip of the horizontal ramus and a small portion of the posterodorsal suture to the prefrontal (Fig. 2.4F,G). The horizontal ramus is deep, and the dorsal margin is arcuate, with a ridge extending its length (seen best on the medial side, Fig. 2.4G); there is no lachrymal horn. Unfortunately, the length of the ramus is unknown, although about half of the anterior portion is believed to be missing. The dorsal surface is relatively wide, indicating that the lachrymal was broadly exposed on the skull roof. The lateral surface of the horizontal ramus contains a large oval depression, the lachrymal fossa, which contains the smaller lachrymal fenestra (Fig. 2.4F). A short ridge occupies the middle of the fossa, effectively dividing the fossa horizontally. The ridge connects the floor of the fossa with the posterior rim of the fossa. The lachrymal fenestra pierces the lachrymal just posteroventral to this ridge. On the medial side, the horizontal ramus has sutures along the posterior portion for the prefrontal, and along the anterodorsal portion for the nasal (Fig. 2.4G). The prefrontal suture extends upward on a short spur of damaged bone. A slight notch along the ventral margin of the horizontal ramus coincides with a very shallow depression on the medial side. These features probably denote the suture for the posterodorsal-most portion of the dorsal ramus of the maxilla (Fig. 2.4F,G; compare with Fig. 2.5). The ventral, or vertical, ramus of the lachrymal is constricted in the

(continued from page 30)
Maxillary teeth (U, V). Scale in mm. Abbreviations:
(A) f—foramen at the base of the ascending or nasal process;
(C) f—large foramen, idp— interdental plates; (F) f—lacrimal foramen; (I) ow—orbital wall;
(K) q—suture for the quadrate;
(L) qj—suture for the quadratojugal; (O) f—part of the quadrate foramen; (R)—small splenial foramen;
(S, T) r—retroarticular process.

(Above)
Figure 2.5. Restoration of the skull of Tanycolagreus. *Parts of the squamosal, articular, and splenial hidden by other (missing) bones are shown faintly. Scale in mm.*

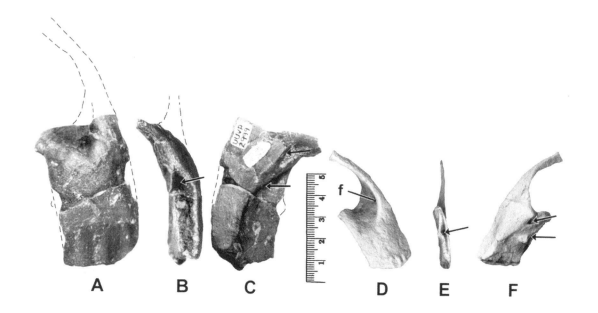

A   B   C   D   E   F

Figure 2.6. Comparison of the premaxillae of Tanycolagreus topwilsoni. *(A–C) specimen referred to "Stokesosaurus clevelandi" (UMNH 7821) with that of the holotype of* Tanycolagreus topwilsoni *(TPII 2000-09-29) (reversed). Lateral (A, D), posterior (B, E), and medial (C, F) views. In the posterior views (B, E), note the deep fossa within the sutural surface for the maxilla (arrow). Note the paired foramina (arrows) on the medial surface (F). The medial ala, which receives the medial premaxillary process from the maxilla, is damaged in UMNH 7821. Scale in mm. Abbreviation: (D) f—foramen.*

middle, giving it an hourglass shape; it does not have an orbital projection. The ramus is pierced anteroposteriorly by a foramen at its narrowest point, thus connecting the antorbital fenestra with the orbit. The lachrymal is thickest along the orbital border.

Postorbital. The postorbital is a T-shaped bone and has a shallow fossa on the external surface (Fig. 2.4H) that may be due to crushing. The horizontal process is incomplete posteriorly and on its medial side, whereas the vertical process is complete (Fig. 2.4I). The horizontal process apparently excluded all except perhaps a small portion of the frontals from the orbital rim. The external surface of the process is smooth, lacking any prominences. Medially, there is a small posterior orbital wall at the junction between the horizontal and vertical processes. The vertical process has a slight orbital projection that would form a slight constriction of the orbit (Fig. 2.5). This feature has previously been suggested as characteristic of large-headed theropods by Chure (1998). The vertical process is longer anteroposteriorly than it is wide. Distally, the ventral process has an oblique suture for the jugal.

Squamosal. Most of the squamosal is damaged (Fig. 2.4P,Q), but enough remains to show that in dorsal view it tapered into a V shape posteriorly, rather than a broad U shape as in *Allosaurus* and other large theropods.

Quadratojugal. The left quadratojugal is complete, L-shaped, and as tall as it is long (Fig. 2.4J,K); it is not fused to the quadrate. The horizontal jugal process is vertically narrow and projects nearly straight. In contrast, the horizontal squamosal process is slightly arcuate posteriorly, and it slightly widens dorsally. The sutural surface for the squamosal is broad and sloped posteriorly at about 50° relative to the vertical axis. The quadratojugal forms the lateral wall of the quadrate foramen.

Quadrate. The right quadrate is nearly complete, lacking only a portion of the pterygoid process and squamosal condyle (Fig. 2.4L–O); the quadrate is not pneumatic. It is slightly arced posteriorly to accommodate the tympanic membrane. On the lateral surface, the facet for the quadratojugal is very prominent and is slightly expanded anteriorly to provide an immoveable suture. The dorsal rim delineates the ventral border of the quadrate foramen (Fig. 2.4O). The foramen is relatively large and walled laterally by the quadratojugal. The pterygoid process apparently arced ventrally, as indicated by the preserved portions (Fig. 2.4N). The posterior face of the quadrate (Fig. 2.4O) is triangular, with a notch for the quadrate foramen located about mid-height on the lateral margin. Distally, the medial mandibular condyle is expanded medially and posteriorly (Fig. 2.4O,N). The medial and lateral condylar surfaces are slightly damaged, the medial one more so. Nevertheless, it is clear that they have an hourglass shape in ventral view (Fig. 2.4M). The two condyles are separated by a slight, oblique groove as in other theropods.

Splenial. The splenial is slightly damaged, lacking most of the dorsoposterior process (Fig. 2.4R). Overall, it is shaped like an isosceles triangle. A small foramen pierces the splenial near its anteroventral border.

Articular. The articular is a small, rectangular block in dorsal view (Fig. 2.4S). The facets for the quadrate condyles are separated from each other by a low, oblique ridge, which accommodates the groove between the lateral and medial condyles of the quadrate. The retroarticular process is short and vertical (Fig. 2.4T).

Cheek Teeth. Two crushed maxillary or dentary teeth are present (Fig. 2.4U,V) but are so damaged that not even the serrations are preserved.

Vertebrae

The vertebrae consist of two anterior dorsal centra, four complete posterior-most dorsals, the first sacral centrum, two anterior caudal centra, two mid-caudal centra, and three distal caudals (Fig. 2.7).

Dorsals. The anterior dorsals consist only of the centra, and the neural arches are not fused, indicating that the individual was not fully mature. One of the centra is a parallelogram, with the posterior articular surface lower than the anterior. All of the centra have slightly amphicoelous articular surfaces. The centra are also strongly constricted in the middle, so as to have an hourglass shape in ventral view. In cross-section, the centra are taller than they are wide. The lateral surfaces do not have a pleurocoel; instead they have a shallow, anteroposteriorly elongated fossa, termed pleurofossa ("side fossa") by Carpenter and Tidwell (in press). A representative complete dorsal is presented in Figure 2.7A–D. The suture between the neural arch and centra is visible on most of the dorsals, or else is visible as a faint line. Ventrally, the centra have short parallel ridges near their ends, produced by the ventral horizontal ligament. The capitular facets for the ribs are located just ventrolateral to the prezygapophyses and are connected to the transverse process by a thin paradiapophyseal lamina.

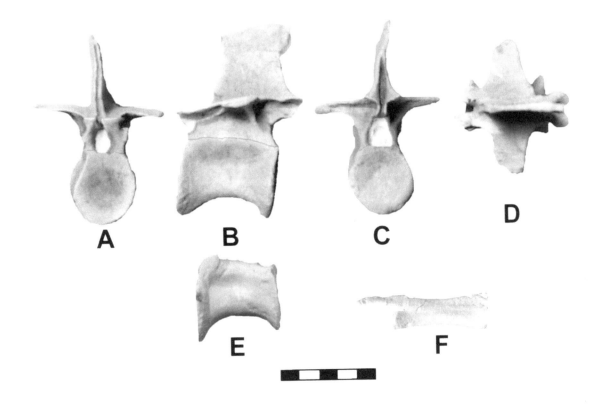

A   B   C   D

E   F

Figure 2.7. Vertebrae of Tanycolagreus topwilsoni *(TPII 2000-09-29) Representative posterior dorsal in anterior (A), left lateral (B), posterior (C), and dorsal (D) views. First sacral centrum (E) and distal caudal (F). Scale in cm.*

This lamina is almost non-existent on the anterior-most dorsal (9? 10?) and is most prominently developed on the last dorsal (13? 14?). The neural spines are about the same height as the centra and are simple vertical blades. The groove for the interspinous ligaments extends almost to the top of the spines. The transverse processes vary considerably. The anterior-most complete dorsal (9? 10?) has triangular, posteriorly directed processes that are angled dorsally about 10°. The other dorsals all have laterally directed, horizontal processes that are sub-rectangular in dorsal view. A well-developed centrodiapophysis is present, connecting the transverse process with the centrum. The postzygapophyses overhang the posterior end of the centra. Well-developed hyposphene-hypantrum accessory articulations are present.

Sacral. The sacral (Fig. 2.7E) is slightly procoelus, with the anterior surface concave and the center of the posterior surface very slightly convex. The height of the posterior articular surface is about three-quarters the height of the anterior surface as measured from the neural canal. The floor of the neural canal is hourglass-shaped and has a deep trough. Like the dorsals, the centrum has a pleurofossa, is hourglass-shaped in ventral view, and has scars for the ventral horizontal ligament.

Caudals. The anterior and mid-caudal centra are constricted laterally, so are hourglass-shaped in ventral view. They also have pleurofossa, although these are not as prominent as those on the dorsals or sacral; ventral horizontal ligament scars are also present. The posterior caudals are elongated, being about 140 percent longer than the anterior

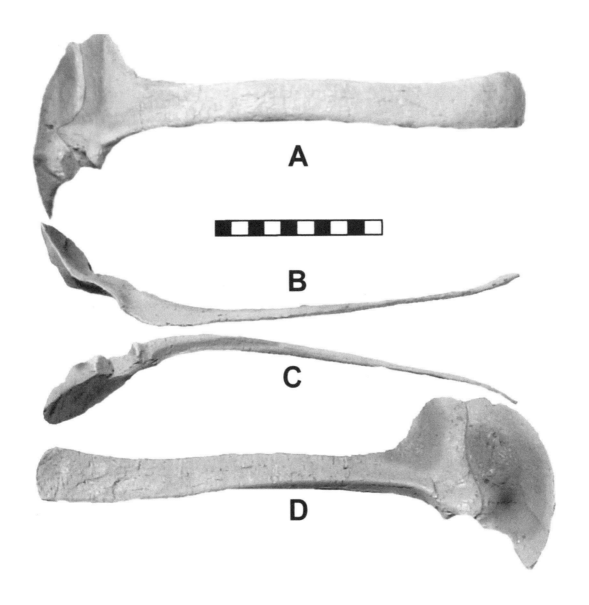

centra. The centra are laterally compressed and have a pair of parallel ridges connecting the anterior and posterior chevron facets. Although the prezygapophyses are long, they do not extend beyond half a centrum length.

Ribs (not illustrated). Fourteen complete and partial ribs are present. They are long and slender but otherwise lack any distinctive features.

Chevrons (not illustrated). Seven complete or nearly complete anterior and mid-caudal chevrons are present. They are long, slender blades, but they lack any unique features.

Gastralia (not illustrated). Several rod-like fragments are thought to be those of gastralia. Unfortunately, they are too incomplete to be diagnostic. A furculum has not been recognized among the fragments.

Figure 2.8. Left scapulocoracoid of Tanycolagreus topwilsoni *(TPII 2000-09-29) in lateral (A), dorsal (B), ventral (C), and medial (D) views. Sutural contact between the scapula and coracoid has been digitally enhanced. Scale in cm.*

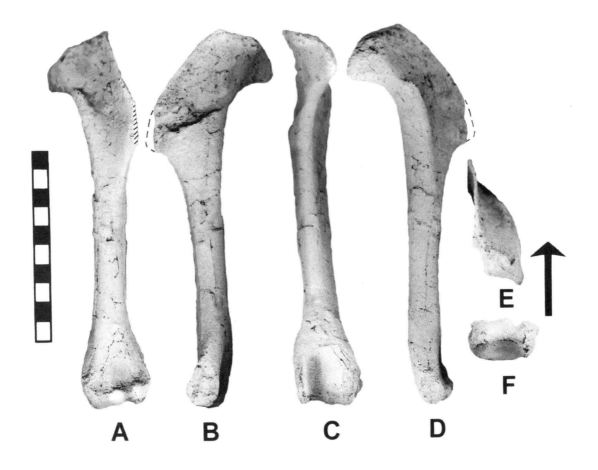

A B C D E F

Figure 2.9. Left humerus of
Tanycolagreus topwilsoni (TPII
2000-09-29) in anterior
(A), lateral (B), posterior
(C), medial (D), proximal
(E), and distal (F) views. Arrow
indicates anterior for E and F.
Scale in cm.

Forelimb

Scapula (Fig. 2.8). The scapula is long, and strap-like, with a length almost ten times the mid-shaft vertical height; the distal end is very slightly expanded. The blade is bowed medially just below the posterior edge of the acromion. The acromion is very large, and its height is almost twice the vertical height of the scapular blade at mid-shaft. Its posterior margin is sloped to the scapular blade. There is a slight notch developed between the dorsal margin of the acromion and the coracoid. The length of the scapular portion of the glenoid is twice the length of the coracoid portion, and the glenoid faces ventroposteriorly.

Coracoid (Fig. 2.8). The coracoid is large and attached to the scapula, although the suture between them is visible. In profile, the coracoid is over twice as tall as it is anteroposteriorly long. The ventral process is long, although only about as long as the glenoid. The coracoid protuberance (wrongly called the biceps tubercle) is well developed and may be an artifact caused by the coracobrachialis and supercoracoideus (Carpenter 2002). This feature is developed opposite a fossa on the medial side (compare Figs. 2.8A and D). In ventral view, the angle between the coracoid and scapula at the glenoid is about 30°.

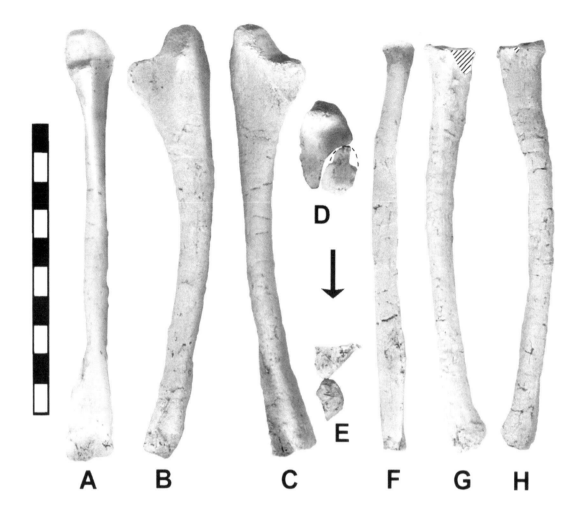

Figure 2.10. Forelimb elements of Tanycolagreus topwilsoni (TPII 2000-09-29). Left ulna in anterior (A), lateral (B), and medial (C) views. Ulna and radius in articulation proximal (D) and distal (E) views (arrow points anteriorly). Left radius in anterior (F), lateral (G), and medial (H) views. Scale in cm.

Humerus (Fig. 2.9). The humerus is straight shafted, although humeral torsion is present, so that the distal end is about 45° relative to the proximal end. The diameters of the proximal and distal ends are over twice the mid-shaft diameter. The humeral head is not well developed, and it is confluent with the medial (=internal) tuberosity and the deltopectoral crest. The medial tuberosity is a triangular process that projects ventromedially, and slightly posteriorly. It is separated from the humeral head by a shallow groove. The deltopectoral crest is expanded anteriorly (Fig. 2.9E) and offset from the shaft in profile (Fig. 2.9D). On the basis of the right humerus, the deltopectoral crest is triangular in profile. The distal condyles are developed on the anterior and ventral surfaces of the shaft. The medial condyle, or entepicondyle, is well developed and is as prominent as the lateral condyle. A shallow ulnar fossa is present dorsal of the lateral condyle, and it extends ventrally toward the notch between the two condyles. On the posterior side, the olecranon fossa is broad and triangular; it is deepest near the lateral margin and shallows medially. Ventrally, it merges with the medial condyle.

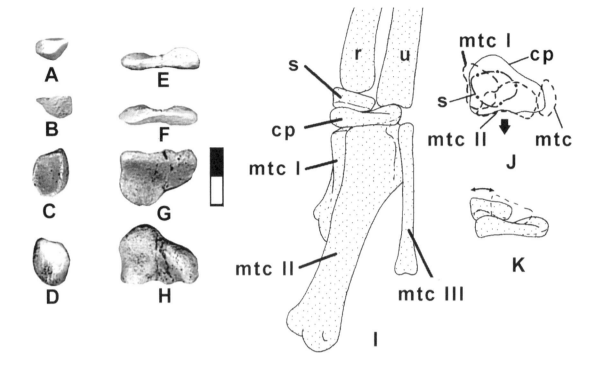

Figure 2.11. Carpal elements of Tanycolagreus topwilsoni *(TPII 2000-09-29). Semilunate in anterior (A), posterior (B), proximal (C), and distal (D) views. Distal carpal block in posterior (E), anterior (F), distal (G), and proximal (H) views. Scale in cm. Restored wrist in anterior (I) and proximal (J) views (arrow points anteriorly). Amount of movement possible for the semilunate carpal on the distal carpal block (K). Abbreviations: cp—carpal block; mtc I–III—metacarpals I–III; r—radius; s—semilunate; u—ulna.*

Ulna (Fig. 2.10A–C). The ulna is bowed posteriorly, with most of the arching occurring proximally. The olecranon is short and rounded, and it grades anteriorly into the humeral, or medial condylar, notch. The radial notch is well developed, transversely expanded, and concave. The diameter of the radial shaft is about the same as that of the radius. The distal end is triangular in ventral view (Fig. 2.10E) and set at about 60° relative to the proximal end. The contact surface for the radius is small.

Radius (Fig. 2.10F–H). The radius is bowed anteriorly. Its proximal end is ellipsoidal, being slightly wider anteriorly than posteriorly (Fig. 2.10D). The distal end is sloped about 50° relative to the axis of the shaft, and its contact with the ulna is small and rounded. The shaft is laterally compressed proximally and is subtriangular at mid-shaft, with the apex on the posterior side.

Carpals (Fig. 2.11). The proximal carpal (radiale) is modified into a boat-shaped semilunate carpal (Fig. 2.11A–D) that has a concave dorsal surface for the radius and an elongated convex ventral surface. In end view, it is wedge-shaped, with the tapered edge palmar. The distal carpals (Fig. 2.11E–H) are fused and are functionally a single carpal. The dorsal side has two low, elongate prominences that act as guides to constrain the direction of movement of the semilunate carpal (Fig. 2.11K). The ventral side is slightly concave, with a lip along the palmar side (Fig. 2.11F). The carpal overlies metacarpals I and II and barely overlies metacarpal III (Fig. 2.11J). Movement at the wrist has been discussed elsewhere (Carpenter 2002).

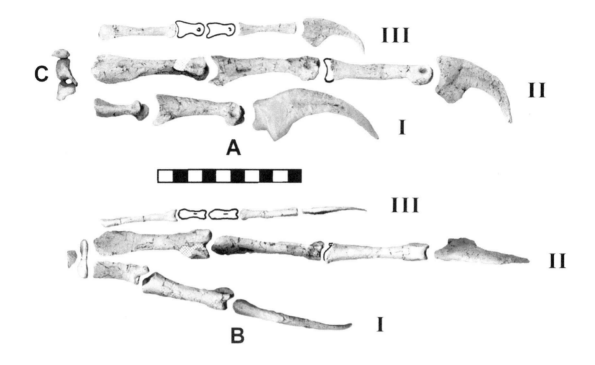

Manus (Fig. 2.12). The manus consists of three digits, all with phalanges and unguals. Digit II is the longest in the manus, followed by digit I and then III. Metacarpal I is slightly less than half as long as metacarpal II. In proximal view it somewhat V-shaped, with a corresponding notch on the proximal end of metacarpal II (Fig. 2.12B). When articulated, the extensor (outer) surfaces of metacarpals I and II form an 80° angle in proximal view (Fig. 2.12B), so that metacarpal I is mostly hidden by metacarpal II (Fig. 2.11I). Distally, metacarpal I is twisted about 35° relative to the extensor surface of the proximal end. The medial and outer lengths are considerably unequal so that the distal condyle is not in the same horizontal plane. The result is that digit I is canted away from digit II. The length of metacarpal II is only 41 percent of humeral length. Metacarpal III is shorter than metacarpal II and has a straight shaft that is significantly narrower than that of metacarpal II. Metacarpal III articulates below the proximal end of metacarpal II, but not on the palmar side. The proximal articulation of metacarpal III is subquadrilateral. The metacarpal-phalangeal joints are hyperextensible (Carpenter 2002), although the extensor pits are not well developed. Digit I extends to the middle of phalanx II-2, which is the longest non-ungual phalanx. The longest ungual is on digit I, and it is five times as long as the height of the articular surface. It is trenchant and rather deep, with an elliptical articular surface. The unguals are moderately curved, and their cross-section at mid-shaft is ovoid, being over twice as deep as it is wide. None of the unguals have a pronounced lip above the articular surface as seen in oviraptorids. On the palmar side, the flexor tubercles of the unguals are well developed and proximally placed. The

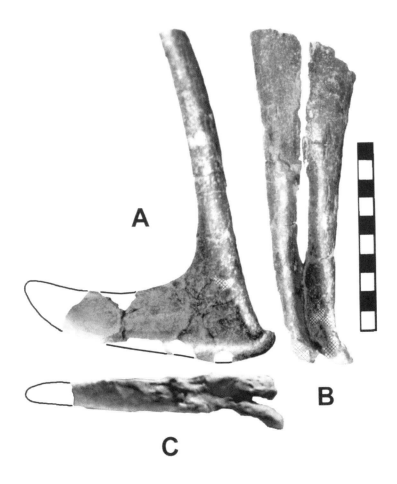

Figure 2.13. Pubis of
Tanycolagreus topwilsoni
(TPII 2000-09-29) in right lateral
(A), anterior (B), and ventral
(C) views. Scale in cm.

ungual grooves divide the unguals equally into palmar and extensor regions.

Previously, Osborn (1916) described a partial manus (AMNH 587) from Bone Cabin Quarry as that of *Ornitholestes hermanni* (it was not found with that skeleton). The specimen was collected about 300 m east of the holotype of *Tanycolagreus topwilsoni* and at approximately the same stratigraphic level. Although we cannot rule out that the manus is that of *Ornitholestes* (the manus is otherwise unknown), we also note that the specimen is indistinguishable from that of *Tanycolagreus*. We therefore consider AMNH 587 as the manus of *Tanycolagreus* until such time as the manus of *Ornitholestes* is known.

Pelvis

Pubis (Fig. 2.13). The pubes are represented by their distal half. In lateral view, the shaft is bowed anteriorly, but it joins the foot at a right angle. The dorsal surface of the foot, where it joins the pubic shaft, is more gently sloped than in *Coelurus*, where the transition is almost abrupt. The pubic foramen is located in the distal half of the pubic shaft and terminates just proximal to the pubic foot (Fig. 2.13B). The foot has a short anterior portion, which is not completely fused (Fig. 2.13C), probably because the individual was not fully mature. The foot is incomplete posteriorly and, in ventral view, is a narrow triangle. The ventral surface of the pubic foot is flat, rather than arcuate or boat-shaped as in *Coelurus* (see Carpenter et al., Chapter 3).

A second pubis from Felch Quarry 1 was previously referred to *Coelurus* by Gilmore (1920), but the ventral surface is flat and the pubic foramen terminates just above the pubic foot as in *Tanycolagreus*, rather than at midway as in *Coelurus* (see Carpenter et al., Chapter 3). We therefore refer this specimen to *Tanycolagreus*.

Hindlimb

Femur (Fig. 2.14A–F). The femoral shaft is bowed anteriorly and is very slightly sigmoidal (Fig. 2.14B). The femoral head is at right angles to the shaft and is transversely elongated. The dorsal surface of the head is flat and extends laterally toward the greater trochanter, whereas in proximal view (Fig. 2.14E), the head tapers laterally to the greater trochanter. We digress briefly to note that the dorsal surface, lateral to the femoral head, has frequently been misidentified as the greater trochanter. Sometimes there may be a slight crest in that region. However, as Gregory (1918, p. 535) has noted, "the greater trochanter . . . remains on the outer side of the shaft more or less near the proximal end. The outer portion of the head itself has sometimes been wrongly called the 'great trochanter.'" He adds in a footnote, "the proximal surface of the so-called great trochanter was covered by bursa, as is the similar smooth surface above the great trochanter in the Ostrich. . . ; and that the gluteal muscles (i.e., M. iliofemoralis externus) were attached on the outer surface of the great trochanter and not upon its top." Thus, the greater trochanter is on the proximal, lateral surface of the femoral shaft, where this region is frequently marked by faint vertical rugosities or ridges that mark the insertion of the M. ilio-trochantericus and M. iliofemoralis externus. The crest that is some-times developed on the dorsal surface is the trochanteric crest (i.e., crista trochanteris), and in extant birds, it serves to restrict abduction of the femur (see further discussion in Carpenter and Kirkland 1998, p. 249–250).

The anterior trochanter (incorrectly called the lesser trochanter by some; see Carpenter and Kirkland 1998, p. 250), is separated from the proximal end of the femoral shaft by a cleft. Furthermore, the tro-chanter projects anteriorly as an alariform structure, and its proximal-most point is below the femoral head (Fig. 2.14A,B). On the femoral shaft, the fourth trochanter is little more than a slight, irregular ridge located proximally, beginning opposite the distal end of the anterior trochanter (Fig. 2.14B). Distally on the anterior surface, the origin scar for the tibial extensor, the M. femorotibialis, is elliptical shallow, and bounded medially by the medial distal crest; no extensor groove is present for the tendon of the M. femoro-tibialis, suggesting that the tibial aponeurosis extended dorsally onto the distal end of the femur. On the medial side of the distal end, there is no medial epicondyle, and on the lateral side, there is a groove in the lateral condyle. On the posterior side, the vertical fibular crest (crista tibiofibularis) is short and

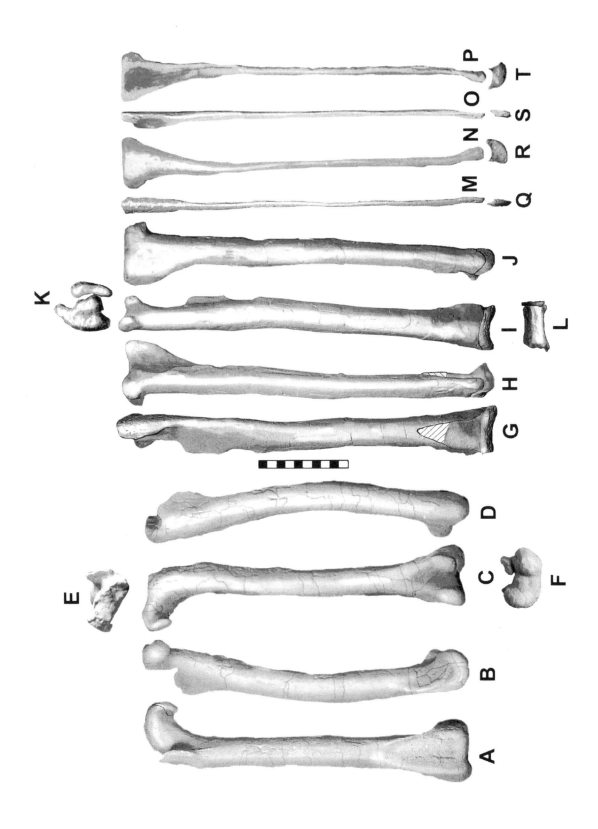

is separate from the lateral condyle; lateral to it is a shallow groove. The poptileal fossa is moderately well developed.

Tibia (Fig. 2.14G–K). The tibial shaft is long and straight. The cnemial crest is short and projects anterolaterally. The proximal fibular crest is subrectangular and almost as long as the cnemial crest (Fig. 2.14G). In proximal view, the lateral and medial condyles are subequal in size, and the lateral condyle has shifted anteriorly to a lateral position (Fig. 2.14K). There is a conspicuous waist between this condyle and the main body of the tibia. The notch (incisura tibialis) between the cnemial crest and lateral condyle is small relative to the overall size of the femoral platform. Laterally, the fibular crest is low on the shaft and subrectangular, but it is otherwise well developed (Fig. 2.14G). The distal end of the tibia is only slightly expanded relative to the shaft width (Fig. 2.14G), but it does back the calcaneum.

Fibula (Fig. 2.14M–P). The fibula is long and slender and is broadly separate from the tibia. The length of the expanded proximal end is almost 75 percent of the anteroposterior length of the femoral platform. Medially, the proximal end is slightly concave, with a sulcus extending ventrally. A slight protuberance is present on the anterior surface of the shaft. Distally, the fibula slightly overlaps the ascending process of the astragalus. The distal terminus is not very expanded.

Tarsals (Fig. 2.14G,L,Q–T). The astragalus is anteroposteriorly short, and the distal condyles are more developed ventrally than anteriorly; a transverse groove extends across the anterior face of the astragalus, just proximal to the condyles. The shallow fossa present at the base of the ascending process is not very prominent. The ascending process is not complete, although how much is missing is unknown. The calcaneum is not fused to the astragalus, and is reduced to a small, narrow disk. Medially, the astragalar tuberosity is small.

Pes (Fig. 2.15). The metatarsals are elongate, with the length of metatarsal III being about 60 percent of femur length. The mid-shaft cross-sections of metatarsals II–IV are as wide as they are long. Metatarsal I is reduced but still retains phalanges; it is placed posteromedially near the mid-shaft of metatarsal II. Metatarsal III is not reduced proximally, and it separates metatarsals II and IV. Metatarsals II and IV are subequal in length, and both are only a little shorter than metatarsal III. In proximal view, metatarsal III is hourglass-shaped. Metatarsal V is reduced to a slender, laterally compressed structure.

The combined length of the phalanges on pedal digit I is greater than the length of the first phalanx of pedal digit III. Pedal digits II and IV are subequal in length, and the first phalanx of digit II allows some hyperextension of phalanx 2. In cross-section, the pedal unguals are similar in overall shape, laterally compressed, and subtriangular. Ungual II is subequal to ungual IV.

Discussion

The holotype of *Tanycolagreus* is of a submature individual, about 3.3 m long, whereas the premaxilla from the Cleveland-Lloyd Quarry

Figure 2.14. Right hindlimb elements of Tanycolagreus topwilsoni *(TPII 2000-09-29). Femur in anterior (A), medial (B), posterior (C), lateral (D), proximal (E), and distal (F) views. Tibia and astragalus in anterior (G), lateral (H), posterior (I), and medial (J) views. Tibia and fibula in articulation in proximal view (K). Astragalus and calcaneum in distal view (L). Fibula in anterior (M), lateral (N), posterior (O), and medial (P) views. Calcaneum in anterior (Q), lateral (R), posterior (S), and medial (T) views. Scale in cm.*

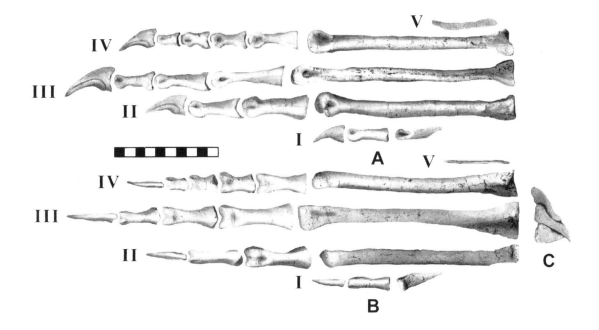

Figure 2.15. Pes of Tanycolagreus topwilsoni *(TPII 2000-09-29) showing digits I–V in medial (A) and extensor (B) views. Proximal view of metatarsals (C). Scale in cm.*

suggests an individual about 4 m long (Fig. 2.16). Whether the Cleveland-Lloyd specimen represents a fully mature individual is not known. When initially discovered, the holotype skeleton of *Tanycolagreus* was identified as that of *Coelurus fragilis* (Miles et al. 1998). However, direct comparison with the holotype makes such an identification untenable; nor can it be referred to *Ornitholestes* (Table 2.1). Comparison with another medium-size theropod from the Morrison Formation, *Stokesosaurus clevelandi,* is not possible, because the ilium is unknown for *Tanycolagreus,* and the referred premaxilla is shown above to be that of *Tanycolagreus.* The referred premaxilla of *Marshosaurus,* another medium-size Morrison theropod, differs from that of *Tanycolagreus* in being long and low, rather than short and deep; and the referred pubis lacks the short "toe" at the anterior part of the pubic boot.

Of all the known theropods from the Morrison Formation, *Tanycolagreus* is most similar to *Coelurus,* although the former is more primitive in a number of points, including the absence of pleurocoels on the anterior dorsals, the relatively long prezygapophyses on the caudals, and the straight humeral shaft. Its identification has increased the diversity of theropods in the Morrison and demonstrates that small to medium-size theropods were considerably more diverse than previously thought.

Acknowledgments. We thank the following people for access to specimens in their care: Dan Chure (Dinosaur National Monument), Jacques Gauthier and Lyndon Murray (Yale Peabody Museum), Charlotte Holton and Mark Norell (American Museum of Natural History), Robert Purdy (National Museum of Natural History), and Kenneth

TABLE 2.1.

Comparison between *Tanycolagreus, Coelurus,* and *Ornitholestes*

	Tanycolagreus	*Coelurus*	*Ornitholestes*
Anterior edge of premaxilla	Straight	Unknown	Rounded
Premaxillary body	Deep	Unknown	Shallow
Jugal process of lachrymal	Angled sharply downward	Unknown	Horizontal
Orbital projection on postorbital	Present	Unknown	Absent
Posterior process of quadratojugal	Moderately developed	Unknown	Absent
Posterior dorsal neural spine/centrum length	>75%	>75%	<75%
Dorsal neural spine	Long, overhangs centrum posteriorly	Long, overhangs centrum posteriorly	Short, does not overhang centrum posteriorly
Centrodiapophyseal lamina	Present	Present, with two very deep fossa on each side	Absent
Pleurocoel in anterior dorsal centra	Absent	Present	Absent
Dorsal centrum height/length	>75%	<75%	<75%
Posterior caudal prezygapophysesnot	Moderately long, extending to about 1/3 preceding centrum length	Short, not extending much beyond end of centrum	Long, to about half preceding centrum
Lateral side of distal caudal centrum	Smooth	Smooth	Ridge
Humeral shaft in lateral view	Straight	Strongly sigmoidal	Straight
Proximal internal tuberosity	Sharp corner, angled down below level of humeral head	Sharp corner, nearly level with humeral head	Rounded, lower than humeral head
Ulna	Strongly bowed posteriorly	Strongly bowed posteriorly	Unknown
Radius	Bowed anteriorly, slender	Straight, slender	Straight very robust for length
Semilunate carpal	Present	Present	Unknown
Pubic foot	Flat bottomed	Arced bottom (boat shape)	Unknown
Femur shaft in anterior view	Straight	Sigmoidal	Straight
Medial end of femoral head	Level with greater trochanter	Below greater trochanter (angled ventromedially)	Unknown
Metatarsal IV/humeral length	1.01	1.72	0.91

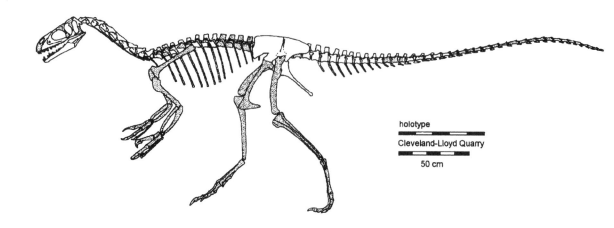

holotype

Cleveland-Lloyd Quarry

50 cm

Figure 2.16. Skeletal reconstruction of Tanycolagreus topwilsoni. Shaded bones are known. Scale = 50 cm.

Stadtman (Brigham Young University Earth Science Museum). Thanks to Ben Creisler for the moniker *Tanycolagreus*. Finally, we thank an anonymous benefactor for donating the holotype skeleton, which is now on display at the North American Museum of Ancient Life.

References Cited

Carpenter, K. 2002. Forelimb biomechanics of nonavian theropod dinosaurs in predation. *Senckenbergiana Lethaea* 82: 59–76.

Carpenter, K., and J. I. Kirkland. 1998. Review of Lower and Middle Cretaceous ankylosaurs from North America. In S. G. Lucas, J. I. Kirkland, and J. W. Estep (eds.), *Lower and Middle Cretaceous Terrestrial Ecosystems*, pp. 249–270. New Mexico Museum of Natural History and Science Bulletin, no. 14. Albuquerque: New Mexico Museum of Natural History and Science.

Carpenter, K., and V. Tidwell. In press. Reassessment of the Early Cretaceous sauropod *Astrodon johnsoni* Leidy 1865 (Titanosauriformes). In V. Tidwell and K. Carpenter (eds.), *The Thunder-Lizards*. Bloomington: Indiana University Press.

Chure, D. J. 1994. *Koparion douglassi*, a new dinosaur from the Morrison Formation (Upper Jurassic) of Dinosaur National Monument: The oldest troodontid (Theropoda: Maniraptora). *Brigham Young University Geology Studies* 40: 11–15.

Chure, D. J. 1998. On the orbit of the theropod dinosaurs. In B. P. Pérez-Moreno, T. Holtz Jr., J. L. Sanz, and J. Moratalla (eds.), *Gaia: Aspects of Theropod Biology*, vol. 15, pp. 233–240. Lisbon: Museu Nacional de História Natural.

Chure, D. J. 2001. The second record of the African theropod *Elaphrosaurus* (Dinosauria, Ceratosauria) from the Western Hemisphere. *Neues Jahrbuch für Geologie und Paläontologie, Monatshefte* 2001: 565–576.

Chure, D. J., B. B. Britt, J. R. Foster, J. H. Madsen, and C. A. Miles. 2000. New records of *Ceratosaurus*, *Torvosaurus*, *Coelurus*, and *Stokesosaurus* and their implication for theropod community structure and biozonation in the Late Jurassic of the Western Interior. *Journal of Vertebrate Paleontology* 20 (3 suppl.): 35A.

Currie, P. J., and Zhao X.-J. 1993. A new carnosaur (Dinosauria: Theropoda) from the Jurassic of Xinjiang, People's Republic of China. *Canadian Journal of Earth Sciences* 30: 2037–2081.

Foster, J. R., T. R. Holtz, Jr., and D. J. Chure. 2001. Contrasting patterns of diversity and community structure of the Late Jurassic and Late Cretaceous of western North America. *Journal of Vertebrate Paleontology* 21 (3 suppl.): 51A.

Galton, P. M. 1982. *Elaphrosaurus*, an ornithomimid dinosaur from the Upper Jurassic of North America and Africa. *Paläontologische Zeitschrift* 56: 265–275.

Gilmore, C. W. 1909. Osteology of the Jurassic reptile *Camptosaurus*. *U.S. National Museum Proceedings* 36: 197–332.

Gilmore, C. W. 1914. *Osteology of the armoured Dinosauria in the United States National Museum, with special reference to the genus* Stegosaurus, pp. 1–143. U.S. National Museum Bulletin 89. Washington, D.C.: Government Printing Office.

Gilmore, C. W. 1920. *Osteology of the Carnivorous Dinosauria in the United States National Museum, with Special Reference to the Genera* Antrodemus (Allosaurus) *and* Ceratosaurus. U.S. National Museum Bulletin no. 110. Washington, D.C.: Government Printing Office.

Gregory, W. K. 1918. Note on the morphology and evolution of the femoral trochanters in reptiles and mammals. *Bulletin of the American Museum of Natural History* 38: 528–538.

Janensch, W. 1925. The coelurosaurs and theropods of the Tendaguru Formation, German East Africa. *Palaeontographica* 1 (suppl. 7): 1–100.

Madsen, J. H., Jr. 1974. A new theropod dinosaur from the Upper Jurassic of Utah. *Journal of Paleontology* 48: 27–31.

Madsen, J. H., Jr. 1976. A second new theropod from the Late Jurassic of east-central Utah. *Utah Geology* 3: 51–60.

Makovicky, P. J. 1997. A new small theropod from the Morrison Formation of Como Bluff, Wyoming. *Journal of Vertebrate Paleontology* 17: 755–757.

Marsh, O. C. 1879. Notice of new Jurassic reptiles. *American Journal of Science*, ser. 3, 18: 501–505.

Marsh, O. C. 1881. A new order of extinct Jurassic reptiles (Coeluria). *American Journal of Science* 21: 339–341.

Marsh, O. C. 1884. Principal characters of American Jurassic dinosaurs. Part 8: The Order Theropoda. *American Journal of Science* 27: 29–40.

Marsh, O. C. 1888. Notice of a new genus of Sauropoda and other dinosaurs from the Potomac Formation. *American Journal of Science* 135: 89–94.

Marsh, O. C. 1896. The dinosaurs of North America. In *Sixteenth Annual Report of the U.S. Geological Survey*, pp. 133–230.

Miles, C. A., and D. W. Hamblin. 1999. Historical update. In J. H. Ostrom and J. S. McIntosh, *Marsh's Dinosaurs*, pp. vii–xiv. New Haven, Conn.: Yale University Press.

Miles, C. A., K. Carpenter, and K. Cloward. 1998. A new skeleton of *Coelurus fragilis* from the Morrison Formation of Wyoming. *Journal of Vertebrate Paleontology* 18 (3 suppl.): 64A.

Osborn, H. F. 1903. *Ornitholestes hermanni*, a new compsognathid dinosaur from the Upper Jurassic. *Bulletin of the American Museum of Natural History* 19: 459–464.

Osborn, H. F. 1916. Skeletal adaptations of *Ornitholestes, Struthiomimus, Tyrannosaurus. Bulletin of the American Museum of Natural History* 35: 733–771.

Ostrom, J. H. 1980. *Coelurus* and *Ornitholestes:* Are they the same? In L. Jacobs (ed.), *Aspects of Vertebrate History,* pp. 245–256. Flagstaff: Museum of Northern Arizona Press.

Russell, D. A. 1972. Ostrich dinosaurs from the Late Cretaceous of western Canada. *Canadian Journal of Earth Sciences* 9: 375–402.

Russell, D. A., P. Béland, and J. S. McIntosh. 1980. Paleoecology of the dinosaurs of Tendaguru (Tanzania). *Mémoires de la Société géologique de France* 139: 169–175.

3. Redescription of the Small Maniraptoran Theropods *Ornitholestes* and *Coelurus* from the Upper Jurassic Morrison Formation of Wyoming

KENNETH CARPENTER, CLIFFORD MILES,
JOHN H. OSTROM, AND KAREN CLOWARD

Abstract

The small tetanuran–basal maniraptoran theropods *Ornitholestes hermanni* and *Coelurus fragilis* have been described only briefly. Owing to their importance to evolutionary studies of theropods, a detailed description is presented here. *Ornitholestes* has a relatively short neck and body (as indicated by the anteroposteriorly short cervical and dorsal centra); cervicals with opisthocoelous centra; prezygapophyses of distal caudals extending about midway on the preceding caudal; ventrally bifurcated chevrons on distal caudals; a long, low ilium; a posteroventrally curved pubic shaft; an ischium with a large obturator notch and a large, triangular obturator process; and a metatarsal IV longer than metatarsal II and slightly shorter than metatarsal III. In contrast, *Coelurus* is characterized by an elongate neck and body (as indicated by the elongated cervical and dorsal centra); a slender, gracile dentary; very thin walled centra; amphicoelous cervical, dorsal, and caudal centra; long, low neural spines on all vertebrae; short prezygapophyses; a sigmoidal humeral shaft; a posteriorly bowed ulna; a semilunate carpal with distinct facets for metacarpals; a posteriorly bowed proximal shaft of the pubis; a pubic foot with a short anterior process that is ventrally arcuate; a femur with a low fourth trochanter; an alariform anterior trochanter; an allosauroid-type astragalus; and a very long metatarsal IV, which is nearly as long as the tibia or femur, so that the hind legs are proportionally very long.

Introduction

The Morrison Formation is a widespread Upper Jurassic non-marine deposit scattered across much of the Western Interior, from northern New Mexico to the Oklahoma panhandle, and north to Montana. The formation has produced a very diverse dinosaur fauna mostly in multi-taxon bone beds (Dodson et al. 1980). Small theropods are rare and include *Coelurus fragilis* Marsh 1879, *Ornitholestes hermanni* Osborn 1903, *Elaphrosaurus?* sp., *Koparion douglassi* Chure 1994, *Tanycolagreus topwilsoni* (Carpenter et al., Chapter 2; previously identified as *Coelurus*, Miles et al. 1998), and another, as yet unnamed theropod reported by Makovicky (1997). Of these, *Coelurus*, *Ornitholestes*, and *Tanycolagreus* are known from partial skeletons. Remarkably, only *Tanycolagreus* has been described in any detail. During our study of *Tanycolagreus*, the material of *Ornitholestes* and *Coelurus* was also studied, and these results are presented below.

The first small theropod from the Morrison Formation to be named was *Coelurus fragilis* by O. C. Marsh (1879). The specimen was apparently widely scattered because it was recovered over a span of several years from Reed's Quarry 13 at Como Bluff, Wyoming, as the quarry was expanded. The locality was an extensive bone bed better known for its numerous specimens of *Stegosaurus* and *Camptosaurus* (Gilmore 1909, 1914). In naming the species, Marsh emphasized the extreme hollowness of the vertebrae and illustrated three of them in 1881. A second species, *Coelurus agilis*, was named a few years later by Marsh (1884) for additional material from Quarry 13, but only a pair of pubes was described and illustrated. As part of our restudy of the *Coelurus* material, it became clear to one of us (JHO), that *C. fragilis* and *C. agilis* are different parts of the same individual (Ostrom 1980). A third species, *Coelurus gracilis*, was erected by Marsh in 1888 for metapodials from the Lower Cretaceous Potomac Formation of Maryland. However, as noted elsewhere, the taxon is a nomen dubium (Ostrom 1980); it is certainly not a species of *Coelurus*.

The other small theropod of our study, *Ornitholestes hermanni*, was named by H. F. Osborn (1903) for a partial skeleton from Bone Cabin Quarry, north of Como Bluff. Osborn figured little of the material at that time, although he did offer a skeletal reconstruction. The skull was figured at a later date, along with a slightly modified rendition of the skeleton (Osborn 1916). In naming this theropod, Osborn noted its small, gracile skeleton and assumed that it was an active, agile hunter. Several years later, Gilmore (1920) synonymized *Ornitholestes hermanni* with *Coelurus fragilis*, arguing that the differences in the vertebrae were minor. However, one of us (JHO) briefly showed later that numerous characters separate the two taxa (Ostrom 1980). Here, we elaborate upon those differences.

Institutional Abbreviations. AMNH, American Museum of Natural History, New York; USNM, United States National Museum (now National Museum of Natural History), Washington D.C.; YPM, Yale Peabody Museum of Natural History, New Haven, Conn.

Systematic Paleontology
Theropoda
Coelurosauria
Maniraptora
Ornitholestes hermanni Osborn 1903
Coelurus fragilis Gilmore 1920 (in part)
Coelurus hermanni Steele 1970

Holotype. AMNH 619, skull with both mandibles, three cervical vertebrae, eleven dorsal vertebrae, four sacral vertebrae, twenty-seven caudal vertebrae, both ischia, left ilium, both pubes missing the distal ends, incomplete left femur, proximal end of left fibula, both humeri, right metatarsals II, III, and IV, left metatarsal IV, four phalanges of the pes, two pedal unguals, one right tarsal, right metacarpal II or III, fragments of two other metacarpals, two fragments of manus phalanges, one ungual of the manus, and numerous fragments. Date received, Summer 1900.

Type Locality. Bone Cabin Quarry, 13 km north of Como Bluff, Wyoming.

Revised Diagnosis. Distal caudals deep and bifurcated ventrally; pubis bowed, distal end directed posteroventrally; pubic apron without pubic fenestra; ischium with broad obturator notch, triangular obturator process.

Coeluridae
Coelurus fragilis Marsh 1879
Coelurus agilis Marsh 1884
Elaphrosaurus agilis Russell, Béland, and McIntosh 1980

Syntypes. YPM 1991, two proximal caudal vertebrae, one proximal caudal centrum, one proximal caudal neural arch (Marsh 1881, plate 10, fig. 2a,b). YPM 1992, eight mid-caudal vertebrae, one partial mid-caudal centrum (Marsh 1881, plate 10, fig. 3a,b). YPM 1993, cervical vertebra, proximal caudal neural arch (Marsh 1881, plate 10, fig. 1 (composite), 1a (composite), 1b).

Type Locality. Quarry 13 East, Como Bluff, Wyoming.

Referred Specimens. Quarry 13 East, Como Bluff, Wyoming: YPM 1992, eight mid-caudal vertebrae, one partial centrum of a mid-caudal vertebra (Marsh 1881, plate 10, fig. 3, 3a). YPM 1993, one cervical vertebra, one proximal caudal neural arch (Marsh 1881, plate 10, fig. 1, 1a). YPM 1994, one caudal centrum. YPM 1995, one caudal vertebra, plus fragments. YPM 2010 (type specimen of *Coelurus agilis*), right dentary?, three cervical vertebrae, two dorsal vertebrae, five dor-

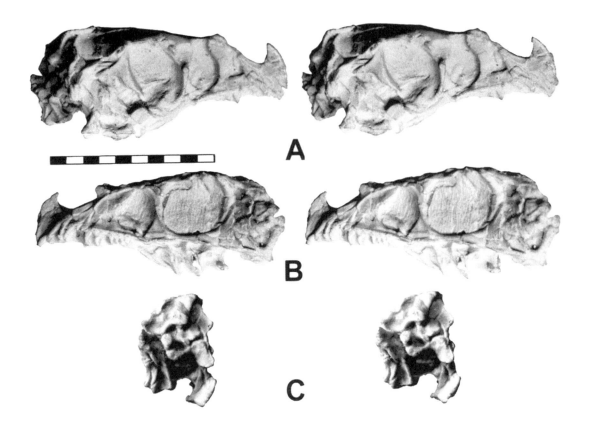

A

B

C

Figure 3.1. Skull of Ornitholestes hermanni *as available to Osborn (1903, 1916) in stereoscopic pairs (YPM 53262 cast): (A) right side showing displacement of bones anteriorly, (B) left side with most bones intact, and (C) posterior view. Scale in cm.*

sal centra, six dorsal neural arches, two indeterminate neural arches, one proximal caudal vertebra, left and right ulnae, distal ends of both radii, left humerus, left femur, proximal end of right femur, both pubes, fragment of an ilium?, distal three-quarters of left tibia, proximal end of left fibula, right scapula, distal end of metatarsal III, metatarsal II or IV, left and right radiale, left and right metacarpal II, right metacarpal III, fragment of right metacarpal IV, seven phalanges of the manus, many unidentified fragments. YPM 9162, one partial sacral vertebra.

Cleveland-Lloyd Quarry, Utah: ?UUVP 11743, 1eft humerus. UMNH 7795, left humerus.

Revised Diagnosis. Very gracile dentary; all centra elongated; all neural spines long and low; all centra very thin walled; paired pleurocoels on some cervicals; triangular diapophyseal ala angled sharply ventrolaterally; centroprezygapophyseal lamina forms fossa lateral to neural canal; centrodiapophyseal lamina present; caudal prezygapophyses very short; pubic foot very arcuate ventrally and projects posterodorsally; pubic fenestra located at middle of pubic apron; very long, gracile metatarsal almost as long as tibia or femur.

Description

Cranial (Figs. 3.1–3.3). The skull of *Ornitholestes* as originally available to Osborn is illustrated in multiple views for the first time in

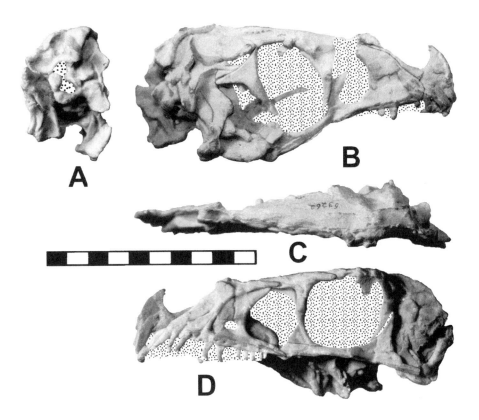

Figures 3.1 and 3.2. The considerable preparation that has been done recently reveals a great deal of detail not previously available to Osborn (Norell, pers. comm.). The skull is presently under study by Mark Norell, and thus little will be written here. The skull is slightly crushed obliquely, resulting in a distortion of the right side. For example, the orbit on the left side is almost circular, whereas it is distinctly ellipsoid on the right. This crushing has displaced the base of the left nasal, which has led to the suggestion that *Ornitholestes* may have had a nasal horn much like *Proceratosaurus* (Paul 1988); clearly it did not (Norell et al. 2001).

Figure 3.2. Skull of Ornitholestes hermanni *(YPM 53262 cast) in (A) occipital, (B) right, (C) dorsal, and (D) left views. Scale in cm.*

Of *Coelurus,* the only possible cranial fragment is a right dentary (Fig. 3.3). The preservation, including color, is the same as that of most of the rest of YPM 2010, suggesting that the fragment belongs with *Coelurus* as determined by Marsh. However, the fragment is very slender and delicate, measuring 7.9 cm long with a mid-length depth of 1.1 cm. This condition has led us to speculate that perhaps the dentary does not belong with the skeleton, even though it was found in the same general area. In dorsal view, the jaw is slightly sigmoid. There are fourteen alveoli present, as well as interdental plates. It has a large Meckelian groove on the medial side that extends the length of the fragment. On the lateral surface there is a prominent groove connecting the nutrient foramina. If this dentary does belong to *Coelurus,* then the taxon may also be characterized by its unique long, slender dentary.

Figure 3.3. Possible right dentary of Coelurus fragilis *in (top) occlussal, (center) medial, and (bottom) lateral views. Scale in cm.*

A very gracile dentary also characterizes the troodontid *Byronosaurus jaffe*, and it also has a prominent groove on the lateral surface (Norell et al. 2000). A slender, gracile dentary also characterizes *Compsognathus* and *Shuvuuia* (Sereno 2001, fig. 12). Further comparisons may reveal additional similarities with these taxa.

Cervicals (Fig. 3.4). Only three cervicals of *Ornitholestes* are known, and these are crushed or distorted (Fig. 3.4A–C). The neural spines are damaged on the more anterior two, but they were apparently anteroposteriorly longer and lower than the spine on the third cervical. The beveled centra suggest that these are all anterior and mid-cervicals, possibly CV3, CV4, and CV6. The beveling does indicate that the neck was somewhat sigmoidal (contrary to Ostrom 1969, p. 48), as it also was in *Coelurus*. The articular faces of the *Ornitholestes* cervicals are opisthocoelous, whereas they are amphicoelous in *Coelurus* as first observed by Marsh (1881). Marsh (1881, p. 339) also reported that "The first three of four behind the axis [of *Coelurus*] had the front articular face of the centrum somewhat convex, and the posterior one deeply concave." The only vertebrae that match this description are those described by Makovicky (1997) as an unnamed small theropod from the Morrison Formation (see further discussion below). Previously, one of us (JHO) had come to a similar conclusion and had excluded these vertebrae from *Coelurus* (Ostrom 1980, Appendix A). In addition, none of the cervical vertebrae of *Coelurus* can be matched to that figured by Marsh (1881, plate 10, fig. 1). It would appear, therefore, that the figured vertebra is actually a composite of two cervicals (Ostrom 1980, p. 253), including one of the non-*Coelurus* vertebrae described by Makovicky.

The neural spine is preserved long and somewhat low on the anterior two cervicals of *Ornitholestes*, whereas it is anteroposteriorly short and moderately tall in the third cervical. In contrast, the frag-

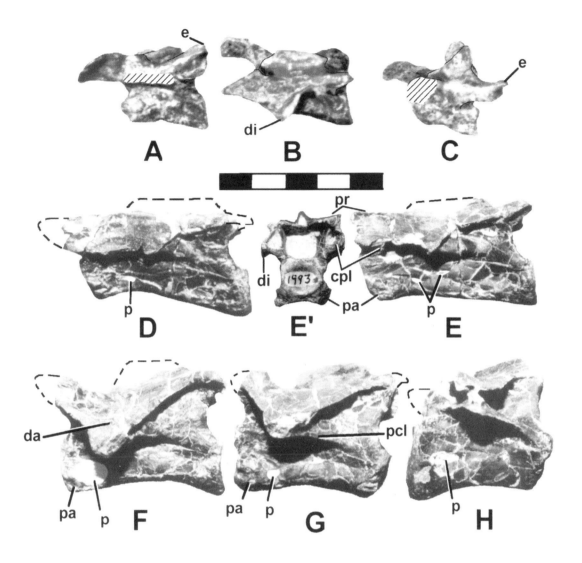

ments of the neural spines of *Coelurus* suggest that they were all very low and long, extending almost the entire length of the neural arches. Admittedly most of the spines were damaged in collecting; thus it is difficult to know if any were tall. There is a small, triangular depression at the base of the neural spine on the anterior side of the neural spine, and a vertically elliptical one on the posterior side. In addition, there is a peculiar, moderately deep fossa at the base of the prezygapophysis. It is formed by the centroprezygapophyseal lamina. The diapophyseal lamina is a triangular ala that projects ventrolaterally from the neural arch and connects the pre- and postzygapophyses. A posterior centrodiapophyseal lamina is also present connecting the posterior part of the centrum to the diapophysis. In *Ornitholestes*, the diapophysis extends subhorizontally (although it is distorted in the cervical preserving this feature, Fig. 3.4B). The prezygapophyses of *Ornitholestes* project

Figure 3.4. Comparison of cervical vertebrae. Ornitholestes hermanni (A–C) and Coelurus fragilis (D–H). (E') is the anterior view of cervical E showing the matrix filled fossa below the prezygapophyses formed by the centroprezygapophyseal lamina. Abbreviations: cpl—centroprezygapophyseal lamina; da—diapophyseal ala; di—diapophysis; e—epipophysis; p—pleurocoel; pa—parapophysis; pcl—posterior centrodiapophyseal lamina; pr—prezygapophysis. Scale in cm.

anterodorsally and are slightly flexed in lateral view. In dorsal view, the pre- and postzygapophyses are strongly divergent, forming an X shape. In contrast, the prezygapophyses are horizontal in the anterior cervicals of *Coelurus* and slightly angled anterodorsally in the mid- and posterior cervicals; in dorsal view the pre- and postzygapophyses are moderately divergent. However, the full extent of the prezygapophyses is unknown for most of the vertebrae because they are incomplete. The postzygapophyses overhang the posterior surface of the centrum in *Ornitholestes*, especially in the mid-cervical (contrast Fig. 3.4A,B with 3.4C), but not in *Coelurus*. In addition, the articular facets of the zygapophyses are large in *Ornitholestes* but very small in *Coelurus*. Small, horizontal, prong-shaped epipophyses are present on the cervicals of *Ornitholestes* but apparently not *Coelurus*, although these are probably broken off because most postzygapophyses are incomplete.

None of the centra of *Ornitholestes* or *Coelurus* have a hypapophysis, or keel, ventrally. The centrum of *Ornitholestes* is short relative to the centrum width, whereas in *Coelurus* it is more than four times as long as the width. The anterior articular surface is wider than it is deep in *Ornitholestes*, whereas the articular face is almost circular in *Coelurus*. Pleurocoels are absent on the cervicals of *Ornitholestes* but are present in *Coelurus*. These pleurocoels in *Coelurus* vary considerably in size and number on each side of the centra. Cervical B (Fig. 3.4E) has two small, lenticular pleurocoels on the left side but only one on the right. One cervical, C, has an enormous pleurocoel on one side (Fig. 3.4F), but a considerably smaller one on the other. This asymmetry appears to be real. Another cervical, E (Fig. 3.4H), has a large pleurocoel on the left side, but three smaller pleurocoels on the right side. All of the *Coelurus* vertebrae are very pneumatic. Not only are the centra camellate as noticed by Marsh (1881), but the broken surfaces of the diapophyses suggest that pneumaticity also extended into the diapophyseal ala. Contrary to Marsh (1881), cervical ribs are not fused to the vertebrae in *Coelurus*.

Dorsals (Fig. 3.5A–U). Ten dorsals are preserved in the holotype of *Ornitholestes hermanni*, and nine centra and neural aches in *Coelurus fragilis*. Some of the dorsals of *Ornitholestes* are distorted, and many others have lost their transverse processes (Fig. 3.5A). The neural spines of the anterior dorsals are anteroposteriorly short, and they rise vertically and have a rounded anterior border, whereas those of the mid- and posterior dorsals extend nearly the full length of the neural arch and are almost twice as tall as those of the anterior dorsals. In addition, the neural spine of the last dorsal leans forward. In *Coelurus*, the anterior neural spines are only slightly shorter than the mid- and posterior dorsals. In addition, the spine occupies the length of the neural arch in all of the vertebrae. The apices of the neural spines in both *Ornitholestes* and *Coelurus* are not expanded into a spinal table. Along the anterior and posterior margins, the scars for the interspinous ligaments extend to near the tops of the neural spines in *Coelurus* but terminate well below the tops in *Ornitholestes*.

The transverse processes of *Ornitholestes* are simple horizontal, laterally directed, and subrectangular in dorsal view. In contrast, the

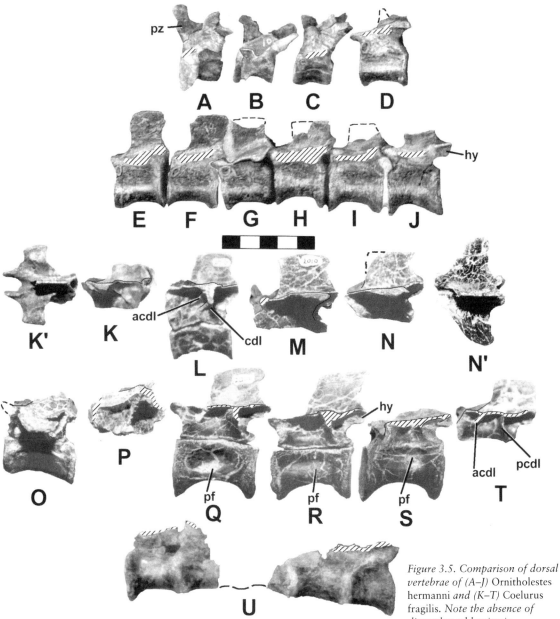

Figure 3.5. Comparison of dorsal vertebrae of (A–J) Ornitholestes hermanni *and* (K–T) Coelurus fragilis. *Note the absence of diapophyseal lamina in* Ornitholestes. *Sequence of vertebrae in* Coelurus *is based upon transverse process shape (compare K' with N'). (U) Sacral vertebrae of* Ornitholestes hermanni. *Abbreviations: acdl— anterior centrodiapophyseal lamina; cdl—centrodiapophyseal lamina; hy—hyposphene; pcdl— posterior centrodiapophyseal lamina; pf—pleurofossa; pz— prezygapophysis. Scale in cm.*

transverse processes of the anterior dorsals in *Coelurus* are horizontal, laterally directed, and subrectangular in dorsal view (Fig. 3.5K') but become strongly back turned and triangular in dorsal view beginning with the mid- and extending to the posterior dorsals (Fig. 3.5N'); a similar change is seen in *Tanycolagreus* (see Carpenter et al., Chapter 2). In addition, there is a well-developed centrodiapophyseal lamina on the ventral side of the transverse process, and this separates two deep diapophyseal fossa. Some transverse processes also have anterior and posterior centrodiapophyseal lamina as well.

The prezygapophyses of *Coelurus* are small, slightly inclined structures that merge laterally into the transverse processes. In contrast, those of *Ornitholestes* are distinct, separate structures. The postzygapophyses in the anterior dorsals of *Coelurus* do not extend beyond the posterior surface of the centrum, whereas they do in the mid- and posterior dorsals. In *Ornitholestes*, the postzygapophyses overhang the centrum of all the dorsals. A hyposphene is developed only on the posterior dorsals of *Ornitholestes* (Fig. 3.5J) and on the mid- and posterior dorsals of *Coelurus* (Fig. 3.5L–T).

In both *Ornitholestes* and *Coelurus,* the centra are taller than they are wide, although the amount of compression varies. In *Ornitholestes*, the centrum is slightly compressed, or spool-shaped, in ventral view, and the cross-section of the mid-section is still over 60 percent the height of the articular face. In contrast, the centrum of *Coelurus* is much more constricted in the middle, being hourglass-shaped in ventral view. The articular faces are amphiplatyan in *Ornitholestes* and somewhat amphicoelous in *Coelurus*. No pleurocoels are present on the dorsals of either taxon.

Sacrals (Fig. 3.5U). Five sacral vertebrae are preserved in *Ornitholestes*, although there is a gap dividing them in half. The neural spines are unfortunately missing on all of them. None of the centra have pleurocoels. The articular face of the first sacrum is amphiplatyan, and its width is subequal to that of the last sacral. There is a damaged centrum that may be a sacral centrum of *Coelurus*.

Caudals (Fig. 3.6). Caudals are the most abundant vertebrae in both specimens. None of them have a pleurocoel, although the centra do have an elongated, shallow groove termed a pleurofossa (Carpenter and Tidwell, in press). In addition, the lateral sides of the centra in *Ornitholestes* have a ridge extending nearly the length of the centra. Ventrally, the centra have a groove separating the paired ridges that connect the chevron facets. The articular faces of the centra are rather boxy, implying less mobility rather than more as stated by Holtz (1998, Appendix 1). The anterior caudals, especially those of *Ornitholestes*, are short. The length of these anterior caudals is about 75 percent the length of the posterior ones in *Ornitholestes* and over 80 percent in *Coelurus*. In addition, the centra faces of these anterior caudals are beveled so that the posterior articular surface is lower than the anterior. Unfortunately, because so many caudals are missing, the transition point between the anterior and posterior portions of the tail cannot be determined in either specimen. The prezygapophyses of the mid- and posterior caudals are proportionally longer in *Ornitholestes* than in *Coelurus,* but even the longest do not extend to mid-centrum (Fig. 3.6B'); in *Coelurus* the prezygapophyses are shorter than in *Tanycolagreus*. The neural spines are damaged in both taxa, but from their remnants they appear to have been longer and lower in *Coelurus* than in *Ornitholestes;* neither show the bifid spines seen in *Sinosauropteryx* (Currie and Chen 2001). The distal chevrons of *Ornitholestes* are elongate, being about as long anteriorly as posteriorly. Structurally, they are unusual in that some of them are bifurcated laterally as well as ventrally (Fig. 3.6B").

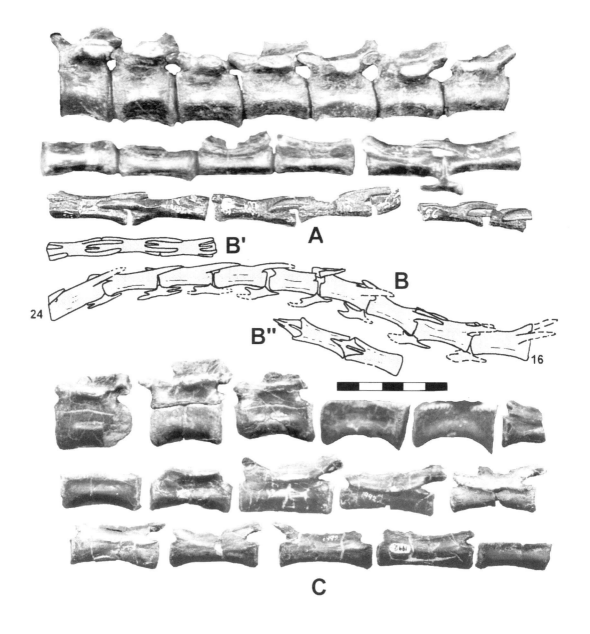

There is one odd caudal cataloged with YPM 2010 that we believe does not belong to *Coelurus* and may actually belong to the cervical vertebrae described by Makovicky (1997). In addition, it has the same black color and preservation, suggesting that it also came from the same quarry. This vertebra is amphicoelous but with a thick rim around the articular face. The neural spine is a tall, spike-like structure, as are the horizontal caudal ribs. Ventrally, there are very faint ridges connecting the chevron facets.

Scapula (Fig. 3.7). The only scapula is that of *Coelurus*, and unfortunately, even that is considerably damaged. Nothing about the development of the acromion or about whether the distal end of the

Figure 3.6. Comparison of the caudal vertebrae of Ornitholestes hermanni *(A, B) and* Coelurus fragilis *(C). The distal caudals of* Ornitholestes *shown in articulation in B, two vertebrae in dorsal view in B' and in ventral view in B'' (adapted from an unpublished sketch by Erwin Christman). Note the peculiar, boat-shaped chevrons in B''. Scale in cm.*

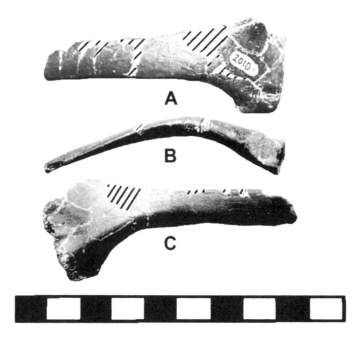

(Opposite page)

Figure 3.7. Scapular fragment of Coelurus fragilis in lateral (A), ventral (B), and medial (C) views. Scale in cm.

Figure 3.8. Forelimb elements of Ornitholestes hermanni and Coelurus fragilis: (A) right humerus of Ornitholestes hermanni in anterior and lateral views; (B) left humerus of Coelurus in anterior, medial (showing prominent scar near the deltopectoral crest), lateral, and posterior views, as well as proximal and distal views; (C) right and (D) left ulnas of Coelurus in proximal, lateral, anterior, medial, and posterior views; (E) right radius of Ornitholestes; and (F) left radius of Coelurus in proximal, lateral, anterior, medial, and posterior views. Scale in cm.

scapular blade was expanded can be determined. On the medial side, there is no prominent horizontal ridge as in *Tanycolagreus*. In ventral view (Fig. 3.7B), the scapula is sharply bent medially.

Coracoid. The coracoid is unfortunately missing in both specimens.

Humerus (Fig. 3.8A–B). A right humerus is known for *Ornitholestes* and a left for *Coelurus*. That of *Ornitholestes* (12.4 cm long) is rather straight shafted, whereas that of *Coelurus* is markedly sigmoidal in lateral view (11.9 cm long). The proximal end is rounded in *Ornitholestes* and markedly flat in *Coelurus*. The humeral head is low and confluent with the deltopectoral and bicipital crests in *Ornitholestes* but is offset and emarginated ventrally by a groove in *Coelurus*. The internal tuberosity is well differentiated and angular in both taxa, but it is conical in *Ornitholestes* and elongated and tapering in *Coelurus*. What these differences mean biomechanically is an area in need of research. In lateral view, the internal tuberosities of both taxa project posteriorly and are well separate from the humeral head. The diameters of both the proximal and distal ends are well expanded to over 150 percent of the mid-shaft diameter, and the ends are slightly offset relative to one another; that is, there is slight torsion of the humeral shaft. The deltopectoral crest projects anteriorly and is expanded and offset from the humeral shaft in both taxa. A prominent scar for the M. pectoralis superficialis is present laterally near the base of the deltopectoral crest in *Coelurus*. In *Tyrannosaurus* (Carpenter and Smith 2001) and apparently in *Ornitholestes*, this scar is a slight tuberosity. Distally, the condyles are more anteriorly than ventrally developed, indicating that the relaxed position of the arms was slightly flexed. The entepicondyle is prominent in both taxa also. Ventrally, the ulnar facet is expanded and merges with the entepicondyle.

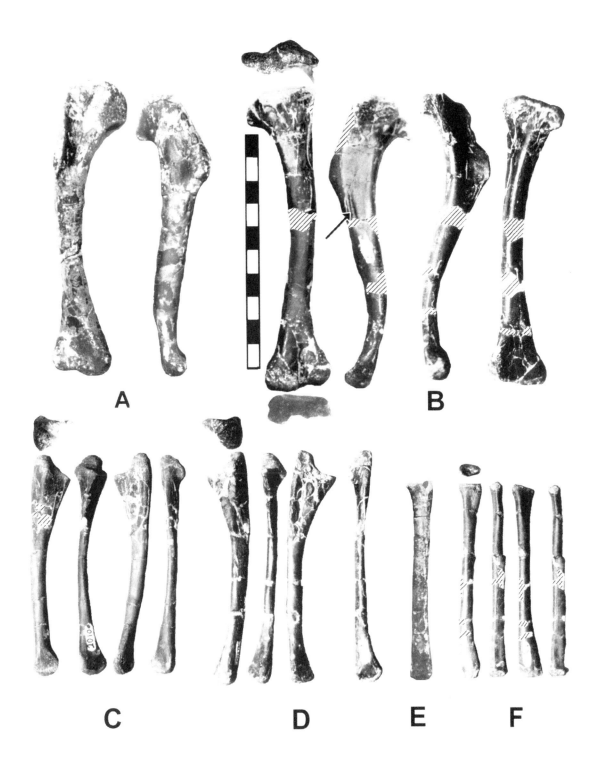

A

B

C

D

E

F

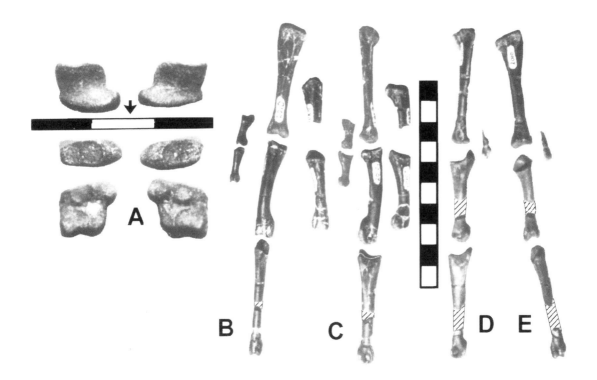

The humerus/ulna ratio is greater than 100 percent in both taxa,
and the radius/ulna ratio is less than 75 percent, but greater than 50
percent (Holtz 1998).

Ulna (Fig. 3.8C,D). Both ulnas are known for *Coelurus*, although
the left one is slightly crushed laterally (left, 9.6 cm long; right, 9.1 cm).
The shafts are bowed posteriorly; and in anterior view, the shafts are
bowed laterally. The humeral notch is better developed in the left ulna
than in the right; the olecranon is equally developed in both. The radial
notch is shallow and is slightly more prominent in the left ulna than in
the right. The distal end is arcuate but lacks the anteriorly developed
process seen in some theropods, such as *Deinonychus*.

Radius (Fig. 3.8E,F). A single right radius is present for *Ornitholestes* (8.4 cm long), and a left for *Coelurus* (~8.1 cm long), as well as
the proximal end of the right. Their shafts are straight, and they differ
only in that the shaft is slightly more slender in *Coelurus* than in
Ornitholestes. The distal ends of the radius and ulna have a loose joint.
The distal ends are not expanded as they are in *Deinonychus*.

Carpal (Fig. 3.9A). Both semilunate carpals are known for
Coelurus (greatest width: left, 1.2 cm; right, 1 cm). The carpal is deeper
than in *Tanycolagreus* and has a more prominent trochlear surface (see
Carpenter et al., Chapter 2). It has well-developed articular facets on its
ventral side, and it broadly overlaps metacarpal II. The carpal differs
from that of the contemporary large theropod *Allosaurus* in that it is
not restricted to capping metacarpal I, and it lacks the ventral process

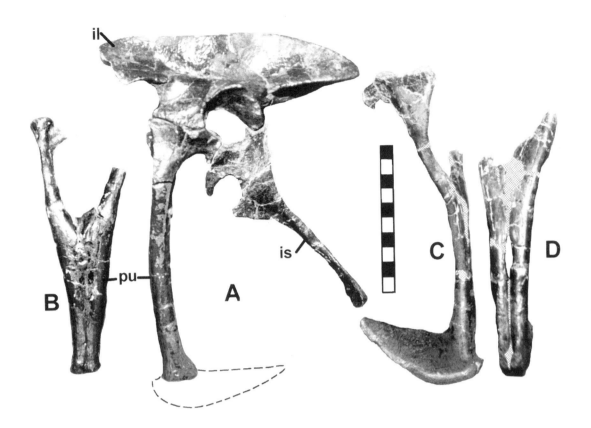

seen in *Allosaurus* (see Carpenter 2002). It is similar to the carpal of *Deinonychus* in overall shape, although considerably smaller. As noted elsewhere, the manus of theropods could not fold up against the forearm in avian fashion (Carpenter 2002).

Manus (Fig. 3.9B–E). The isolated manus attributed to *Ornitholestes* by Osborn (1916) has been referred to *Tanycolagreus* because of the great similarities between the two, especially in the curvature of the first ungual (Carpenter et al., Chapter 2); the holotype of *Tanycolagreus topwilsoni* comes from a few hundred meters west of the holotype of *Ornitholestes hermanni*. The manual elements of *Coelurus* are long and slender, and parts of both hands are represented. Unfortunately, metacarpal I and IV are missing; metacarpal III is present, and it retains its phalanges. Collateral tendon facets are well developed. Overall, the manual elements resemble those of *Tanycolagreus*, although they are much smaller.

Ilium (Fig. 3.10A). The ilium of *Ornitholestes* is long and low in profile. It has a moderately long preacetabular process that is notched anteriorly, and a much longer, tapering postacetabular process. Ventrally beneath the postacetabular process, the brevis fossa is long and tapering; and anteriorly, a well-developed M. cuppedicus fossa is present along the ventral edge of the preacetabular process. The lateral surface above the acetabulum is smooth, lacking the vertical ridge seen in *Stokesosaurus*. The dorsal lip of the acetabulum flares laterally into

Figure 3.10. Pelvic elements of Ornitholestes hermanni *(A, B)* and Coelurus fragilis *(C, D).* (A) Pelvis of Ornitholestes *in left lateral view showing long, low ilium, anteriorly bowed pubis (foot shown with dashed lines), and ischium with very large obturator notch; (B) pubis of* Ornitholestes *in anterior view show the absence of a pubic fenestra; (C) pubis of* Coelurus *in right lateral view (note bend in shaft in the proximal portion); and (D) pubis of* Coelurus *in anterior view showing slit-like pubic fenestra. Abbreviations: il—ilium; is—ischium; pu—pubis. Scale in cm.*

a supra-acetabular crest. The distal end of the ischial peduncle is considerably smaller than that of the pubic one. The pubic peduncle projects slightly anteriorly and is anteroposteriorly longer than it is wide. It also extends ventrally to the level of the ischial peduncle. No ilium is known for *Coelurus*.

Pubis (Fig. 3.10A–D). The pubis is known for both *Ornitholestes* and *Coelurus*. The shaft is bowed in *Ornitholestes* so that the pubic foot is posterior to the vertical plane (Fig. 3.10A), as if it were becoming the precursor to the opisthopubic condition of dromaeosaurids. The shaft is relatively straight in *Coelurus*, except proximally, where it is bowed posteriorly; this bowing appears to be natural and not distortion (Fig. 3.10C). Distally, the left and right pubic feet are in contact in *Coelurus* (Fig. 3.10D). Unfortunately, the distal end for *Ornitholestes* is incomplete, and the damaged end is plastered. It is possible, although doubtful, that the distal end of the pubis retained a rounded, cerato-saur-type pubic expansion. A small part of what is probably the pubic foot is present on the posterior and anterior sides at the distal end (Fig. 3.10A). The pubic foot in *Coelurus* is considerably longer posteriorly than the short anterior projection. The length of the foot is about 46 percent of the length of the pubic shaft as measured just above the foot. The ventral side of the pubic foot is arcuate, whereas it is flat in *Tanycolagreus* (Carpenter et al., Chapter 2) and *Aristosuchus* (Naish et al. 2001, fig. 9.28). The obturator foramen is ventrally open to form an obturator notch in the pubis of both *Ornitholestes* and *Coelurus*; no other foramen is present in the pubis ventral to the notch in lateral view. The pubic apron is relative wide and ventrally long in both taxa. Furthermore, it is pierced about mid-length by a lenticular-shaped pubic foramen in *Coelurus* but not in *Ornitholestes* (compare Figs. 3.10B and D). This foramen is located just above the pubic foot in *Tanycolagreus*, and the pubis from Cañon City previously referred by Gilmore (1920) as *Coelurus*; we therefore transfer this pubis to *Tany-colagreus* (see Carpenter et al., Chapter 2). The contact with the is-chium is dorsoventrally narrow.

Ischium (Fig. 3.10A). The ischium for *Ornitholestes* is slightly curved posteroventrally. Proximally, there is no antitrochanter along the posterior margin of the acetabulum. The obturator notch is very wide, and the obturator process is triangular and distally placed; no foramen pierces this process. The anterior margin of the obturator notch is a long, tapering process that extends ventrally from the pubic articulation. A small scar on the posterolateral side of the ischial shaft is present. The distal end of the ischium is slightly expanded but does not have a foot. If we are correct that a *Coelurus-Tanycolagreus*-like foot was present at the end of the pubis, then the ischium/pubis length ratio was 66–68 percent.

Femur (Fig. 3.11A–C). A partial left femur is known for *Orni-tholestes* (>20 cm long), and a complete left (~21 cm long) and partial right are known for *Coelurus*. The femur of *Ornitholestes* is antero-posteriorly crushed and is missing its head (Fig. 3.10C). Nevertheless, the shaft is bowed anteriorly and has a slight sigmoid curvature in

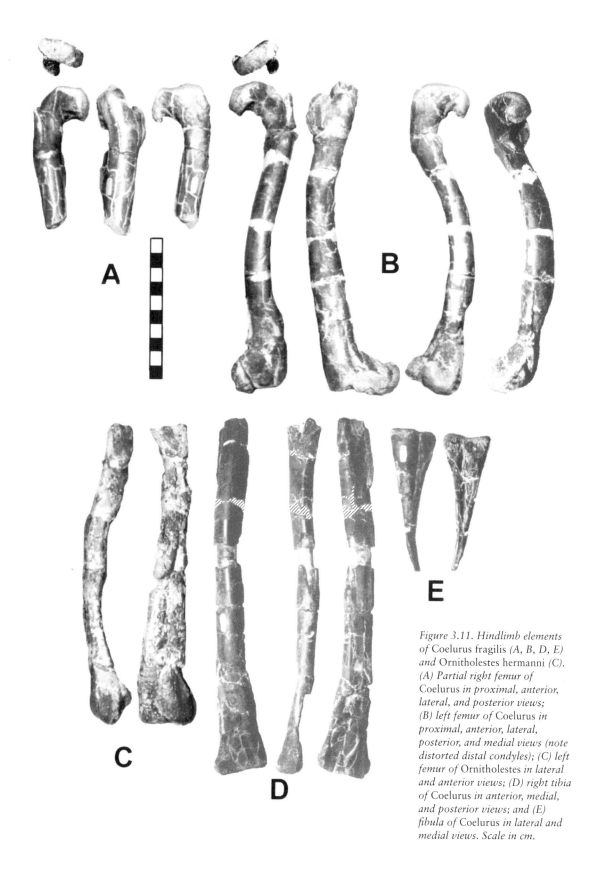

Figure 3.11. Hindlimb elements of Coelurus fragilis *(A, B, D, E)* and Ornitholestes hermanni *(C)*. *(A) Partial right femur of* Coelurus *in proximal, anterior, lateral, and posterior views; (B) left femur of* Coelurus *in proximal, anterior, lateral, posterior, and medial views (note distorted distal condyles); (C) left femur of* Ornitholestes *in lateral and anterior views; (D) right tibia of* Coelurus *in anterior, medial, and posterior views; and (E) fibula of* Coelurus *in lateral and medial views. Scale in cm.*

anterior view. In marked contrast, the femur of *Coelurus* is strongly bowed anteriorly and more strongly curved sigmoidally in anterior view. The femoral head is transversely elongated and rather blocky, or squared; the head is slightly angled ventrally, but less so than in *Nedcolbertia* (Kirkland et al. 1998). The greater trochanter is a mound-like structure in the right femur and less mound-like in the left. In the left femur, the greater trochanter is confluent with the head, giving the proximal end an almost rectangular appearance. The anterior trochanter is alariform, being a plate-like structure that projects antero-dorsally below, and separate from, the greater trochanter. The top of the anterior trochanter is well below the top of the greater trochanter, but less so than in *Nedcolbertia*. The fourth trochanter is apparently absent in *Ornitholestes*, although the bone is poorly preserved. It is, however, present in *Coelurus* as a long, low ridge on the right femur (it is broken in the left). The medial epicondyle is short, its length being less than one-quarter the femoral length in *Ornitholestes* but about one-quarter the femur length in *Coelurus*. The distal end of the femur is pathological in *Coelurus*, as evidenced by the filigree on the shaft just above the condyles; the two condyles appear to be approximately the same size. The "patellar" sulcus, or groove, at the anterior surface of the distal end is very shallow in both taxa.

Tibia (Fig. 3.11D). Only the right tibia for *Coelurus* is known, and this is lacking the proximal head. Thus, nothing can be said about the shape of the cnemial crest. The fibular crest (crista fibularis) is present, however, and this is long and low. The presence of the crest and the missing proximal end including the cnemial crest suggest that the crest was distally placed. The shaft is fractured and is missing some of the cortical bone at mid-shaft. Nevertheless, the tibia appears to be laterally bowed. The distal end is fractured in numerous places, as well as slightly crushed. For these reasons, entirely delineating the sutural scar for the ascending process of the astragalus is difficult. The fibular process, or flange, is not as well developed as in *Tanycolagreus*.

Fibula (Fig. 3.11E). The proximal end of the right fibula is known for *Coelurus*. It is triangular and tapers distally. The proximal end is slightly concave. On its medial side, the sulcus, or fossa, is not very well developed, unlike in many theropods.

Tarsal (Fig. 3.12). The only tarsal element available is the right astragalus of *Coelurus*. An astragalus purportedly belonging to *Coelurus* was described by Welles and Long (1974) under its access number, YPM 1252, rather than its catalog number, YPM 9163 (Ostrom 1980). They referred to it as of the allosauroid type, noting that the ascending process does not extend the entire width of the astragalus. Although this is true, it is clear that the ascending process is incomplete and that it projected higher than they acknowledged. In fact, their reported width of 7.4 cm is at odds with our measured width of 3.2 cm. It is unfortunate that they did not figure the specimen. However, their overall description matches the astragalus in Figure 3.12; we can only assume that there is an error in their reported measurements or that there is another, larger astragalus in the Yale collections.

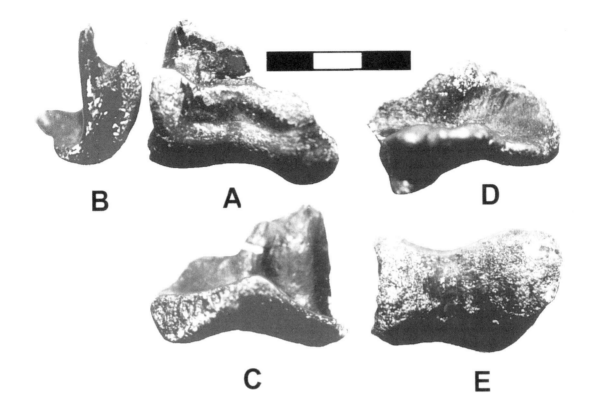

B   A   D

C   E

The ascending process occupies the lateral half of the astragalar body (Fig. 3.12A), and it has a prominent, wide groove separating it from a short dorsal projection from the lateral distal condyle (Fig. 3.12B); a fossa is not present at the base of the ascending process. The anterior face of the distal condyles has a shallow horizontal groove, or sulcus, extending across it (Fig. 3.12A). The facet for the calcaneum is not well developed, but it was slightly interdigitating. The condyles are oriented anteroventrally (Fig. 3.12B).

Pes (Fig. 3.13). The foot of *Ornitholestes* is represented by right metatarsals II–IV and phalanges (Fig. 3.13A,B), and that of *Coelurus* only by a left metatarsal IV and the distal end of metatarsal III (Fig. 3.13C,D). Metatarsal IV of *Ornitholestes* is distorted so that it is not appressed against metatarsal III. It is clear, however, from metatarsals II and III that originally the metatarsals were closely appressed, although not fused together. Proximally, metatarsal III is somewhat hourglass-shaped, and it separates metatarsals IV and II; metatarsal IV backs metatarsal III. Metatarsal IV (11.3 cm long) is also longer than metatarsal II (10.9 cm long) but not as long as metatarsal III (11.9 cm long). The width of the cross-section of the metatarsals is about the same as the anteroposterior length. The phalanges and unguals resemble those of other theropods.

In contrast, the only complete metatarsal of *Coelurus* is extremely long and slender and is about the same length as the femur, whereas

Figure 3.12. Right astragalus of Coelurus fragilis *in (A) anterior, (B) lateral, (C) posterior, (D) proximal, and (E) ventral views. Scale in cm.*

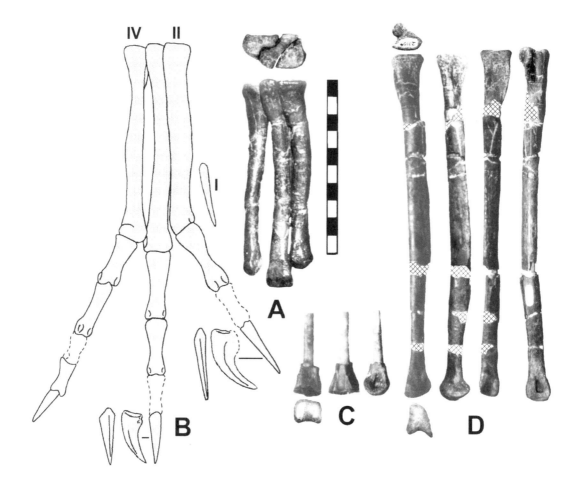

Figure 3.13. Pes elements of Ornitholestes hermanni *(A, B) and* Coelurus fragilis *(C, D). (A) Right metatarsals II–IV of* Ornitholestes *in proximal and anterior views; (B) reconstructed pes of* Ornitholestes *(adapted from an unpublished sketch by Erwin Christman); (C) distal end of metatarsal III of* Coelurus *in anterior, posterior, and lateral views; and (D) left metatarsal IV of* Coelurus *in proximal, anterior, distal, lateral, posterior, and medial views. Scale in cm.*

the length of metatarsal IV is less than 50 percent of the femur length in *Ornitholestes*. In addition, the cross-section of the metatarsal in *Coelurus* is wider than it is anteroposteriorly long.

Discussion

Small theropod remains are known from various sites in the Morrison. Some of these have been referred to *Ornitholestes* or *Coelurus* more because of size than morphology. Makovicky (1997) has identified two vertebrae in the Marsh collection that were apparently considered by Marsh as anterior *Coelurus* cervicals (we note that his "fourth cervical" is dorsoventrally crushed, as indicated by the fracture patterns on the diapophyseal lamina, and that the centrum would normally be visible in lateral view). A posterior cervical, otherwise matching these vertebrae, is also known from the Small *Stegosaurus* locality (Carpenter 1998), indicating that the geographic range of this small theropod was wider than the Como Bluff region. A long slender tibia is also known from the Small Quarry, which size-wise, seems reasonable as belonging to this as yet unnamed theropod. Because the original specimens described by Makovicky came from Quarry 9 (mam-

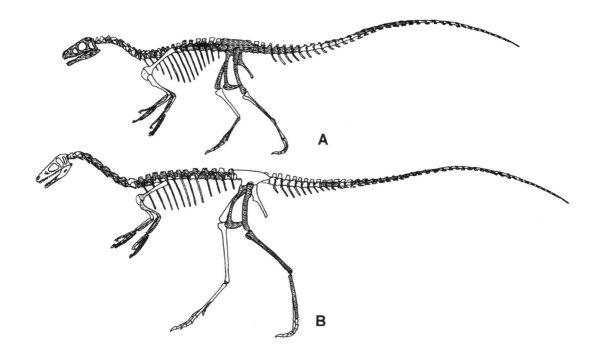

Figure 3.14. Skeletal reconstruction of Ornitholestes hermanni (A) and Coelurus fragilis (B) to same femoral length. Note differences in relative proportions.

mal quarry), it is possible that some of the small theropod bones from Quarry 9 at the National Museum of Natural History actually belong to this particular individual: USNM 5810 cervical vertebra, USNM 5809 sacral vertebra, USNM 6624 distal caudal, USNM 6625 distal caudal, USNM 6626 mid-caudal, USNM 6627 distal caudal, USNM 6628 anterior caudal, USNM 162447 right femur, and uncatalogued distal end of metatarsal.

Skeletal reconstructions for *Ornitholestes* and *Coelurus* are presented in Figure 3.14 to the same femoral length, which is actually close to being similar in the two specimens. Several important differences are readily apparent. First is that on the basis of the dentary section, the skull of *Coelurus* must have been more slender or gracile than that of *Ornitholestes*. In this, the former is more similar to that of *Mononykus* and *Compsognathus* than to the proportionally more robust skull seen in other small theropod skulls, such as *Deinonychus*. In addition, it may have also been proportionally small relative to body size, as in *Mononychus* as well. Second, the neck and body are proportionally longer in *Coelurus* than in *Ornitholestes*. This difference is due to the longer vertebrae of *Coelurus* as compared with those of *Ornitholestes* (Figs. 3.4 and 3.5). Finally, the hind leg of *Coelurus* is very long and gracile, and it more closely resembles that of *Mononykus*. This extreme long-leggedness is due to the very long metatarsal.

Acknowledgments. We thank the following people for access to specimens in their care: Donald Burge (College of Eastern Utah), Jacques Gauthier and Lyndon Murray (Yale Peabody Museum), Charlotte Holton (American Museum of Natural History), Robert Purdy (National Museum of Natural History), and Kenneth Stadtman (Brigham Young University). In addition, we thank Gregory S. Paul for review comments.

References Cited

Carpenter, K. 1998. Vertebrate biostratigraphy of the Morrison Formation near Cañon City, Colorado. In K. Carpenter, D. Chure, and J. I. Kirkland (eds.), *The Morrison Formation: An Interdisciplinary Study. Modern Geology* 23: 407–426.

Carpenter, K. 2002. Forelimb biomechanics of nonavian theropod dinosaurs in predation. *Senckenbergiana Lethaea* 82: 59–76.

Carpenter, K. and M. Smith. 2001. Forelimb osteology and biomechanics of *Tyrannosaurus rex*. In Tanke, D. H. and Carpenter, K. (eds.) *Mesozoic Vertebrate Life*, pp. 90–116. Bloomington: Indiana University Press.

Chure, D. J. 1994. *Koparion douglassi*, a new dinosaur from the Morrison Formation (Upper Jurassic) of Dinosaur National Monument: The oldest troodontid (Theropoda: Maniraptora). *Brigham Young University Geology Studies* 40: 11–15.

Currie, P. J., and Chen, P. J. 2001. Anatomy of *Sinosauropteryx primia* from Liaoning, northeastern China. *Canadian Journal of Earth Sciences* 38: 1705–1727.

Dodson, P., A. K. Behrensmeyer, R. T. Bakker, and J. S. McIntosh. 1980. Taphonomy and paleoecology of the dinosaur beds of the Jurassic Morrison Formation. *Paleobiology* 6: 208–232.

Gilmore, C. W. 1909. Osteology of the Jurassic reptile *Camptosaurus. U.S. National Museum Proceedings* 36: 197–332.

Gilmore, C. W. 1914. *Osteology of the armoured Dinosauria in the United States National Museum, with special reference to the genus* Stegosaurus, pp. 1–143. U.S. National Museum Bulletin 89. Washington, D.C.: Government Printing Office.

Gilmore, C. W. 1920. *Osteology of the Carnivorous Dinosauria in the United States National Museum, with Special Reference to the Genera* Antrodemus (Allosaurus) *and* Ceratosaurus. U.S. National Museum Bulletin no. 110. Washington, D.C.: Government Printing Office.

Holtz, T. R., Jr. 1998. A new phylogeny of the carnivorous dinosaurs. B. P. Pérez-Moreno, T. Holtz Jr., J. L. Sanz, and J. Moratalla (eds.), *Gaia: Aspects of Theropod Paleobiology*, vol. 15, pp. 5–61. Lisbon: Museu Nacional de História Natural.

Kirkland, J. I., B. Britt, C. H. Whittle, S. K. Madsen, and D. L. Burge. 1998. A small coelurosaurian theropod from the Yellow Cat Member of the Cedar Mountain Formation (Lower Cretaceous, Barremian) of eastern Utah. In S. G. Lucas, J. I. Kirkland, and J. W. Estep (eds.), *Lower and Middle Cretaceous Terrestrial Ecosystems*, pp. 239–248. New Mexico Museum of Natural History and Science Bulletin, no. 14. Albuquerque: New Mexico Museum of Natural History and Science.

Makovicky, P. J. 1997. A new small theropod from the Morrison Formation of Como Bluff, Wyoming. *Journal of Vertebrate Paleontology* 17: 755–757.

Marsh, O. C. 1879. Notice of new Jurassic reptiles. *American Journal of Science,* ser. 3, 18: 501–505.

Marsh, O. C. 1881. A new order of extinct Jurassic reptiles (Coeluria). *American Journal of Science* 21: 339–341.

Marsh, O. C. 1884. Principal characters of American Jurassic dinosaurs. Part 8: The Order Theropoda. *American Journal of Science* 27: 29–40.

Marsh, O. C. 1888. Notice of a new genus of Sauropoda and other dinosaurs from the Potomac Formation. *American Journal of Science* 35: 89–94.

Miles, C. A., K. Carpenter, and K. Cloward. 1998. A new skeleton of *Coelurus fragilis* from the Morrison Formation of Wyoming. *Journal of Vertebrate Paleontology* 18 (3 suppl.): 64A.

Naish, D., S. Hutt, and D. M. Martill. 2001. Saurischian dinosaurs 2: Theropods. In D. M. Martill and D. Naish (eds.), *Dinosaurs of the Isle of Wight*, pp. 242–309. Palaeontological Association Field Guides to Fossils, no. 10. London: Palaeontological Association.

Norell, M. A., P. J. Makovicky, and J. M. Clark. 2000. A new troodontid theropod from Ukhaa Tolgod, Mongolia. *Journal of Vertebrate Paleontology* 20: 7–11.

Norell, M. A., J. M. Clark, and P. J. Makovicky. 2001. Phylogenetic relationships among coelurosaurian theropods. In J. Gauthier and L. F. Gall (eds.), *New Perspectives on the Origin and Early Evolution of Birds: Proceedings of the International Symposium in Honor of John H. Ostrom*, pp. 29–67. New Haven, Conn.: Peabody Museum of Natural History, Yale University.

Osborn, H. F. 1903. *Ornitholestes hermanni*, a new compsognathid dinosaur from the Upper Jurassic. *Bulletin of the American Museum of Natural History* 19: 459–464.

Osborn, H. F. 1916. Skeletal adaptations of *Ornitholestes, Struthiomimus, Tyrannosaurus. Bulletin of the American Museum of Natural History* 35: 733–771.

Ostrom, J. H. 1969. Osteology of *Deinonychus antirrhopus*, an unusual theropod from the Lower Cretaceous of Montana. *Bulletin of the Peabody Museum of Natural History* 30: 1–165.

Ostrom, J. H. 1980. *Coelurus* and *Ornitholestes:* Are they the same? In L. Jacobs (ed.), *Aspects of Vertebrate History*, pp. 245–256. Flagstaff: Museum of Northern Arizona Press.

Paul, G. S. 1988. The small predatory dinosaurs of the mid-Mesozoic: The horned dinosaurs of the Morrison and Great Oolite—*Ornitholestes* and *Proceratosaurus*—and the sickle-claw theropods of the Cloverly, Djadokhta and Judith River—*Deinonychus, Velociraptor* and *Saurornitholestes. Hunteria* 2: 1–9.

Russell, D. A., P. Béland, and J. S. McIntosh. 1980. Paleoecology of the dinosaurs of Tendaguru (Tanzania). *Mémoires de la Société géologique de France* 139: 169–175.

Sereno, P. C. 2001. Alvarezsaurids: Birds or ornithomimosaurs. In J. Gauthier and L. F. Gall (eds.), *New Perspectives on the Origins and Early Evolution of Birds*, pp. 69–98. New Haven, Conn.: Peabody Museum of Natural History.

Steele, R. 1970. Teil 14, Saurischia. *Handbuch der Paläoherpetologie.* Stuttgart: Verlag Fischer.

Welles, S. P., and R. A. Long. 1974. The tarsus of theropod dinosaurs. *Annals of the South African Museum* 64: 191–218.

4. The Enigmatic Theropod Dinosaur *Erectopus superbus* (Sauvage 1882) from the Lower Albian of Louppy-le-Château (Meuse, France)

Ronan Allain

Abstract

The remains of the theropod from Louppy-le-Château, France, described as *Megalosaurus superbus* by Sauvage in 1882, were made the type of *Erectopus sauvagei* Huene 1932. For a long time, the Pierson collection, which includes the type of *Erectopus superbus*, was considered lost. However, casts of the type skeleton have been rediscovered in the collections of Muséum National d'Histoire Naturelle, Paris. Moreover, the original anterior portion of a maxilla described by Sauvage as a dentary has been retrieved from a Parisian fossil dealer. These remains are redescribed here. *Erectopus superbus* is the valid name to which the whole of the material should be assigned. Most of the unusual characters of *Erectopus* listed in the literature are due to Sauvage's misinterpretations or to incorrect restorations of some specimens. A few derived characters suggest that *Erectopus* is more closely related to Carnosauria than to any other clade of theropod.

Introduction

The presence of carnivorous dinosaurs in the Gault (Lower Cretaceous, Aptian-Albian) of the northeastern part of the Paris Basin,

France, was established in the second half of the nineteenth century as a result of the works of Barrois (1875) and Sauvage (1876). The first described remains consisting of a single vertebra and three isolated teeth found in Grandpré (Ardennes) and Louppy-le-Château (Meuse). They were referred to the genus *Megalosaurus* (Barrois 1875; Sauvage 1876) known from the Middle and Upper Jurassic of England (Buckland 1824; Mantell 1827; Owen 1842, 1856). However, the teeth of *Megalosaurus bucklandi*, the type species of the genus from the Bathonian (Middle Jurassic), have serrations only on the apical part of the anterior carinae (Buckland 1824), whereas serrations are present on the entire length of the anterior and posterior carinae of the tooth crowns of the Grandpré and Louppy specimens. This difference led Barrois (1875) and Sauvage (1876) to consider the megalosaur from Meuse as a new species, but considering the scarcity of the material, neither erected a new name for it. In 1882, Sauvage, who had access to the private collection of Louis Pierson, described the remains of a new theropod for which he coined the name *Megalosaurus superbus*. The Pierson collection comes from the phosphate-bearing beds of La Penthiève, near Louppy-le-Château, which has also yielded ichthyosaur, plesiosaur, and crocodile remains (Sauvage 1882).

Sauvage (1882) reported that *M. superbus* included postcranial material, as well as cranial material, the latter being represented by a fragment of a dentary bearing a coronoid process. Such a process is unknown in other theropods and makes *M. superbus* enigmatic, but it unfortunately was never figured by Sauvage (1882). In addition, the unusual posteriorly offset head of the femur described by Sauvage has long been recognized by various authors as a diagnostic feature of the Louppy theropod (Huene 1923; Molnar 1990; Buffetaut et al. 1991; Rauhut 2000), but it has never been observed except in the figures of Sauvage (1882). Thus, some bones referred by Sauvage to *Megalosaurus superbus* are quite different from those of other theropods, and the taxon is clearly in need of revision (Molnar 1990; Buffetaut et al. 1991). Although Huene (1926, 1932) attempted a revision, it was based on Sauvage's description and figures, and the Pierson collection has for a long time been considered lost (Molnar 1990; Buffetaut et al. 1991; Buffetaut 1994).

Casts of some bones of the type series, mentioned by Piveteau (1923, p. 122), have been recently rediscovered in the collections of the Muséum National d'Histoire Naturelle (MNHN). These casts did not appear, until recently, in the Catalogue d'Anatomie Comparée du Muséum National d'Histoire Naturelle, although their labels state that they were given to the museum by Louis Pierson. Among these casts is the anterior part of a left maxilla, which has never been figured and which had been described by Sauvage (1882, p. 9) as the posterior part of a dentary with a coronoid process. Interestingly, this maxilla was found through a Parisian fossil dealer some years ago and purchased by Christian de Muizon, paleomammalogist at the museum. It is now the only original bone of the Pierson collection, which was probably dispersed after his death (Fig. 4.1).

Although the scientific literature discussing the osteology and the

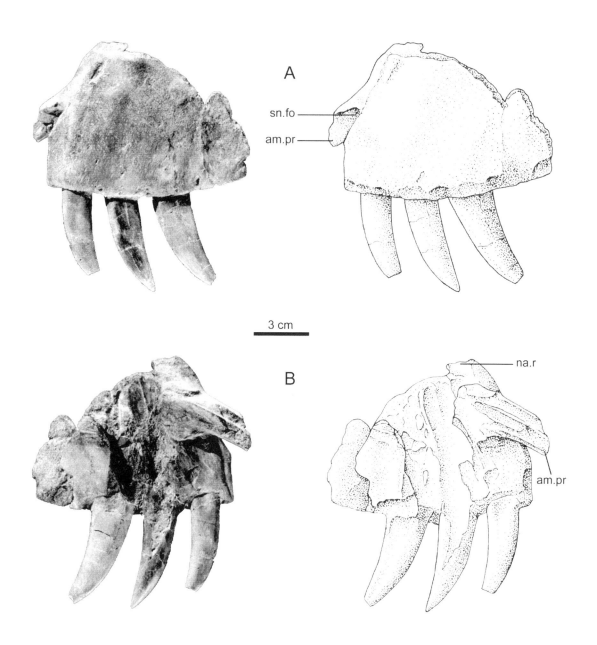

A

sn.fo

am.pr

3 cm

B

na.r

am.pr

Figure 4.1. Anterior part of the left maxilla of Erectopus superbus, *lectotype*, MNHN 2001-4, *from the lower Albian of Louppy-le-Château: (A) lateral view and (B) medial view. Abbreviations: Am.pr— anteromedial process; na.r—nasal ramus; sn.fo—subnarial foramen.*

affinities of the theropod from Louppy is limited (Sauvage 1882; Huene 1923, 1926, 1932), some problems surrounding its exact scientific name still persist (Molnar 1990; Buffetaut et al. 1991). The taxonomic name *Megalosaurus superbus* was coined by Sauvage in 1882, after he studied the collection of Louis Pierson. Noting differences between the postcranial material referred to the Megalosauridae (*Megalosaurus, Poekilopleuron,* and *Streptospondylus*) and the French theropod, Huene (1923, 1936) created the new genus *Erectopus.*

The lectotype of *Megalosaurus bucklandii,* the type species of the genus, is a rostral portion of a right dentary (OUM-J13505) (Allain and Chure 2002). Since there is no dentary material of the theropod from

Louppy, it is impossible to refer the theropod to *Megalosaurus*. Moreover, it seems difficult to believe that this genus survived the 60 million years separating the Bathonian from the Albian. The generic name *Erectopus* Huene 1923 is thus retained for the material described by Sauvage (1882).

Huene (1932) later abandoned the specific name *superbus*, arguing that the teeth described by Barrois (1875) and Sauvage (1876) were too large to be associated with the other bones subsequently described (Sauvage 1882), and that Sauvage erected the name *superbus* on the basis of the teeth before having studied the Pierson collection (Huene 1932, p. 238). Thus, Huene retained the specific name for the teeth, which he called "Gen. indeterm. *superbus*," and erected the name *Erectopus sauvagei* for the material described in 1882. In doing this, Huene misunderstood Sauvage. Even though Sauvage referred the teeth of Louppy to *Megalosaurus* as early as 1876, he carefully waited for more material before coining the specific name (Sauvage 1882). Contrary to what is claimed by Huene, Sauvage did not designate any holotype among the material described in 1882. Moreover, Huene overlooked that the isolated tooth collected in the Albian phosphate-bearing beds of Louppy and described in 1876 has the same size and morphology as the teeth of the maxilla described in 1882. *Erectopus sauvagei* is thus considered a junior synonym of *Erectopus superbus*, the maxilla of which is here designated as the lectotype of the species.

Institutional Abbreviations. ML, Museum of Lourinhã, Lourinhã; MNHN, Muséum National d'Histoire Naturelle, Paris; OUM, Oxford University Museum, Oxford.

Systematic Paleontology
Saurischia Seeley 1887
Theropoda Marsh 1881
Tetanurae Gauthier 1986
Allosauroidea Marsh 1878
Erectopus Huene 1923

Type species—*Erectopus superbus* (Sauvage 1882), by original description.

Erectopus superbus (Sauvage 1882)
Megalosaurus superbus Sauvage 1882 (p. 9; plate 1, figs.
1–5; plate 2, figs. 1–2; plate 3, fig. 1; plate 4, figs. 1–4).
Erectopus sauvagei Huene 1932, p. 239–240.

Lectotype. MNHN 2001-4, anterior part of the left maxilla.
Plastotype. MNHN 2001-4, anterior part of the left maxilla, partial right manus, left femur, proximal and distal ends of the left tibia, left calcaneum, right metatarsal II.
Type Locality. Louppy-le-Château (Meuse, eastern France).
Horizon and Age. Phosphate-bearing beds of La Penthiève, *Mammilatum* Zone, Lower Cretaceous, lower Albian.
Diagnosis. Rounded anterior ramus of maxilla; slender neck of

femur; posterior curvature of proximal half of femur; anterodorsal edge of calcaneum dorsally projected; calcaneum twice as long as deep vertically; posteromedial process for tibia on articular surface of astragalus; length of second metatarsal equal to half the length of femur; lateral margin of proximal end of second metatarsal regularly concave.

Description

Left Maxilla. The lectotype corresponds to the anterior-most part of a left maxilla broken through the fourth alveolus (Fig. 4.1A,B). The lateral surface of the maxilla is vertical and flat. Contrary to the condition in the Abelisauridae and Carcharodontosauridae, it is smooth except on its ventral part, immediately above the tooth row, where it is pierced by the superior labial foramina. The anterior margin, which contacts the premaxilla, is 8 cm high. It is vertical for 3 cm and then inclined posterodorsally at about 45° relative to the maxillary tooth row. It forms about 2 cm of the posteroventral boundary of the subnarial foramen before reaching the base of the nasal ramus of the maxilla. The latter is broken but is in a very anterior position on the maxillary, being confluent with the anterior rim of the maxilla body and the sutural contact with the premaxilla. In contrast, it is offset from the anterior rim in some theropods such as *Poekilopleuron? valesdunensis* (Allain 2002a), *Eustreptospondylus* (Huene 1926), *Torvosaurus* (Britt 1991), and *Neovenator* (Hutt et al. 1996). The anteromedial process of the maxilla considered by Sauvage (1882) as the coronoid process of the dentary is prominent and protrudes anteriorly from the body of the maxilla. In medial view, the process is higher in position than that of *Sinraptor* (Currie and Zhao 1993) and slopes anteroventrally at an angle of about 40° relative to the maxillary tooth row. The medial surface of the maxilla is badly damaged. Only the fourth interdental plate is preserved. It is taller than it is long and was not fused with the adjacent plates. The first three teeth are preserved in the lectotype, whereas the second one is absent in the cast of the lateral side of the maxillary given by Pierson to the MNHN. In fact, during preparation of the lectotype, the second replacement tooth was extracted from its alveolus (Christian de Muizon, pers. comm.). The teeth are transversally compressed, tapered, and symmetrical in anterior view. The entire length of the anterior and posterior carinae of each tooth bears small serrations. There are fifteen serrations per 5 mm on the anterior carina and thirteen on the posterior one. Anterior serrations are perpendicular to the long axis of the tooth, whereas posterior serrations slope slightly toward the apex of the tooth.

Manus. The elements of the right manus were cast from the block containing them. Only one side of each phalanx or metacarpal is thus observable (Fig. 4.2). Manual elements include the distal end of metacarpal III, and phalanges I-1, I-2, II-1, II-2, II-3, III-1, and III-2. All are in semi-articulation. The distal end of the third metacarpal is well expanded in contrast to the shaft of the bone, and the discontinuity of the extensor surface thus observed is reminiscent of that of *Allosaurus*

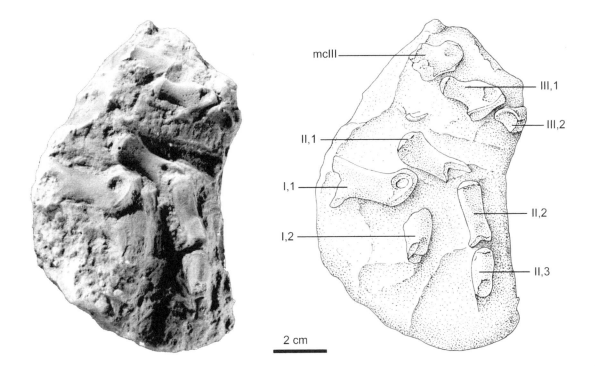

Figure 4.2. Cast of some elements of the right hand of Erectopus superbus *enclosed in matrix, MNHN 2001-4, from the lower Albian of Louppy-le-Château. McIII, third metacarpal; I, 1–2, phalanges of the first digit; II, 1–3, phalanges of the second digit; III, 1–2, phalanges of the third digit.*

(Madsen 1976, plate 43) and *Sinraptor* (Currie and Zhao 1993, fig. 20G). The collateral ligament fossae are shallow on the third metacarpal and absent on phalanges III-1 and III-2 as in *Sinraptor* (Currie and Zhao 1993), whereas they are deep on the phalanges of digits I and II. As in *Allosaurus*, phalanx I-1 is the longest (35 mm) of the preserved phalanges, and the ungual phalanx of digit I is longer than that of digit II.

Left Femur. The left femur is 48 cm long (Fig. 4.3A–C). The maximum width of the proximal head is 10 cm (Fig. 4.3B), and the distal width is 9.5 cm (Fig. 4.3D). The shaft below the fourth trochanter is 5.5 cm anteroposteriorly by 5.3 cm mediolaterally, dimensions that suggest an individual of about 200 kg (Anderson et al. 1985). The dimensions contrast with those of *Sinraptor,* in which the mediolateral width is greater than the anteroposterior one (Currie and Zhao 1993). In medial view, the distal third of the shaft of the femur is straight, whereas the proximal two-thirds is concave posteriorly (Fig. 4.3C). The proximal end of the femur exhibits certain unusual features, such as a posteriorly offset head and an anterior trochanter that faces anteriorly (Fig. 4.3A). These features have long been recognized as autapomorphies of *Erectopus* (Molnar 1990; Buffetaut et al. 1991; Buffetaut 1994; Rauhut 2000) but are in fact due to bad preparation and restoration of the original femur by Louis Pierson. The head of the femur was originally broken under the anterior trochanter, as is clearly seen on the cast (Fig. 4.3A–B). During the restoration, this part was caudally displaced and rotated clockwise, giving the femur its unusual aspect.

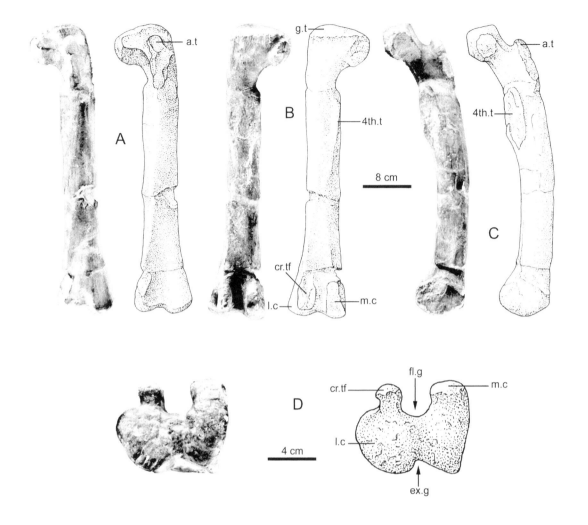

Figure 4.3. Cast of the left femur of Erectopus superbus, *MNHN 2001-4, from the lower Albian of Louppy-le-Château in (A) anterior, (B) posterior, (C) medial, and (D) distal views. Abbreviations: A.t—anterior trochanter; cr.tf—crista tibiofibularis; ex.g—extensor groove; fl.g—flexor groove; g.t—great trochanter; l.c—lateral condyle; m.c—medial condyle; 4th.t—fourth trochanter.*

Correctly oriented, the head of the femur is medially directed, and the anterior trochanter is more lateral in position as in most theropods. The greater trochanter is continuous with the head of the femur. The anterior trochanter is broken. Contrary to a previous report (Molnar 1990), it is wing-like as in *Allosaurus* and *Sinraptor*, extending proximally at least as far as the mid-height of the femoral head, and is separated from the main body of the femur by a deep slot (Fig. 4.3C). The fourth trochanter is a 10-cm-long pronounced crest proximal to the mid-shaft (Fig. 4.3B–C). Its medial side is concave for insertion of the caudofemoralis longis muscle.

The distal end of the femur is divided into a longitudinally expanded and transversally narrow medial condyle and a more or less rounded lateral condyle. This latter is not as stout as in *Sinraptor* or *Streptospondylus* (Allain 2001). As in all theropods except *Baryonyx* (Charig and Milner 1997), the medial condyle extends below the lateral one. The crista tibiofibularis (the ectocondylar tuber of Welles 1984)

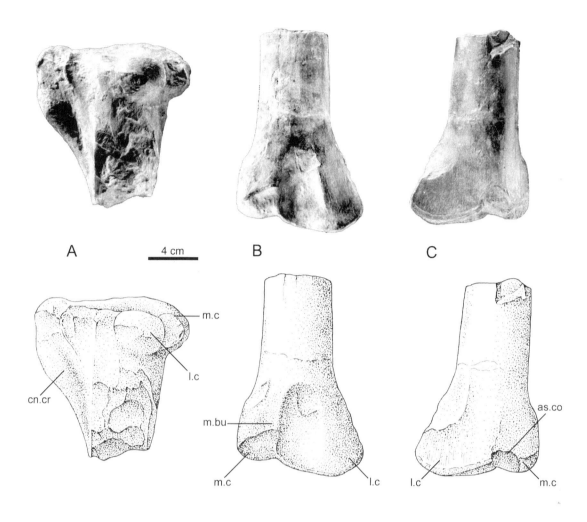

Figure 4.4. Cast of the left tibia of Erectopus superbus, MNHN 2001-4, from the lower Albian of Louppy-le-Château: (A) proximal end in lateral view; (B) distal end in anterior view; and (C) distal end in posterior view. Abbreviations: As.co—contact with the astragalus; cn.cr—cnemial crest; l.c—lateral condyle; m.bu—medial buttress; m.c—medial condyle.

arises on the posteromedial surface of the lateral condyle, parallel to the posterior expansion of the medial condyle. Distally, the crista tibio-fibularis is slightly deflected laterally. It bounds medially the trochlea fibularis (Currie and Zhao 1993) into which the proximal end of the fibula fitted, and laterally the 3-cm-deep U-shaped flexor groove (Fig. 4.3D). The longitudinal ridge on the floor of the flexor groove that is present in *Allosaurus* and *Sinraptor* is absent. The distal condyles are also separated anteriorly by an extensor groove. This groove is asymmetrical and deeper than that observed in *Poekilopleuron? valesdunensis* (Allain 2002b), *Streptospondylus* (Allain 2001), *Eustreptospondylus* (Huene 1926), and in *Ceratosaurus* (Gilmore 1920).

Left Tibia. The remains of the left tibia consist of the proximal and distal ends. Sauvage (1882, plate 4, fig. 1) misidentified this latter as the distal end of the radius. The proximal width of the tibia is 13 cm anteroposteriorly and 7.7 cm mediolaterally (Fig. 4.4A). In proximal view, the tibia is comparable with that of *Allosaurus*. The paired articulations on the proximal end is not as deeply separated by the posterior intercondylar groove as the articulations of *Allosaurus,* but more than those of *Sinraptor.* The medial condyle is higher than the lateral (Fig. 4.4A). The cnemial crest is well developed. It arises on the anteromedial surface of the shaft and extends anterolaterally. The shaft of the proximal part of the tibia is broken at the level of the crista fibularis, with the latter distally placed. In contrast to the condition in *Ceratosaurus* (Madsen and Welles 2000), the crista fibularis arises abruptly from the shaft of the tibia and is not connected with the proximal end of the tibia.

The distal half of the left tibia is broken 15.5 cm above its distal articulation (Fig. 4.4B–C). At that level, the shaft diameter of the tibia is 4 cm anteroposteriorly and 5.6 cm mediolaterally. The distal width is 4.1 cm anteroposteriorly and 10.1 cm mediolaterally; and laterally, the tibia backs the calcaneum. The anterior surface of the distal end of the tibia is depressed where it contacts the astragalus. The astragalar overhang (medial buttress of Molnar et al. 1996), which accommodates the ascending process of the astragalus proximomedially, extends from the mediodistal corner of the tibia proximolaterally (Fig. 4.4B). The angle formed between the astragalar overhang and the horizontal level is about 30° for 3.5 cm, before the slope becomes nearly vertical. As in *Allosaurus* and *Streptospondylus,* the astragalar overhang extends only as far as the median line of the anterodistal surface of the tibia laterally, whereas in *Torvosaurus* (Britt 1991), *Eustreptospondylus* (Huene 1926), and *Poekilopleuron bucklandii* (Allain and Chure 2002), it has a more pronounced lateral expansion. The distal articular surface that contacts the astragalus extends slightly onto the posterior surface of the tibia (Fig. 4.4C). This suggests that the astragalus of *Erectopus* should have a posteromedial process as in *Allosaurus* and *Sinraptor* (Currie and Zhao 1993, fig. 23E–F).

Calcaneum. The calcaneum of *Erectopus* is more slender that the calcanea of *Allosaurus* and *Sinraptor,* its maximum anteroposterior length (5.6 cm) being twice its median height (2.7 cm) (Fig. 4.5). It is wider ventrally than dorsally. The lateral surface of the calcaneum is flat, whereas the proximal part that contacts the fibula is markedly concave, with a higher lateral margin than in *Torvosaurus, Eustreptospondylus,* and *Streptospondylus.* The fibular articulation is shared with the astragalus. In lateral and medial view, the anterodorsal corner of the calcaneum is dorsally projected. This feature is present in *Sinraptor* (Currie and Zhao 1993, fig. 23G–H) and *Allosaurus* (Gilmore 1920, fig. 48B). In *Ceratosaurus* (Gilmore 1920, fig. 65) and in spinosauroids such as *Baryonyx* (Charig and Milner 1997, fig. 43), *Torvosaurus* (Britt 1991, fig. 23H), and *Streptospondylus* (Allain 2001, fig. 6A), the dorsal margin of the calcaneum is only slightly concave and is devoid of any anterodorsal projection. The medial articular surface of the calcaneum is divided into three distinct depressions (Fig. 4.5). The posterior one is the largest. It faces posteromedially and articulates with the tibia. The medial concavity is much smaller than in *Allosaurus* and *Sinraptor.* Associated with the elongated anterior concavity, the medial concavity should result in a tight-fitting articulation with the

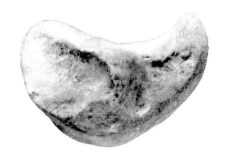

2 cm

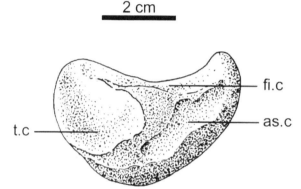

astragalus, allowing no movement between the two bones. The rounded distal articular surface of the calcaneum tapers both anteriorly and posteriorly.

Metatarsal II. The second right metatarsal is the only preserved element of the foot that was cast (Fig. 4.6). It is longer relative to the femur than in most other large theropods: its length (23 cm) is 48 percent the length of the femur, compared with 32–39 percent in *Allosaurus* (Gilmore 1920), 32 percent in *Acrocanthosaurus* (Currie and Carpenter 2000), 41 percent in *Sinraptor,* and 44 percent in *Neovenator* (Hutt et al. 1996; Naish et al. 2001). This feature can be used to diagnose *Erectopus;* however, it is difficult to assess whether the ratio reflects phylogeny rather than ontogeny, because immature theropods have relatively longer metatarsals than adults do (Russell 1970). The shaft of the metatarsal is longitudinally straight and more or less rounded in cross-section. The proximal end is slightly different from the proximal ends of *Allosaurus* and *Sinraptor:* the lateral margin, which contacts metatarsal III, is slightly concave along its entire length (Sauvage 1882, fig. 3b), whereas it is convex in *Sinraptor* (Currie and Zhao 1993, fig. 26C) and sigmoid in *Allosaurus* (Gilmore 1920, fig. 51). Unlike the second metatarsal of *Torvosaurus* (Britt 1991), the metacarpal II of *Erectopus* bears deep pits at the point of origin of the collateral ligaments.

Figure 4.5. Cast of the left calcaneum of Erectopus superbus, *MNHN 2001-4, from the lower Albian of Louppy-le-Château, in medial view. Abbreviations: As.c—contact with the astragalus; fi.c—contact with the fibula; t.c—contact with the tibia.*

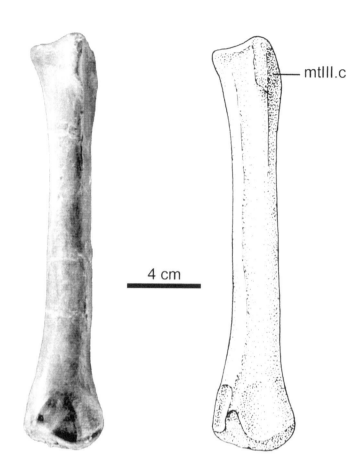

mtIII.c

Figure 4.6. Cast of the second metatarsal of Erectopus superbus, *MNHN 2001-4, from the lower Albian of Louppy-le-Château, in posterior view. Abbreviation: MtIII.c—contact with the third metatarsal.*

Discussion

The limited and fragmentary nature of the only known specimen of *Erectopus superbus* makes phylogenetic analysis difficult. It was originally described as a Megalosauridae (Sauvage 1882) and was later placed by Huene (1932) within Carnosauria in its own family, the Erectopodidae. Waiting for a thorough revision, Molnar (1990) failed to find any carnosaurian characters in Sauvage's figures, and he classified *Erectopus* as a problematic Carnosauria incertae sedis. If the phylogeny of carnivorous dinosaurs of Holtz (2000) is used, *Erectopus* is clearly a tetanuran theropod. The tetanuran synapomorphies of *Erectopus* and the synapomorphies found at nodes H to L in Holtz's phylogeny include the femoral head directed horizontally; the femoral head transversally elongated; the anterior trochanter of the femur placed at approximately the mid-height of the femoral head; a tibia distal end that backs the calcaneum; and an astragalar ascending process that is mediolaterally reduced, craniocaudally wide, and proximodistally low, as shown by the contact surface for astragalus on the distal end of the tibia.

The relationships of *Erectopus* within tetanuran theropods are more difficult to assess. None of the synapomorphies listed by Holtz (2000) in avetheropods or by Sereno (1998, 1999) in neotetanuran theropods is recognized in *Erectopus*, except for the extensor groove in the craniodistal region of the femur, which is deep and conspicuous but not as deep as in basal Tetanurae such as *Poekilopleuron? valesdunensis* (Allain 2002b; MNHN, 1998-13), *Streptospondylus* (Allain 2001; MNHN, 9645), *Eustreptospondylus* (Huene 1926; OUM-J13558), or *Lourinhanosaurus* (Mateus 1998; ML 370). Although the astragalus of *Erectopus* is not preserved, the morphology of the distal end of the tibia clearly shows that the astragalus had a posterodorsally directed medial process on the posterior margin of its articulating surface. Such a process, which wraps around the distal articular surface of the tibia, is known only in allosauroids such as *Allosaurus* and *Sinraptor* (Currie and Zhao 1993, fig. 23E–F). *Erectopus superbus* has never been included in a cladistic analysis, except in a recent work that corroborates its inclusion within Carnosauria (Allain 2002b, 2002c, in prep.). *Erectopus* is thus regarded here as a Carnosauria. It is the third and youngest carnosaur reported in the Lower Cretaceous of Europe, along with the Montmirat theropod from the Valangian of southern France (Pérez-Moreno et al. 1993) and *Neovenator salerii* from the Barremian of the Isle of Wight (Hutt et al. 1996).

Cretaceous carnosaurs are more diversified than previously believed. Other Cretaceous carnosaurs include the Gondwanan carcharodontosaurids *Carcharodontosaurus* (Stromer 1931; Sereno et al. 1996) and *Giganotosaurus* (Coria and Salgado 1995), and the North American allosaurid *Acrocanthosaurus* (Stovall and Langston 1950; Currie and Carpenter 2000). The establishment of a biogeographic pattern to explain this distribution will remain uncertain as long as the phylogenetic relationships of carnosaurs and allosauroids are not ascertained, but the post-separation vicariance from a Middle to Late Jurassic global endemic carnosaur fauna is likely (Hutt et al. 1996).

Acknowledgments. I am grateful to Professor P. Taquet, who entrusted me with the material, and to K. Padian, J.-M. Mazin, A. de Ricqlés, and D. Gouget for critical comments on an earlier French draft of the manuscript. Thanks to P. Powell and O. Mateus for access to material in their care in Oxford and in Lourinhã. Thanks also to F. Pilard for drawings and to D. Serrette and P. Loubry for photographs.

References Cited

Allain, R. 2001. Redescription de *Streptospondylus altdorfensis,* le dinosaure théropode de Cuvier, du Jurassique de Normandie. *Geodiversitas* 23: 349–367.

Allain, R. 2002a. Discovery of megalosaur (Dinosauria, Theropoda) in the middle Bathonian of Normandy (France) and its implications for the phylogeny of basal Tetanurae. *Journal of Vertebrate Paleontology* 22: 548–563.

Allain, R. 2002b. Les Megalosauridae (Dinosauria, Theropoda). Nouvelle découverte et révision systématique: Implications phylogénétiques et paléobiogéographiques. Ph.D. thesis, Muséum National d'Histoire Naturelle, Paris.

Allain, R. 2002c. The phylogenetic relationships of Megalosauridae within basal tetanurine theropods. *Journal of Vertebrate Paleontology* 22 (3 suppl.): 31A.

Allain, R., and D. J. Chure. 2002. *Poekilopleuron bucklandii*, the theropod dinosaur from the Middle Jurassic (Bathonian) of Normandy. *Palaeontology* 45: 1107–1121.

Anderson, J. F., A. Hall-Martin, and D. A. Russell. 1985. Long-bone circumference and weight in mammals, birds and dinosaurs. *Journal of Zoology, London* 207: 53–61.

Barrois, C. 1875. Les reptiles du terrain Crétacé du nord-est du Bassin de Paris. *Bulletin scientifique, historique et littéraire du Nord* 6: 1–11.

Britt, B. B. 1991. Theropods of Dry Mesa Quarry (Morrison Formation, Late Jurassic), Colorado, with emphasis on the osteology of *Torvosaurus tanneri*. *Brigham Young University Geology Studies* 37: 1–72.

Buckland, W. 1824. Notice on the *Megalosaurus* or great fossil lizard of Stonesfield. *Transactions of the Geological Society of London* 2: 390–396.

Buffetaut, E. 1994. The significance of dinosaur remains in marine sediments: An investigation based on the French record. *Berliner Geowissenschaften Abhandlungen* 13: 125–133.

Buffetaut, E., G. Cuny, and J. Le Loeuff. 1991. French dinosaurs, the best record in Europe? *Modern Geology* 16: 17–42.

Charig, A. J., and A. C. Milner. 1997. *Baryonyx walkeri*, a fish-eating dinosaur from the Wealden of Surrey. *Bulletin of the Natural History Museum, Geology Series*, 53: 11–70.

Coria, R. A., and L. Salgado. 1995. A new giant carnivorous dinosaur from the Cretaceous of Patagonia. *Nature* 377: 224–226.

Currie, P. J., and K. Carpenter. 2000. A new specimen of *Acrocanthosaurus atokensis* (Theropoda, Dinosauria) from the Lower Cretaceous Antlers Formation (Lower Cretaceous, Aptian) of Oklahoma, USA. *Geodiversitas* 22: 207–246.

Currie, P. J., and Zhao X.-J. 1993. A new carnosaur (Dinosauria, Theropoda) from the Jurassic of Xinjiang, People's Republic of China. *Canadian Journal of Earth Sciences* 30: 2037–2081.

Gilmore, C. W. 1920. *Osteology of the Carnivorous Dinosauria in the United States National Museum, with Special Reference to the Genera* Antrodemus (Allosaurus) *and* Ceratosaurus. U.S. National Museum Bulletin no. 110. Washington, D.C.: Government Printing Office.

Holtz, T. R., Jr. 1998. A new phylogeny of the carnivorous dinosaurs. B. P. Pérez-Moreno, T. Holtz Jr., J. L. Sanz, and J. Moratalla (eds.), *Gaia: Aspects of Theropod Paleobiology,* vol. 15, pp. 5–61. Lisbon: Museu Nacional de História Natural.

Huene, F. 1923. Carnivorous Saurischia in Europe since the Triassic. *Bulletin of the Geological Society of America* 34: 449–458.

Huene, F. 1926. The carnivorous Saurischia in the Jura and Cretaceous formations, principally in Europe. *Revista del Museo de la Plata* 29: 35–167.

Huene, F. 1932. Die fossile Reptil-Ordnung Saurischia, ihre Entwicklung und Geschichte. *Monographien zur Geologie und Palaeontologie,* ser. 1, 4: 1–361.

Hutt, S., D. M. Martill, and M. J. Barker. 1996. The first European allosaurid dinosaur (Lower Cretaceous, Wealden Group, England). *Neues Jahrbuch für Geologie und Paläontologie, Monatshefte* 10: 635–644.

Madsen, J. H., Jr. 1976. Allosaurus fragilis: *A revised osteology*. Utah Geological and Mineral Survey Bulletin, no. 109. Salt Lake City: Utah Geological and Mineral Survey, Utah Department of Natural Resources.

Madsen, J. H., Jr., and S. P. Welles. 2000. *Ceratosaurus* (Dinosauria, Theropoda), a revised osteology. *Utah Geological Survey Miscellaneous Publication* 00-2: 1–80.

Mantell, G. 1827. *Illustrations of the geology of Sussex: Containing a general view of the geological relations of the south-eastern part of England; with figures and descriptions of the fossil of Tilgate Forest*. London: Relfe L.

Mateus, O. 1998. *Lourinhanosaurus antunesi*, a new Upper Jurassic Allosauroid (Dinosauria: Theropoda) from Lourinhã, Portugal. *Memorias da Academia das Ciencias de Lisboa* 37: 111–124.

Molnar, R. E. 1990. Problematic Theropoda: "Carnosaurs." In D. B. Weishampel, P. Dodson, and H. Osmólska (eds.), *The Dinosauria*, pp. 306–317. Berkeley: University of California Press.

Molnar, R. E., A. Lopez Angriman, and Z. Gasparini. 1996. An Antarctic Cretaceous theropod. *Memoirs of the Queensland Museum* 39: 669–674.

Naish, D., S. Hutt, and D. M. Martill. 2001. Saurischian dinosaurs 2: Theropods. In D. M. Martill and D. Naish (eds.), *Dinosaurs of the Isle of Wight*, pp. 242–309. Palaeontological Association Field Guides to Fossils, no. 10. London: Palaeontological Association.

Owen, R. 1842. Report on British fossil reptiles. Pt. II. *Report of the British Association for the Advancement of Science* 11: 60–204.

Owen, R. 1856. *Monograph on the Fossil Reptilia of the Wealden Formations*. Part III: *Megalosaurus bucklandi*, pp. 1–26. Palaeontographical Society Monographs, no. 9. London: Palaeontographical Society.

Pérez-Moreno, B. P., J. L. Sanz, J. Sudre, and B. Sigé. 1993. A theropod dinosaur from the Lower Cretaceous of southern France. *Revue de Paléobiologie*, vol. spéc., 7: 173–188.

Piveteau, J. 1923. L'arrière crâne d'un dinosaurien carnivore de l'Oxfordien de Dives. *Annales de Paléontologie* 12: 115–123.

Rauhut, O. W. M. 2000. The interrelationships and evolution of basal theropods (Dinosauria, Saurischia). Ph.D. thesis, University of Bristol.

Russell, D. A. 1970. *Tyrannosaurs from the Late Cretaceous of Western Canada*. National Museums of Canada, National Museum of Natural Sciences, Publications in Palaeontology, no. 1. Ottawa.

Sauvage, H. E. 1876. De la présence du type dinosaurien dans le Gault du nord de la France. *Bulletin de la Société géologique de France* 4: 439–442.

Sauvage, H. E. 1882. Recherches sur les reptiles trouvés dans le Gault de l'est du Bassin de Paris. *Mémoires de la Société géologique de France* 2: 1–41.

Sereno, P. C. 1998. A rationale for phylogenetic definitions, with application to the higher-level taxonomy of Dinosauria. *Neues Jahrbuch für Geologie und Paläontologie, Abhandlungen* 210: 41–83.

Sereno, P. C. 1999. The evolution of dinosaurs. *Science* 284: 2137–2147.

Sereno, P. C., D. B. Dutheil, M. Iarochene, H. C. E. Larsson, G. H. Lyon, P. M. Magwene, C. A. Sidor, D. J. Varricchio, and J. A. Wilson 1996. Predatory dinosaurs from the Sahara and Late Cretaceous faunal differentiation. *Science* 272: 986–991.

Stovall, J. W., and W. Langston. 1950. *Acrocanthosaurus atokensis*, a new genus and species of Lower Cretaceous Theropoda from Oklahoma. *American Midland Naturalist* 43: 696–728.

Stromer, E. 1931. Ergebnisse der Forschungsreisen Prof Stromers in den Wüsten Ägyptens II Wirbeltier-Reste der Baharije-Stufe (unterstes Cenoman). 10. Ein Skelett-rest von *Carcharodontosaurus* nov gen. *Abhandlungen der Bayerischen Akademie der Wissenschaften, Mathematisch, Naturwissenschaftliche, Abteilung (Neue Folge)* 9: 1–23.

Welles, S. P. 1984. *Dilophosaurus wetherilli* (Dinosauria, Theropoda): Osteology and comparisons. *Palaeontographica Abt. A* 185: 85–180.

5. Holotype Braincase of *Nothronychus mckinleyi* Kirkland and Wolfe 2001 (Theropoda; Therizinosauridae) from the Upper Cretaceous (Turonian) of West-Central New Mexico

James I. Kirkland, David K. Smith, and Douglas G. Wolfe

Abstract

The occipital braincase of *Nothronychus mckinleyi* Kirkland and Wolfe 2001 from the lower middle Turonian Moreno Hill Formation of west-central New Mexico is that of an advanced therizinosaurid. It is ventrally expanded and highly pneumatic, as is the braincase of *Erlicosaurus andrewsi* Perle from Mongolia, the only other described therizinosaurid braincase; this similarity supports the close relationship of these two taxa.

Introduction

The Moreno Hill Formation in west-central New Mexico has recently yielded the remains of several dinosaurian taxa, including the first therizinosaurid outside of Asia to be represented by an associated

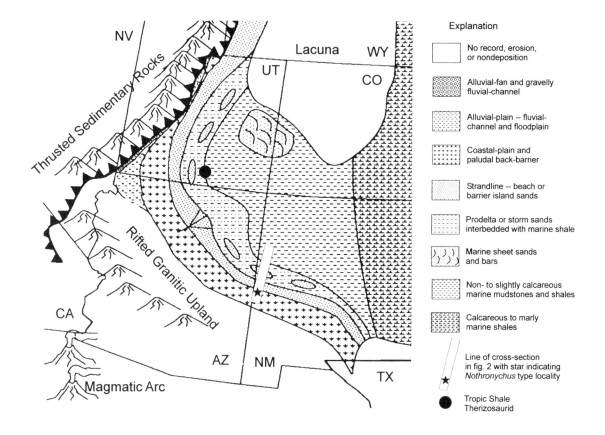

Explanation

	No record, erosion, or nondeposition
	Alluvial-fan and gravelly fluvial-channel
	Alluvial-plain -- fluvial-channel and floodplain
	Coastal-plain and paludal back-barrier
	Strandline -- beach or barrier island sands
	Prodelta or storm sands interbedded with marine shale
	Marine sheet sands and bars
	Non- to slightly calcareous marine mudstones and shales
	Calcareous to marly marine shales
	Line of cross-section in fig. 2 with star indicating *Nothronychus* type locality
	Tropic Shale Therizosaurid

Figure 5.1. Middle Turonian paleogeographic map showing line of cross-section in Figure 5.2, Nothronychus type locality, and site of lower Turonian Tropic Shale therizinosaurid (modified after Elder and Kirkland 1993).

skeleton. The newly recovered posterior portion of the skull can now be added to the holotype specimen (MSM P-2117) of *Nothronychus mckinleyi* Kirkland and Wolfe 2001.

The type locality of *Nothronychus* within the Moreno Hill Formation is well established as being "lower" middle Turonian on the basis of stratigraphic relationships with the underlying Atarque Sandstone, which preserves both *Collignoniceras woollgari* and elongate specimens of the inoceramid bivalve *Mytiloides* to the north of the locality (Figs. 5.1 and 5.2). These index fossils occur together only at the base of the middle Turonian in the basal *Collignoniceras woollgari* Zone (Kirkland 1991, 1996; Kauffman et al. 1993; Kennedy et al. 2000).

The type specimen of *Nothronychus* occurs as disarticulated bones that represent the only theropod in a small bone bed preserving a maximum of six individuals of *Zuniceratops christopheri* (Wolfe and Kirkland 1998; Wolfe 2000). Subsequently, a second therizinosaurid was recovered from a lower Turonian horizon approximately 1 million years older than the type locality, in the marine Tropic Shale of southern Utah (Gillette and Albright 2001; Alan Titus, pers. comm.). Although this second specimen appears to be similar to *Nothronychus* and is considerably more complete, it lacks any skull material (Gillette et al. 2002; David Gillette, pers. comm.).

The holotype of *Nothronychus* consists of a partial disarticulated skeleton, including isolated teeth, the new braincase, several cervical

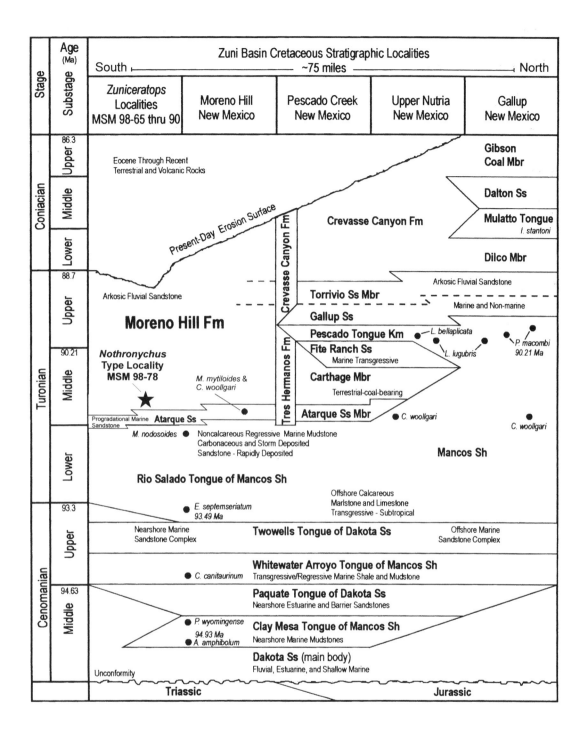

Figure 5.2. Stratigraphic and biostratigraphic synthesis of Cretaceous strata in the southern Zuni Basin showing Nothronychus type locality (modified after Hook et al. 1983; Wolfe and Kirkland 1998).

Holotype Braincase of *Nothronychus mckinleyi* Kirkland and Wolfe 2001 • 89 vertebrae, the first? dorsal vertebra, ribs, fused gastralia, medial caudal vertebrae, the scapula, the right humerus, complete right and partial left ulnae, metacarpals, manual phalanges, two ischia, two tibiae, the right fibula, partial metatarsals, pedal phalanges, and four pedal unguals. Discussions with David Gillette regarding the new Utah therizinosaurid specimen lead us to conclude that the identification of the two thinner claws as manual unguals was incorrect (Kirkland and Wolfe 2001). We now believe they represent pedal unguals similar to those recognized in *Erlicosaurus* (Barsbold and Maryańska 1990; Maryańska 1997), although the pedal unguals of the Utah therizinosaurid are described as blunt (Gillette et al. 2002).

Institutional Abbreviation. MSM, Mesa Southwest Museum, Mesa, Arizona.

Systematic Paleontology
Theropoda Marsh 1881
Tetanurae Gauthier 1986
Coelurosauria von Huene 1914
Therizinosauridae Barsbold 1976
Nothronychus mckinleyi Kirkland and Wolfe 2001

Holotype. Paleontological collections of Mesa Southwest Museum (MSM), Mesa, Arizona, MSM P-2117, partial disarticulated skeleton.

Type Locality. Haystack Butte, locality (MSM 98-78) southern Zuni Basin, Catron County, New Mexico.

Amended Diagnosis. Braincase narrow with supraoccipital above foramen magnum oriented nearly horizontally and lacking distinct nuchal crest; teeth with coarse serrations extending close to constriction with circular root; anterior dorsal vertebrae with long pedicle and large pleurocoel encasing multiple separate pneumatic foramina; scapula slender with laterally facing glenoid; slender, straight humerus with short deltopectoral crest and lacking spur on humeral shaft; thin ischium nearly excluded from acetabulum with large, rectangular, medially situated obturator process; fibula with M. iliofibularis tubercle approximately at mid-shaft; pedal unguals two, three, and four thin and straight with thicker pedal ungual one.

Description of Braincase

For an animal with a tibia length of 614 mm, the skull is small (as is typical for therizinosaurids), measuring only 80 mm high and 79 mm across the paroccipital processes (Fig. 5.3). The braincase preserves the supraoccipital, exoccipital, basioccipital, opisthotic, prootic, basisphenoid, and partial laterosphenoid. It is completely co-ossified, so that the sutures, except the one between the basioccipital and basisphenoid, are not visible. It is similar in most aspects to that of the only other therizinosaurid for which the braincase is known, *Erlicosaurus andrewsi* (Perle 1981), from the upper part of the Baynshirenskaya Svita of Mongolia (Clark et al. 1994). This similarity is most apparent

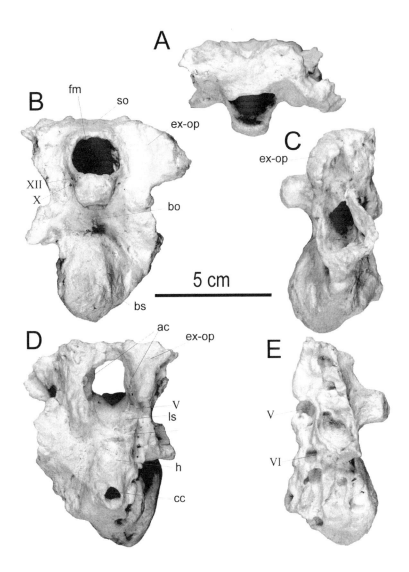

Figure 5.3. Partial braincase of Nothronychus mckinleyi in (A) dorsal, (B) caudal, (C) right lateral, (D) rostral view, and (E) left lateral views. Abbreviations: ac—fossa auriculae cerebelli; bo—basioccipital; bs—basisphenoid; cc—ostium canalium caroticorum; ex-op—exoccipital—opisthotic; fm—foramen magnum; h—fossa hypophysialis; ls—laterosphenoid; so—supraoccipital; V, VI, X, and XII—foramina for cranial nerves V, VI, X, and XII. Scale = 5 cm.

in the great ventral expansion of the extensively pneumatized basisphenoid, which is likewise subdivided into a large medial space and lateral areas for middle ear cavities (Fig. 5.3). Additionally, there are numerous similarities to the described braincases of *Troodon* (Currie and Zhao 1993) and *Chirostenotes* (Sues 1997) from the Dinosaur Park Formation of southern Alberta, Canada, which are particularly useful in comparisons with the anterior view.

The supraoccipital is a solid, horizontal plate above the circular foramen magnum; the nuchal crest is reduced. In *Erlicosaurus*, the supraoccipital extends dorsally beyond the paroccipital processes. As in *Erlicosaurus* and *Troodon*, the anterior margin at the mid-line of the supraoccipital of *Nothronychus* appears to form the posterior edge of a small dorsal venous foramen penetrating the skull roof. The missing parietal would form the anterior margin of this foramen.

The exoccipital is completely fused with the opisthotic. The exoccipitals form a raised edge of the foramen magnum. A faint line on the dorsal surface of the occipital condyle just within the foramen magnum suggests that the exoccipitals might meet above the basioccipital; they extend laterally as short paroccipital processes. The left process is broken, and the right is complete. Separated from the foramen magnum by a deep depression, the hollow paroccipital processes are flattened posteriorly, are inclined back at 10–15°, and are shorter than in *Erlicosaurus*, indicating a more narrowly constructed skull. The bases of the pneumatic paroccipital processes extend dorsally to the level of the skull roof as in *Chirostenotes*. They are pendant in *Erlicosaurus* and *Chirostenotes*, but in *Nothronychus*, there is a convex dorsal margin and straight ventral margin. The exoccipital forms a raised lateral margin of the foramen magnum. The base of this margin extends posteriorly to meet the occipital condyle, forming a dorsolateral ridge along the neck of the condyle. Small openings for cranial nerves X and XII are recognized near the juncture of the paroccipital processes and the occipital condyle. The opisthotic forms the posterior part of the middle ear, below the paroccipital processes, and is broken on both sides. On the medial side, there is a well-preserved fossa auriculae cerebelli that accommodated a moderate-size flocculus.

The basioccipital makes up most of the occipital condyle and a basicranial fontanelle. As is typical for small theropods, the occipital condyle (15.2 mm wide by 12.4 mm high) is smaller than the foramen magnum (18.5 mm wide). In posterior view, the occipital condyle is oval, with a flat dorsal rim above the neck, rather than a concave margin, as in *Erlicosaurus*. In *Erlicosaurus*, there is a ventral ridge from the occipital condyle to the base of the basioccipital that is accentuated by shallow fossae on either side that extend laterally. In *Nothronychus*, this ridge is reduced and the fossae are not as obvious. The shallow fossae that are present extend more ventrally in this specimen. The basioccipital forms a concave suture with the basisphenoid that is visible. There are no ventrolateral accessory struts along the neck of the occipital condyle as are present in some theropods.

The basisphenoid of *Nothronychus* is large and complexly pneumatized, as in *Erlicosaurus* and *Chirostenotes*. It extends ventrally beyond the basioccipital. The basal tubera and basipterygoid processes are not developed as recognizable structures. The ventral portion of the basisphenoid is convex and smooth. Laterally, there is a large strut that extends from the ventral inflated region to the base of the middle ear.

In anterior view, there is a pronounced fossa hypophysialis extending ventrally from the dorsum sellae and ending above a large ostium canalium caroticorum that enters the enlarged pneumatic sinus; this fossa is similar to that in *Troodon*. Presumably this canal accommodated a combined internal carotid arteris, as in *Troodon* and *Itemerus*. The fossa hypophysialis does not extend ventrally beyond the level of the canal in *Nothronychus*, as it does in *Troodon*.

Only the left laterosphenoid is preserved, and it is fragmentary. It forms the ventral margin of the sella turcica and the base of the foramen for cranial nerve V. The prootic forms the anterior margin of the middle

Figure 5.4. Reconstruction of Nothronychus *based on preserved elements. Scale = 1 m.*

ear cavity. This bone forms the posterior margin of the foramen for cranial nerve V and is penetrated by the foramen for cranial nerve VI lateral to the fossa hypophysialis and ventral to the foramen for cranial nerve V.

Discussion

The braincase of *Nothronychus* is remarkably similar to that of *Erlicosaurus*, except that the supraoccipital above the foramen magnum is oriented nearly horizontally and it lacks a distinct nuchal crest. Additionally, the braincase is relatively narrower. As with *Erlicosaurus*, the pneumatic portion of the braincase did not extend over the mid-line of the foramen magnum, and the basipterygoid processes and basituber are indistinct. *Nothronychus* shares the highly pneumatic and ventrally expanded braincase so distinctive of *Erlicosaurus* (Clark et al. 1994).

Surprisingly, the braincase of *Nothronychus* is also similar to that of *Chirostenotes* (Sues 1997) in being highly pneumatic with broad, anteroposteriorly flattened paroccipital processes and in having a ventrally inflated basisphenoid. This is also the case for the oviraptosaur *Citipati* (Clark et al. 2002), suggesting that these characters may be shared between oviraptosaurs and therizinosaurids.

The braincase of *Nothronychus*, considered together with the teeth and postcranial elements of the skeleton (i.e., ischia, scapula, vertebrae, etc.), supports the hypothesis that the new North American therizinosaurid is most closely related to *Segnosaurus* and *Erlicosaurus* (Kirkland and Wolfe 2001). This strengthens the paleogeographic ties between Asia and western North America during the early part of the Late Cretaceous (Cifelli et al. 1997; Kirkland et al. 1998) and may help resolve the age of the Baynshirenian of central Asia as closer to Turonian than to Campanian (Jerzykiewicz and Russell 1991; Currie and Eberth 1993). Finally, the small size of this braincase indicates that the previous skeletal reconstruction for *Nothronychus* had too large a skull (Kirkland and Wolfe 2001, fig. 1) and that the skull was even smaller relative to the postcranial skeleton (Fig. 5.4).

Acknowledgments. Fieldwork was conducted under Bureau of Land Management permit MSM-8172-RS-1A. We thank the staff and volunteers of the Mesa Southwest Museum and Dinamation International Society for their assistance in collecting the specimens. The specimens were prepared under the direction of Harold and Phyllis Bollan, with touch-up preparation by Don DeBlieux. Robert Gaston generously provided study casts of the material. We thank Greg Paul for Figure 5.4. Thanks are due to Jim Clark, Phil Currie, and R. Kent Sanders for discussions regarding coelurosaur braincases. James Kirkland appreciates discussions with Dave Gillette and Alan Titus regarding the Utah therizinosaurid. Don DeBlieux, Mike Lowe, Mike Hylland, and Kenneth Carpenter read drafts of the manuscript. We thank Sooz Kirkland and Hazel Wolfe for their constant support throughout the project.

References Cited

Barsbold, R., and T. Maryańska. 1990. Segnosauria. In D. B. Weishampel, P. Dodson, and H. Osmólska (eds.), *The Dinosauria,* pp. 408–415. Berkeley: University of California Press.

Cifelli, R. L., J. I. Kirkland, A. Weil, A. L. Deinos, and B. J. Kowallis. 1997. High-precision Ar^{40}/Ar^{39} geochronology and the advent of North America's Late Cretaceous terrestrial fauna. *Proceedings National Academy of Science USA* 94: 11163–11167.

Clark, J. M., A. Perle, and M. A. Norell. 1994. A skull of *Erlicosaurus andrewsi,* a Late Cretaceous "segnosaur" from Mongolia. *American Museum Novitates,* no. 3115: 1–39.

Clark, J. M., M. A. Norell, and T. Rowe 2002. Cranial anatomy of *Citipati osmolskae* (Theropoda, Oviraptosauria), and a reinterpretation of the holotype of *Oviraptor philoceratops. American Museum Novitates,* no. 3364: 1–24.

Currie, P. J., and D. A. Eberth. 1993₅. Paleontology, sedimentology, and paleoecology of the Iren Dabasu Formation (Upper Cretaceous), Inner Mongolia. *Cretaceous Research* 14: 127–144.

Currie, P. J., and Zhao X.-J. 1993. A new troodontid (Dinosauria, Theropoda) braincase from Dinosaur Park Formation (Campanian) of Alberta. *Canadian Journal of Earth Science* 30: 2231–2247.

Gillette, D. D., and B. L. Albright. 2001. Discovery and paleogeographic implications of a therizinosaurid dinosaur from the Turonian (Late Cretaceous) of southern Utah. *Journal of Vertebrate Paleontology* 21 (3 suppl.): 54A.

Gillette, D. D., B. L. Albright, A. L. Titus, and M. H. Graffam. 2002.

Discovery and excavation of a therizinosaurid dinosaur from the Upper Cretaceous Tropic Shale (early Turonian), Kane County, Utah. *Geological Society of America, Rocky Mountain Section, Abstracts with Programs* 34 (4): A5.

Jerzykiewicz, T., and D. A. Russell. 1991. Late Mesozoic stratigraphy and vertebrates of the Gobi Basin. *Cretaceous Research* 12: 345–377.

Kauffman, E. G., B. B. Sageman, J. I. Kirkland, W. P. Elder, P. J. Harries, and T. Villamil. 1993. Molluscan biostratigraphy of the Western Interior Cretaceous basin, North America. In W. G. E. Caldwell and E. G. Kauffman (eds.), *Evolution of the Western Interior Basin,* pp. 397–434. Canadian Association of Geologists, Special Paper, no. 39. St. John's, Newfoundland: Geological Association of Canada.

Kennedy, W. J., I. Walasczyk, and W. A. Cobban. 2000. Pueblo, Colorado, USA, candidate global boundary stratotype section and point for base of the Turonian stage and for the base of the middle Turonian substage, with a revision of the Inoceramidae (Bivalvia). *Acta Geologica Polonica* 50 (3): 295–334.

Kirkland, J. I. 1991. Lithostratigraphic and biostratigraphic framework for the Mancos Shale (late Cenomanian to middle Turonian) at Black Mesa, Northeastern Arizona. In D. Nations and J. G. Eaton (eds.), *Stratigraphy, Depositional Environments, and Sedimentary Tectonics of the Western Margin, Cretaceous Western Interior Seaway,* pp. 85–111. Geological Society of America, Special Paper, no. 260. Boulder, Colo.: Geological Society of America.

Kirkland, J. I. 1996. *Paleontology of the Greenhorn Cyclothem (Cretaceous: Late Cenomanian to Middle Turonian) at Black Mesa, Northeastern Arizona.* New Mexico Museum of Natural History and Science Bulletin, no. 9. Albuquerque: New Mexico Museum of Natural History and Science.

Kirkland, J. I., and D. G Wolfe. 2001. First definitive therizinosaurid (Dinosauria: Theropoda) from North America. *Journal of Vertebrate Paleontology* 21: 410–414.

Kirkland, J. I., S. G. Lucas, and J. W. Estep. 1998. Cretaceous dinosaurs of the Colorado Plateau. In S. G. Lucas, J. I. Kirkland, and J. W. Estep (eds.), *Lower and Middle Cretaceous Terrestrial Ecosystems,* pp. 67–89. New Mexico Museum of Natural History and Science Bulletin, no. 14. Albuquerque: New Mexico Museum of Natural History and Science.

Maryańska, T. 1997. Segnosaurs (Therizinosaurs). In J. O. Farlow and M. K. Brett-Surman (eds.), *The Complete Dinosaur,* pp. 234–241. Bloomington: Indiana University Press.

Perle, A. 1981. [A new segnosaurid from the Upper Cretaceous of Mongolia]. [*Transactions, Soviet-Mongolian Paleontological Expedition*] 15: 50–59. [In Russian.]

Sues, H.-D. 1997. On *Chirostenotes,* a Late Cretaceous oviraptorosaur (Dinosauria: Theropoda) from western North America. *Journal of Vertebrate Paleontology* 17: 698–716.

Wolfe, D. G. 2000. New Information on the skull of *Zuniceratops christopheri,* a neoceratopsian dinosaur from the Cretaceous Moreno Hill Formation, New Mexico. In S. G. Lucas (ed.), *Dinosaurs of New Mexico,* pp. 93–94. New Mexico Museum of Natural History and Science Bulletin, no. 17. Albuquerque: New Mexico Museum of Natural History and Science.

Wolfe, D. G., and J. I. Kirkland. 1998. *Zuniceratops christopheri* n. gen. &

n. sp., A ceratopsian dinosaur from the Moreno Hill Formation (Cretaceous, Turonian) of west-central New Mexico. In S. G. Lucas, J. I. Kirkland, and J. W. Estep (eds.), *Lower and Middle Cretaceous Terrestrial Ecosystems,* pp. 303–317. New Mexico Museum of Natural History and Science Bulletin, no. 14. Albuquerque: New Mexico Museum of Natural History and Science.

6. Anatomy of *Harpymimus okladnikovi* Barsbold and Perle 1984 (Dinosauria; Theropoda) of Mongolia

YOSHITSUGU KOBAYASHI AND
RINCHEN BARSBOLD

Abstract

Harpymimus okladnikovi, a primitive ornithomimosaur from the Early Cretaceous of Mongolia, was originally described on the basis of select material. The holotype is a nearly complete skeleton, and re-examination reveals a great deal of new anatomical information. *Harpymimus okladnikovi* is more derived than *Pelecanimimus polyodon* but is basal to the clade of *Garudimimus brevipes* plus Ornithomimidae. The arrangement of a rhamphotheca on the upper jaw and small teeth in the lower suggests that the anterior portions of the jaws were used for grabbing and holding. Some features in the manus (the short metacarpal I with laterally rotated and ginglymoid articulation at the distal end) and caudal vertebrae (short prezygapophyses and vertical articular surfaces of prezygapophyses) of *Harpymimus okladnikovi* are primitive conditions in ornithomimosaurs. In derived ornithomimosaurs, the manus was probably adapted for grasping and raking, and the posterior caudal vertebrae had less flexibility in dorsoventral movement than in *Harpymimus*. Ornithomimosaurs originated in either eastern Asia or Europe prior to the Barremian, and derived ornithomimosaurs (Ornithomimidae) dispersed to the North American continent during or prior to the Late Cretaceous.

Introduction

Harpymimus okladnikovi, described by Barsbold and Perle (1984) as a basal ornithomimosaurian dinosaur, was discovered in the Shinekhudug Formation (Hauterivian to Barremian; Khand et al. 2000) of Dundgov (southeastern Mongolia) during a joint Soviet-Mongolian paleontological expedition. Barsbold and Perle (1984) established the family Harpymimidae for the single specimen of *Harpymimus okladnikovi* based upon a nearly complete skeleton. They recognized its primitive status within Ornithomimosauria because of the presence of dentary teeth and a short metacarpal I. Barsbold and Osmólska (1990) added anatomical information mainly from the skull and limb elements, and hypothesized the basal phylogenetic position of *Harpymimus* relative to *Garudimimus* and a monophyletic Ornithomimidae in the Ornithomimosauria.

Pelecanimimus polyodon from the Early Cretaceous of Spain was described by Pérez-Moreno et al. (1994) as another toothed ornithomimosaur, but the phylogenetic relationship of *Harpymimus* and *Pelecanimimus* was not discussed. Pérez-Moreno and Sanz (1995) suggest that the manus of *Pelecanimimus* is more derived than that of *Harpymimus*. Phylogenetic analyses by Norell et al. (2001b) and Xu et al. (2002) included *Pelecanimimus, Harpymimus, Garudimimus, Gallimimus,* and *Struthiomimus,* and their analyses agree with Barsbold and Osmólska (1990) that *Harpymimus* is basal to the Ornithomimidae; however, the relationships between *Pelecanimimus, Harpymimus,* and *Garudimimus* remained unresolved. A recent analysis by Kobayashi and Lü (2003) resolved the relationships of basal ornithomimosaurs and suggested that *Harpymimus* is a sister taxon to the clade of *Garudimimus* and Ornithomimidae. Despite the crushed condition of the type specimen of *Harpymimus okladnikovi*, it remains important because it is nearly complete, it provides us with detailed anatomy, and it helps us to understand ornithomimosaur evolution.

Institutional Abbreviations. AMNH, American Museum of Natural History, New York; IGM, Mongolian Academy of Sciences, Ulaan Baatar, Mongolia; ROM, Royal Ontario Museum, Toronto, Ontario; TMP, Royal Tyrrell Museum of Palaeontology, Drumheller, Alberta; UCMZ, Museum of Zoology, University of Calgary, Calgary, Alberta.

Systematic Paleontology

Theropoda Marsh 1881
Ornithomimosauria Barsbold 1976
Harpymimidae Barsbold and Perle 1984
Harpymimus okladnikovi Barsbold and Perle 1984

Holotype. IGM 100/29, nearly complete skeleton, missing parts of the pectoral girdles, pelvic girdles, and hindlimbs.

Type Locality and Horizon. Dundgov (Eastern Gobi Province), the Shinekhudug Formation (Hauterivian to Barremian; Khand et al. 2000).

(Opposite page)
Figure 6.1. Skull, axis, and dorsal vertebrae (fifth to seventh) of Harpymimus okladnikovi *(IGM 100/29) (A and B). Shaded areas indicate the pneumatization of vertebrae. Anterior end of the right dentary in dorsal (C) and medial (D) views and anterior end of the left dentary in dorsal (E) and medial (F) views. Numbers represent alveoli positions from the posterior end. Abbreviations: an—angular; ar—articular; ax—axis; cev—cervical vertebra; cpf—central pneumatic fossa; cr—cervical rib; d—dentary; dr—dorsal rib; dv—dorsal vertebra; f—frontal; idf—infradiapophyseal fossa; iprzf—infraprezygapophyseal fossa; ipzf—infrapostzygapophyseal fossa; j—jugal; l—lacrimal; m—maxilla; n—nasal; ns—neural spine; p—parietal; pa—prearticular; pf—prefrontal; pm—premaxilla; po—postorbital; pp—parapophysis; pa—prearticular; prz—prezygapophysis; pz—postzygapophysis; q—quadrate; qj—quadratojugal; sa—surangular; so—supraoccipital; sp—splenial; sq—squamosal; sur—surangular; tp—transverse process.*

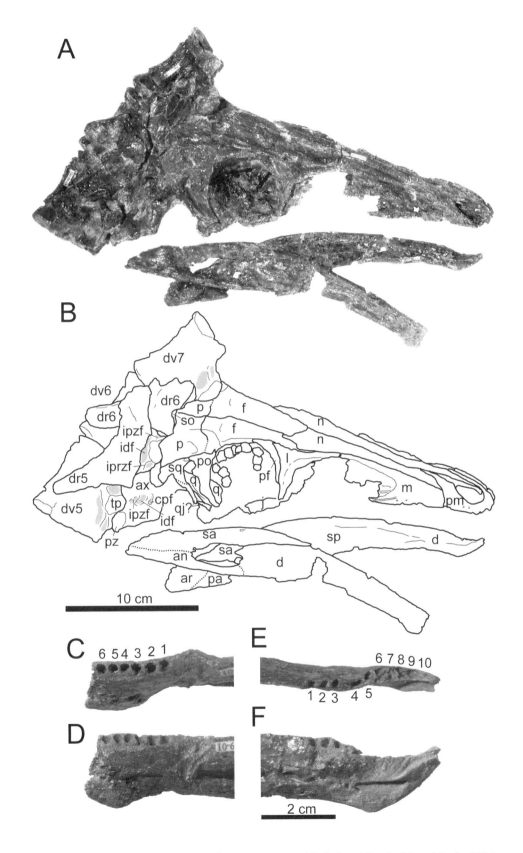

Amended Diagnosis. Eleven dentary teeth, which are anterior in position; transition between anterior and posterior caudal vertebrae at eighteenth caudal; triangular-shaped depression on dorsal surface of supraglenoid buttress of scapula; low ridge dorsal to depression along posterior edge of scapular blade; small but deep collateral ligament fossa on lateral condyle of metacarpal III.

Description

Skull

The skull of *Harpymimus okladnikovi* is nearly complete but crushed laterally (Fig. 6.1). Because it is missing the jugal, the actual height of the skull, excluding the mandible, cannot be measured. However, the anteroposterior length of the skull (262 mm) is roughly more than twice the height and is less than half the neck length (roughly 600 mm). The external narial fossa (28.3 mm long and 11.8 mm high) is teardrop-shaped. The antorbital fenestra is 39.3 mm long, more than half the length of the antorbital fossa (78.5 mm long), but is smaller than the orbit (53.9 mm). The anteroposterior length of the antorbital fossa is roughly same as that of the pre-antorbital region of the skull. The supratemporal fenestra is transversely crushed but is positioned posteriorly within the supratemporal fossa. The dorsal part of the infratemporal fenestra is similar to that of *Struthiomimus* sp. (TMP 90.26.1) in that it is wide and becomes narrower toward the ventral end of the postorbital and is more tightly curved anteriorly than posteriorly. The mandibular fenestra is 30 mm long and 16.69 mm high.

Premaxilla. The right premaxilla is better exposed and preserved than the left but is shifted from its original position dorsally with respect to the nasal. The ventral border of the premaxilla is straight and lacks teeth. The lower premaxilla-maxilla suture on the lateral surface is perpendicular to the ventral border of the upper jaw. The maxillary process is long, but terminates anterior to the anterior border of the antorbital fossa, and excludes the maxilla from the external narial margin. At least two foramina are present on the lateral surface along the ventral edge of the premaxilla, ventral to the external narial opening.

Maxilla. The left maxilla is better preserved than the right but is missing the ventral (jugal) process and is damaged in the antorbital fossa area. The dorsal process meets the anterior process of the lacrimal above the middle of the antorbital fossa. The lateral surface of the ventral process has a shallow groove extending along the ventral border of the element. The ventral border of the maxilla is straight and lacks teeth. Along the ventral edge, the lateral surface lacks the foramina seen in *Struthiomimus* sp. and *Ornithomimus* sp. (TMP 90.26.1, TMP 95.110.1; Osborn 1916, fig. 5). Within the antorbital fossa, two antorbital fenestrae are partially preserved.

Nasal. The unfused nasals are long and flat for their length. These extend from the posterior border of the external narial opening, broaden posteriorly with the maximum width at the level of the lacrimal, and narrow posteriorly toward the frontal. The posterior end of the nasal is slightly anterior to the posterior end of the prefrontal.

Lacrimal and Prefrontal. The lacrimal is triradiate with anterior, posterior, and ventral processes. The ventral process borders the anteroventral portion of the orbital margin. The short posterior process pinches out and fits into a depression on the prefrontal as in other ornithomimosaurs. Unlike the lacrimal of *Pelecanimimus polyodon* (Pérez-Moreno et al. 1994), the lacrimal of *Harpymimus okladnikovi* lacks a dorsal prominence. The area of the prefrontal that is exposed dorsally is slightly less than the dorsally exposed area of the lacrimal. The posterior end of the prefrontal is anterior to the middle of the orbit.

Sclerotic Plates. The right side of the skull preserves eleven in situ sclerotic plates, arranged in a half circle. The diameter of the circle inside the bony elements is 27.6 mm, and the diameter outside is 41.5 mm. Each plate is approximately 9 mm wide.

Jugal, Quadratojugal, and Quadrate. Most of the jugal and quadratojugal are missing on the right side, but a small portion of the jugal ventral to the orbit and the dorsal process and a dorsal tip of the quadratojugal are preserved. The quadrate is partially exposed, and the pterygoid wing extends more anteriorly than the posterior margin of the orbit.

Frontal and Parietal. The frontal is triangular and has a domed structure as in other ornithomimosaurs. The parietal is separated from the orbital rim by the frontal and postorbital. The parietal has a large depression for the supratemporal fossa, extending to the posterodorsal edge of the orbit. Neither element has scars as described in *Pelecanimimus polyodon* (Osmólska et al. 1972).

Postorbital, Squamosal, and Supraoccipital. The anterior process of the postorbital narrows anteriorly, and the posterior process pinches out posteriorly. The ventral process is thick, but it narrows ventrally. The lateral surface of the anterior process of the squamosal has a depression, which is overlain laterally by the posterior process of the postorbital. The ventral and posterior processes of the squamosal are not well preserved. The dorsal portion of the supraoccipital is exposed. Its dorsal edge terminates at the level of the skull table, formed by the parietals, in occipital view. The posterior surface of the dorsal portion is convex as in other ornithomimosaurs. The dorsal portion is U-shaped in dorsal view.

Dentary and Dentition. Both dentaries are laterally crushed, and the anterior tip of the right is missing. The dentary is the longest of the mandibular elements (186 mm, estimated length, whereas the surangular is 135 mm in length). In lateral view, the gently convex dorsal border of the dentary forms a gap between the upper and lower jaws when they articulate. The anterior part of the dentary preserves alveoli, which differ from those in *Pelecanimimus polyodon* in their number and in their more anteriorly restricted distribution (Pérez-Moreno et al. 1994) (Fig. 6.1C–F). The anterior portion of the left dentary preserves ten alveoli, but the total number of dentary teeth could be eleven (the anterior tip of the left dentary is damaged). The alveoli are nearly circular in dorsal view. The five (or six) alveoli of the right dentary extend 19.2 mm anteroposteriorly. The five alveoli in the left dentary extend 15.2 mm. The average distances between consecutive interdental plates are 3.0 and 3.2 mm in anterior and posterior teeth, respectively, which indicates that posterior teeth are slightly larger than anterior ones. The alveoli are present in the anterior 33.9 mm of the dentary, which is only 14 percent of the length (approximately 244 mm) of the mandible.

On the anteroventral edge of the dentary along the symphysis, two foramina are preserved. The lateral surface of the dentary is smooth and lacks the foramen present in *Gallimimus bullatus* (Hurum 2001). The dorsal border of the dentary is rounded in transverse section and shows little development of a cutting edge, unlike the condition seen in Late Cretaceous ornithomimosaurs. The dentary is laterally overlapped by the surangular dorsally. The suture with the angular is not clear. At the anterior part of the external mandibular fenestra, the dentary has a short posterior process. The medial surface of the dentary preserves the Meckelian groove, which narrows anteriorly. The splenial in *Harpymimus okladnikovi* is poorly preserved.

Surangular. The surangular is the second longest mandibular element. The dorsal process of the surangular is low with respect to the length of the process (height at the base of the process is 13.41 mm; process length is 39.3 mm). The process extends further anteriorly than the anterior edge of the mandibular fenestra by 16.1 percent of the total mandibular length, more than in *Struthiomimus altus* (TMP 90.26.1) (approximately 10 percent). There is a foramen at the level of the anterior end of the external mandibular fenestra in the same position as an undescribed pit in *Pelecanimimus polyodon* (Hurum 2001, fig. 4.2A). An anteroposteriorly oriented but laterally curved ridge is present posterior to the external mandibular fenestra for the articular surface with the lateral accessory mandibular condyle of the quadrate. The presence of this curved ridge suggests that the accessory condyle was present in *Harpymimus*, although the quadrate is not well preserved. The lateral surface of the ridge lacks the surangular foramen.

Angular, Articular, and Prearticular. The angular is long and has a convex ventral border. The suture with the surangular is not clear. Most of the articular is missing, but it is visible in lateral view. The posterior part of the prearticular is preserved and shows morphology similar to that in *Garudimimus brevipes* (IGM 100/13).

Vertebral Column

The vertebral series of *Harpymimus okladnikovi* is nearly complete but is badly crushed transversely. Its neck is bent back so that the occipital region and first two cervical vertebrae overlap the fifth to seventh dorsal vertebrae. Therefore, the atlas and axis are not well exposed. The taxon has ten cervical, twelve dorsal, six sacral, and thirty-five (or more) caudal vertebrae. The neurocentral sutures in all vertebrae are fused, indicating the specimen's maturity, if the closure sequence of neurocentral sutures in crocodilian ontogeny can be applied (Brochu 1996).

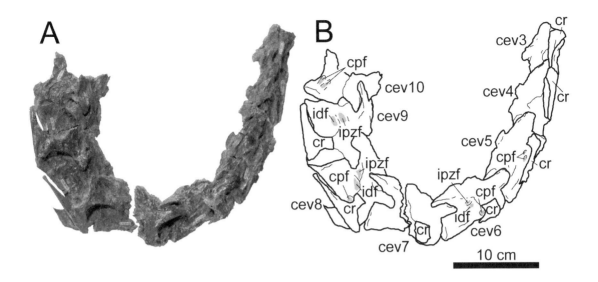

A

B

cr
cev3
cpf
cev10
cr
idf
cev9
cev4
ipzf
cr
cr
cev5
ipzf
cpf
cr
cpf
ipzf
cev8
idf
cpf
cr
cev7
idf
cr
cev6

10 cm

Cervical Vertebrae (Figs. 6.1, 6.2). A partial postzygapophysis of the axis is preserved. The postzygapophysis is short and has a circular articular surface. The lateral side of the axial centrum has infradiapophyseal and infrapostzygapophyseal fossae separated by a lamina. Ventral to the fossae, there is a central pneumatic fossa.

The transition between the anterior and posterior cervicals is between the fifth and sixth. The anterior cervical vertebrae have a straight posterior edge on the neural arch as seen in *Ornithomimus* sp. (Makovicky 1995) and *Gallimimus bullatus* (Osmólska et al. 1972), and the neural arch becomes progressively longer posteriorly (Table 6.1). The neural spines, preserved in the fourth and fifth cervicals, are low and anteroposteriorly long: 35 mm (44 percent of the neural arch length) in the fourth cervical and 30 mm (33 percent) in the fifth. The postzygapophyses are much shorter than the prezygapophyses and do not extend as far as the posterior intervertebral articular surface of the centrum. The diapophysis is near the anterior end of the neural arch. The posterior intervertebral articular surface, exposed in the third and fourth cervical vertebrae, is inclined anteriorly in lateral view. The fifth cervical vertebra has a central pneumatic fossa, slightly posterior to the parapophysis. The latter is divided into two fossae by a thin lamina as in the tenth cervical vertebra.

The lengths of the neural arches are similar in the first three posterior cervical vertebrae (sixth, seventh, eighth), but those in the ninth and tenth cervicals become progressively shorter (Table 6.1). None of the posterior cervical neural arches preserve the neural spines. The prezygapophyses become shorter posteriorly. The postzygapophyses are longer than in the anterior cervical vertebrae. The central pneumatic fossa of the sixth cervical centrum is divided into two fossae by a thin lamina, a condition that is similar to that in the tenth cervical vertebra. The infrapostdiapophyseal laminae in the eighth and ninth cervical centra are much stronger than in the sixth. The parapophyses,

Figure 6.2. Cervical vertebrae (third to tenth) of Harpymimus okladnikovi *(IGM 100/29). Shaded areas indicate the pneumatization of vertebrae. See the caption of Figure 6.1 for abbreviations.*

TABLE 6.1

Measurements of the Cervical, Dorsal, and Sacral Vertebrae of *Harpymimus okladnikovi*

The lengths of cervical vertebrae represent the lengths of neural arches, and those of the dorsal and sacral vertebrae are centrum lengths.

Element	Length (mm)	Element (mm)	Length (mm)
Cervical 3	65	Dorsal 6	59
Cervical 4	79	Dorsal 7	61
Cervical 5	90	Dorsal 8	59
Cervical 6	97	Dorsal 9	65
Cervical 7	94	Dorsal 10	64
Cervical 8	94	Dorsal 11	65
Cervical 9	84	Dorsal 12	67
Cervical 10	75	Sacral 1	71
Dorsal 1	54	Sacral 2	70
Dorsal 2	48	Sacral 3	64
Dorsal 3	50	Sacral 4	57
Dorsal 4	55	Sacral 5	57
Dorsal 5	56	Sacral 6	68

(Opposite page)
Figure 6.3. Dorsal vertebrae of Harpymimus okladnikovi (IGM 100/29): tenth cervical vertebra and first to fourth dorsal vertebrae (A and B) and seventh to twelfth dorsal vertebrae (C and D). Shaded areas indicate the pneumatization of vertebrae. See the caption of Figure 6.1 for abbreviations.

preserved in the sixth to ninth cervical centra, are well developed and located on the anterior part of the ventral surface of the centrum. In lateral view, the posterior intervertebral articular surface of the sixth cervical centrum is tilted anteriorly, but at a much smaller angle than in the third cervical centrum. The posterior intervertebral articular surfaces from the seventh to ninth cervical centra are nearly perpendicular to the main axis of the centrum in lateral view. The tenth cervical is similar in structure to the anterior dorsal vertebrae in the arrangement of fossae in the neural arch. The cervical has well-developed fossae, in which the infradiapophyseal fossa is larger than the infrapostzygapophyseal fossa, and the infraprezygapophyseal fossa is small. The infrapostdiapophyseal lamina is much thicker and longer than the infraprezygapophyseal lamina.

Dorsal Vertebrae (Figs. 6.1, 6.3). The prezygapophyses and postzygapophyses of the anterior dorsal vertebrae are longer than those of the posterior ones. The neural spines become higher in more-posterior dorsal vertebrae and are dorsally expanded in lateral view. The posterior borders of the transverse processes of the fourth and eighth dorsal vertebrae are nearly perpendicular to the sagittal plane and are angled posteriorly in the posterior dorsal vertebrae. The infraprediapophyseal laminae are as prominent as the infrapostdiapophyseal laminae in the first and second dorsal vertebrae (poorly preserved in the third), but are less developed in the fourth dorsal vertebra. The infraprezygapo-

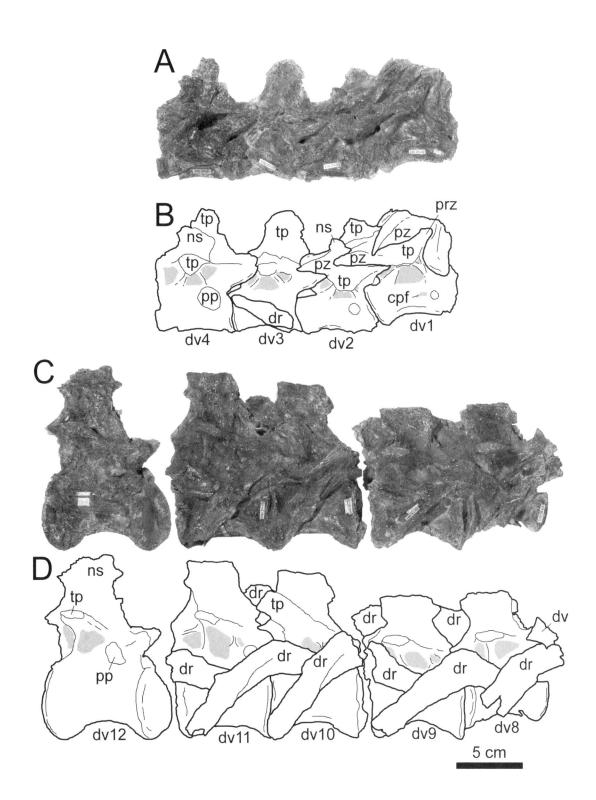

physeal fossae of the dorsal vertebrae become larger posteriorly (at least up to the third dorsal vertebra) because of the posterior shift of the infraprediapophyseal laminae as in *Ornithomimus* sp. (Makovicky 1995). These fossae are small in the ninth and eleventh dorsal vertebrae and absent in the twelfth dorsal. The semicircular parapophysis in the first dorsal vertebra is positioned at the dorsal limit of the centrum, but it shifts more dorsally in more-posterior dorsal vertebrae. The antero-posterior lengths and dorsoventral heights of the centra increase in more-posterior dorsals (Table 6.1). The anterior and posterior intervertebral articular surfaces are slightly concave and are perpendicular to the main axis of the centrum in lateral view. The lateral surface of the centrum lacks pneumatization, except for the first dorsal vertebra. An anteroposteriorly elongated fossa is present posterior to the parapophysis in the first dorsal vertebra. Ventral keels are seen in the first, second, and fourth dorsal centra but are absent in the sixth dorsal centrum (indeterminate in the third and fifth). Paired lateroventral projections known in *Ornithomimus* sp. (Makovicky [1995] noted that the posterior pair is more pronounced than the anterior pair) are absent at least in the sixth and seventh dorsal centra of *Harpymimus okladnikovi.*

Sacral Vertebrae. The sacral vertebrae are crushed and partially covered by the ilia (Fig. 6.8A–D). Most of the fourth sacral centrum is missing. The fourth and fifth have the shortest anteroposterior lengths (Table 6.1). Only the fourth to sixth sacral centra are fused. The first and second sacral centra, positioned anterior to the pubic peduncle of the ilium, are similar to the last dorsal centrum and lack depressions on the lateral surface. The posterior end of the ventral surface of the fifth sacral centrum and the anterior end of the ventral surface of the sixth sacral centrum have paired weak prominences, similar to those in *Garudimimus brevipes* (IGM 100/13). The neural arches of the sacral vertebrae are badly crushed, but the neural spines from the second to fifth sacrals are fused.

Caudal Vertebrae (Fig. 6.4). The caudal series of *Harpymimus okladnikovi* has more than thirty-four vertebrae because the thirty-fourth caudal vertebra has a posterior articular surface for another vertebra. The posterior caudal vertebrae are distinguished from anterior ones in lacking the transverse processes and in having elongated centra. In *Harpymimus okladnikovi*, the position of the first elongated caudal vertebra is not clear, but the first caudal vertebra without the transverse process is the eighteenth; it is the fifteenth in *Ornithomimus* and the twelfth or thirteenth in *Dromiceiomimus* (Russell 1972). The neural spines are taller than they are long (at least up to the sixth caudal vertebra) and are inclined posteriorly. They become lower and anteroposteriorly longer posterior to the fifteenth caudal vertebra. The transverse processes in the first to eighth caudal vertebrae are long and are inclined posteriorly in dorsal view. The processes are reduced in size in more-posterior caudal vertebrae. The transverse process of the seventeenth caudal vertebra is a weak ridge. The articular surfaces of the prezygapophyses in anterior view are tilted more than 45° in the first and second caudal vertebrae and become nearly perpendicular to

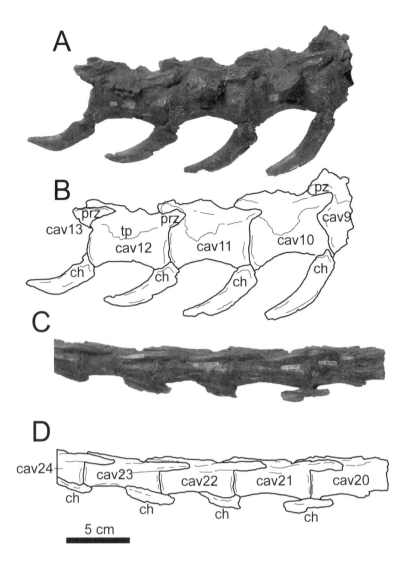

A

B

cav13 prz
tp prz
cav12 cav11 cav10 cav9
ch ch ch ch
pz

C

D

cav24
cav23 cav22 cav21 cav20
ch ch ch ch

5 cm

the horizontal plane in the more-posterior caudal vertebrae. The pre-zygapophyses are short in the first seventeen caudal vertebrae but are longer in those posterior to the eighteenth caudal (Fig. 6.5). The prezygapophyses are nearly equal in length between the eighteenth and thirty-first caudals and become short from the thirty-second. The lengths of the centra are similar from the first to the twentieth caudal vertebra (roughly 60 mm long), but the lengths decrease gradually posteriorly.

The ratios between the lengths of the prezygapophyses, which extend anterior to the intervertebral articular surfaces, and the centrum length increase posterior to the seventeenth caudal. This confirms that the caudal vertebrae can be divided into anterior and posterior regions at the seventeenth or eighteenth caudal vertebra and that the posterior segment is stabilized by longer prezygapophyses than the anterior

Figure 6.4. Caudal vertebrae of Harpymimus okladnikovi (IGM 100/29): tenth to twelfth (and partial ninth and thirteenth) caudal vertebrae (A and B) and twentieth to twenty-fourth caudal vertebrae (C and D). Abbreviations: cav—caudal vertebra; ch—chevron. See the caption of Figure 6.1 for other abbreviations.

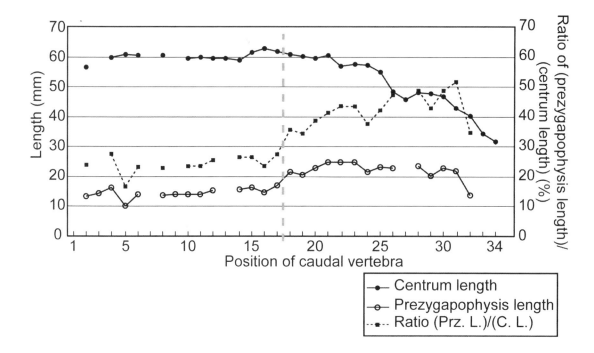

Figure 6.5. Changes in centrum and prezygapophysis lengths and ratios plotted against the caudal vertebra position in Harpymimus okladnikovi (IGM 100/29). Centrum length is anteroposterior length, and prezygapophysis length represents the distance between the anterior tip of the prezygapophyses and the anterior intervertebral articular surface. The dashed vertical line between the seventeenth and eighteenth caudal vertebrae marks a possible transition between the anterior and posterior caudal vertebrae.

segment. The highest prezygapophysis/centrum length ratio is 54 percent at the thirty-first caudal, which is lower than in more advanced ornithomimosaurs (e.g., 75 percent in *Ornithomimus* sp.; Makovicky 1995). The postzygapophyses are fused to each other, forming a single process posterior to the twentieth caudal vertebra. The anterior and posterior intervertebral articular surfaces are semicircular but are higher than wide in anterior caudal vertebrae and lower than wide in posterior caudal vertebrae (the transition point is somewhere between the fourteenth and twentieth). The lateral surface of the caudal centrum has no pneumatization. The ventral surface of each centrum has a sulcus that is narrow up to the sixteenth vertebra. It is dramatically wide in the seventeenth, but the sulci become progressively narrower in more-posterior caudals.

Ribs and Chevrons. Cervical ribs (Fig. 6.2) are fragmentary but have an inverted T-shape with a short anterior process and a long shaft extending anteroposteriorly, and they have a straight ventral edge. These are nearly as long as the corresponding centra. The posterior tips of the cervical ribs do not extend beyond the posterior intervertebral articular surfaces of the centra. In lateral view, the tuberculum is longer than the capitulum and extends perpendicular to the axis of the shaft. Some proximal parts of the dorsal ribs, including the fifth through eleventh ones from the left side and the eighth through eleventh from the right side, are preserved (Figs. 6.1, 6.3). The tuberculum in each anterior dorsal rib is long but becomes shorter in more-posterior dorsal ribs. The angles between tuberculum and capitulum are larger in more-

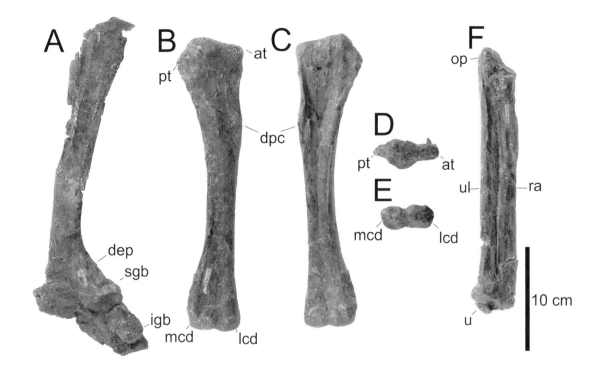

Figure 6.6. Pectoral girdle and forelimb of Harpymimus okladnikovi *(IGM 100/29): left scapula and coracoid (A), right humerus in dorsal (B), ventral (C), proximal (D), and distal (E) views, and articulated left ulna and radius in medial view (F). Abbreviations: at—anterior tuberosity; dep—depression; dpc—deltopectoral crest; igb—infraglenoid buttress; lcd—lateral condyle; mcd—medial condyle; op—olecranon process; pt—posterior tuberosity; ra; radius; sgb—supraglenoid buttress; u—ulnare; ul—ulna.*

posterior ribs. Anterior and posterior expansions are present from the lateral sides of the capitulum, and the expansions extend to the lateral side of the shaft as in other ornithomimosaurs.

The chevrons (Fig. 6.4) are preserved in the fifth to thirty-first caudal vertebrae, with the exception of the seventeenth and nineteenth. The fifth through twelfth chevrons are long, but the more-posterior ones become shorter. The distal half curves posteriorly. The chevrons of the thirteenth through sixteenth caudal vertebrae become dorsoventrally flattened with flat ventral surfaces. The chevrons posterior to the seventeenth caudal vertebra are dorsoventrally flat and anteroposteriorly long, and they decrease in size posteriorly.

Pectoral Girdle and Forelimbs

The nearly complete scapula and partial coracoid from the left side, and a partial right scapular blade, are preserved but are transversely crushed. The left scapula and coracoid are articulated, and the scapula-coracoid suture is straight. The glenoid fossa faces posterolaterally as in other ornithomimosaurs.

Scapula. The scapula is thin and long (Fig. 6.6A). The blade is slightly curved posteriorly and is wider toward the distal end in lateral view. The scapular prominence (attachment for the M. deltoides clavicularis) is low and ends dorsally at the same level as the supraglenoid buttress. This condition is unlike that in *Struthiomimus altus* and *Gallimimus bullatus* (Nicholls and Russell 1985), and the difference may indicate that the area for the muscle attachment is smaller in *Harpymimus okladnikovi* than in the Late Cretaceous forms. The supraglenoid buttress is developed as much as in other ornithomimosaurs, and the flange of the buttress extends onto the anterior surface of the scapula. Dorsal to the buttress, a weak depression for attachment of the M. scapulotriceps is present as in *Gallimimus bullatus* and *Struthiomimus altus* (Osmólska et al. 1972; Nicholls and Russell 1985). The depression is triangular, differing from the oval shape in the Late Cretaceous taxa. There is a low ridge dorsal to the depression along the posterior edge of the blade, which is not present in *Gallimimus bullatus* (IGM 100/10) and *Struthiomimus altus* (UCMZ(VP)1980.1).

Coracoid. The left coracoid is partially preserved but broken posterior to the prominent biceps tubercle. The infraglenoid buttress is more pronounced than the supraglenoid buttress, and its tip is more rounded than in *Gallimimus* sp. (IGM 100/14). The outline of the notch between the infraglenoid buttress and the posterior process is more rounded than in *Gallimimus* sp. (IGM 100/14). The infraglenoid buttress lies along the same line as the posterior process in dorsal view.

Humerus. Both humeri are preserved and are nearly straight (Fig. 6.6B–E). The humerus is as long as the scapula and is more robust than the humerus of other ornithomimosaurs. The ratio of the proximal width to total length is greater than 0.2 as in *Archaeornithomimus asiaticus*, *Gallimimus* sp. (IGM 100/14), *Anserimimus planinychus*, and *Struthiomimus altus*. The deltopectoral crest is weakly developed and extends distally up to one-third of the total length (crest length, 83.6 mm) as in other ornithomimosaurs. The anterior tuberosity is not separated from the head and is slightly lower than the head. The medial condyle is similar in size to the lateral condyle and lacks the entepicondyle. The proximal and distal ends are aligned in the same plane, possibly as an artifact of crushing, and the twist between the ends known in the Late Cretaceous taxa (e.g., *Gallimimus bullatus* and *Struthiomimus altus;* Nicholls and Russell 1981) is absent.

Ulna. Both ulnae and radii are preserved and are similar in general morphology to those of other ornithomimosaurs (Fig. 6.6F). The length of the left ulna is 82 percent of the humerus length (Table 6.2). The olecranon process is well developed as in other ornithomimosaurs, and its length is roughly 10 percent (26 mm) of the ulna length. The distal end is flattened dorsoventrally and has two condyles. The anterior surface of the anterior condyle is flat and smooth for the syndesmotic articulation with the radius. The distal end of the ulna extends slightly more distally than that of the radius when in articulation.

Radius. The straight radius (Fig. 6.6F) is shorter and thinner than the ulna (Table 6.2). Both ends are expanded and oval in proximal and distal views. The proximal and distal ends are rotated relative to each other. The posterior surface of the distal end has a shallow depression for the articulation with the ulna.

Carpals. The right manus of *Harpymimus okladnikovi* preserves three proximal and two distal carpal bones (Figs. 6.6F, 6.7A–D). The ulnare is articulated with the distal end of the ulna and is 17.5 mm long anteroposteriorly and 19.7 mm wide transversely. It is flattened proximodistally and is triangular in distal view. The distal surface is smooth and convex. The shapes of the other two proximal carpals indicate that

TABLE 6.2

Measurements of the Pectoral Girdle and Forelimb
Elements of *Harpymimus okladnikovi*

All elements are from the right side except manual phalanx II-3.
The widths of all elements are measured at the proximal ends,
except for the scapula (width at the supraglenoid buttress).

Element	Length (mm)	Width (mm)
Scapula	303	84
Humerus	294	71
Ulna	242	36
Radius	217	25
Metacarpal I	48	22
Metacarpal II	94	27
Metacarpal III	103	16
Manual phalanx I-1	120	17
Manual phalanx I-2	74	16
Manual phalanx II-1	49	19
Manual phalanx II-2	104	18
Manual phalanx II-3	80	14
Manual phalanx III-1	31	19
Manual phalanx III-2	37	17
Manual phalanx III-3	81	17
Manual phalanx III-4	77	13

they are probably the intermedium and pisiform, but they are displaced
from their original positions and lie on the proximal surface of metacarpal II. The intermedium (14 mm anteroposteriorly, 11 mm in transverse width and 10 mm in height) is triangular in anterior view. Its apex
fits between the ulnare and radiale and has a flat bottom for the
articulation with the distal carpals. The pisiform is roughly the same
size as the intermedium (12.7 mm high, 8.3 mm long, and 6.7 mm
wide). It has a smoothly concave articular surface for the distal end of
the ulna as in *Struthiomimus altus* (Nicholls and Russell 1985), and it
has a flat distal surface. Distal carpal I is slightly larger than the ulnare
and is transversely wider (21.1 mm) than anteroposteriorly long (18.4
mm). It is flat and covers the entire area of the proximal surface of
metacarpal I. The proximal surface of the carpal is concave. Distal
carpal II is incomplete. Its anterior edge is as thick as distal carpal I.

Metacarpals (Fig. 6.7A–E). Metacarpal I is the shortest metacarpal
(Table 6.2). The ratio of metacarpal I length/metacarpal II length is 51
percent, which is the lowest ratio among all ornithomimosaurs. Metacarpal III is longer than metacarpal II, as in *Anserimimus planinychus*
(IGM 100/300) (Table 6.2).

Metacarpal I is triangular at its proximal end. The contact between

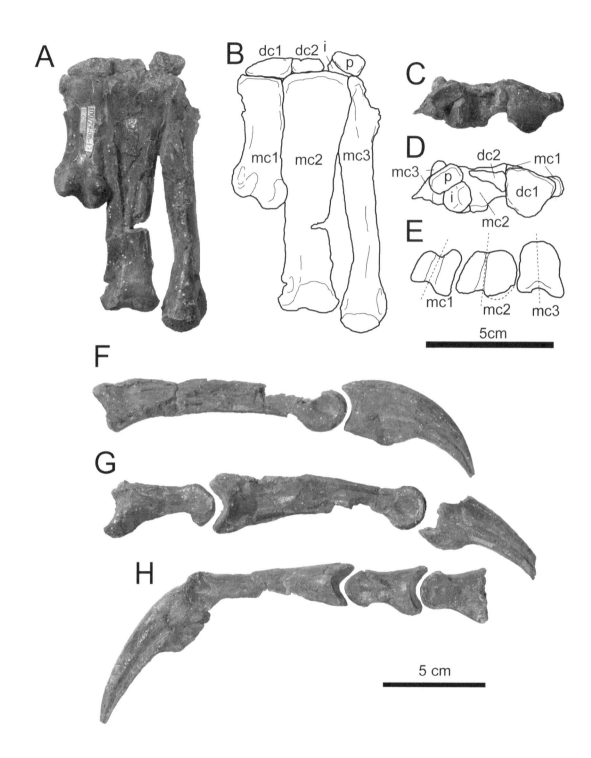

metacarpals I and II is along less than half the length of metacarpal II. The distal end of metacarpal I is rotated medially, indicating that the digits become appressed during hyperextension as in *Allosaurus* (see Nicholls and Russell 1985, fig. 13). The orientation of the intercondylar groove is nearly parallel to that of metacarpal II. The distal end of metacarpal I has a ginglymoidal surface with distinct lateral and medial condyles for articulation with the first phalanx of digit I as in *Pelecanimimus polyodon* (Pérez-Moreno and Sanz 1995) and *Archaeornithomimus asiaticus* (AMNH 21889 and 6569). However, ginglymoidal surface is more prominent than in *Archaeornithomimus asiaticus*. The lateral condyle is larger than the medial one. There is a collateral ligament fossa on the medial side.

The length of the contact surface of metacarpals II and III is roughly one-third of the total length of metacarpal II. Metacarpal II is straight. It is trapezoidal in proximal view where the base of the trapezoid is the anterior surface of the element. The intercondylar groove on the distal end of metacarpal II extends to the dorsal surface, unlike the condition in other ornithomimosaurs. The lateral condyle is slightly larger than the medial one. The collateral ligament fossa on the medial surface is deeper than the lateral one.

Metacarpal III is the thinnest of the metacarpals, and its shaft is straight with a circular cross-section. In proximal view, metacarpal III is triangular. The distal end of metacarpal III is transversely narrower but dorsoventrally taller than the distal end of metacarpal II. The phalangeal surface of metacarpal III is bowl-shaped distally, and condyles are positioned on the posterior side of the distal end. A small, deep collateral ligament fossa is present on the lateral side of the lateral condyle; this features is not present in other ornithomimosaurs.

Manual Phalanges (Fig. 6.7F–H). The phalangeal formula is 2-3-4 as in other ornithomimosaurs. The general morphology of manual phalanges is similar to that of other ornithomimosaurs. The distal ends of all phalanges except the unguals have well-developed ginglymoidal articulations. The first phalanx of digit I is the longest (120 mm) and is longer than metacarpal III. The first phalanx of digit III is the shortest of the manual phalanges. More-proximal phalanges in digits II and III are shorter than more-distal phalanges. The proximal articular surfaces of phalanges I-1 and II-1 are divided into two concavities by medially offset ridges. The ridge of phalanx I-1 is strong and associated with a dorsal intercondylar process from the anterior edge, and the ridge of II-1 is faintly developed. The proximal articular surface of phalanx III-1 is a single depression. The penultimate phalanges II-2 and III-3 are long and similar to phalanx I-1 in having evenly developed distal condyles and proximal surfaces divided into two by ridges. The shafts of penultimate phalanges I-1 and II-2 are slightly curved ventrally, but the shaft of III-3 is straight. The collateral ligament fossa on the lateral side is deeper than the medial fossa in the penultimate phalanges. These fossae are faintly present in II-1, III-1, and III-2. The ungual phalanges are similar in shape to those of other ornithomimosaurs, except for *Anserimimus planinychus* and *Ornithomimus edmontonicus* (Barsbold

(Opposite page)
Figure 6.7. Left carpals and metacarpals and right manual phalanges of Harpymimus okladnikovi *(IGM 100/29): carpals and metacarpals in anterior (A and B) and proximal (C and D) views. Metacarpals in distal view (E), showing the orientations of intercondylar grooves (dotted lines). Digit I (F) and digit II (G) in lateral view and digit III in medial view (H). Abbreviations: dc—distal carpal; i—intermedium; mc—metacarpal; p—pisiform.*

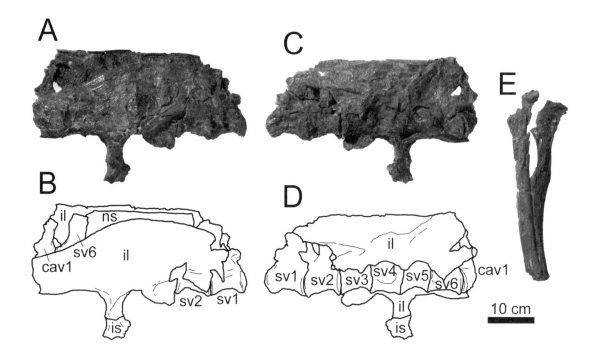

A

C

E

B

D

10 cm

Figure 6.8. Pelvic girdle and sacral vertebrae of Harpymimus okladnikovi *(IGM 100/29) in right (A and B) and left (C and D) lateral views and a photograph of pubis in left lateral view (E). Abbreviations: cav— caudal vertebra; il—ilium; is— ischium; ns—neural spine; sv— sacral vertebra.*

1988; ROM851), in having gently curved main axes in lateral view. The curvature in the ungual of digit I is the strongest. The unguals have a proximally positioned flexor tubercle on the ventral surface. The ungual is shorter than a penultimate phalanx in all digits as in *Ornithomimus* and *Dromiceiomimus* (Nicholls and Russell 1981).

Pelvic Girdle and Hindlimbs

Ilium. Both ilia of *Harpymimus okladnikovi* are preserved but badly crushed (Fig. 6.8A–D). The right ilium is better preserved than the left but is missing the anterior portion of the antilium. The dorsal border of the ilium is concave. It is deep anterior to the acetabulum and becomes shallower posterior to the acetabulum (at its lowest it is 74 mm, roughly half of the maximum depth) so that the sacral neural spines are visible in lateral view. The acetabulum is positioned at the level of the third and fourth sacral centra as in other ornithomimosaurs. The ischial peduncle is strong and transversely wide. The brevis fossa is well developed as in other ornithomimosaurs.

Pubis and Ischium. Both pubes (Fig. 6.8E) are preserved but lack the distal ends, pubic boot, and ischial peduncles. Although incomplete, the total length of the pubis is at least 360 mm. The iliac peduncle is thick transversely (roughly 36 mm). The shaft is nearly straight, unlike that in *Archaeornithomimus asiaticus* (Smith and Galton 1990). The aprons of the pubes meet each other 160 mm from the proximal end. Only a badly crushed portion of the iliac peduncle of the ischium is preserved.

Femur. The proximal ends of both femora are preserved but are

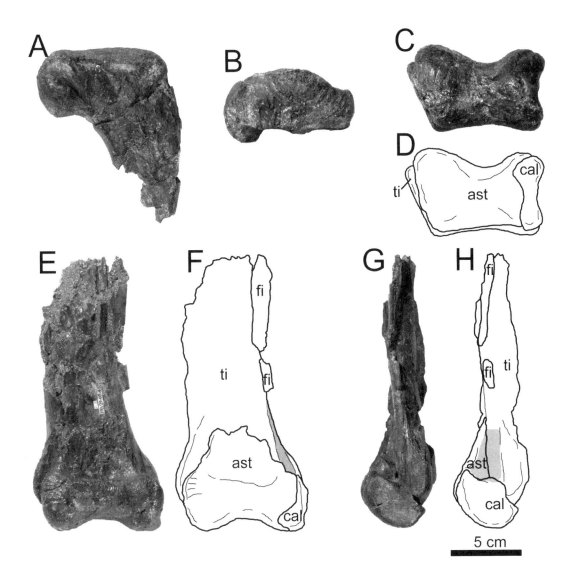

A

B

C

D

ti
cal
ast

E

F

fi
ti
fi
ast
cal

G

H

fi
fi
ti
ast
cal

5 cm

missing the lesser trochanters (Fig. 6.9A,B). The transverse width of the proximal end of the right femur is 91.3 mm. In proximal view, the femur is thickest anteroposteriorly (43.3 mm) at its middle portion, which is slightly less than half the transverse width.

Tibia. The distal ends of both tibiae are preserved. The left is articulated with the calcaneum, astragalus, and the partial shaft of the fibula (Fig. 6.9C–H). The distal end of the tibia has an articular groove for the distal end of the fibula, and the fibula reaches the lateral side of the ascending process of the astragalus and the dorsal edge of the calcaneum. The groove shows that the distal end of the fibula articulates to the lateral side of the tibia, whereas the distal end of the fibula in some derived ornithomimosaurs is placed on the anterior side of the tibia. The shape of the groove indicates that the distal end of the tibia

Figure 6.9. Left hindlimb elements of Harpymimus okladnikovi *(IGM 100/29): femur in anterior (A) and proximal (B) views. Tibia with fragmentary fibula, astragalus, and calcaneum in distal (C and D), anterior (E and F), and lateral (G and H) views. Shaded areas in F and H are articular surfaces for the distal end of the fibula. Abbreviations: ast—astragalus; cal—calcaneum; fi—fibula; ti—tibia.*

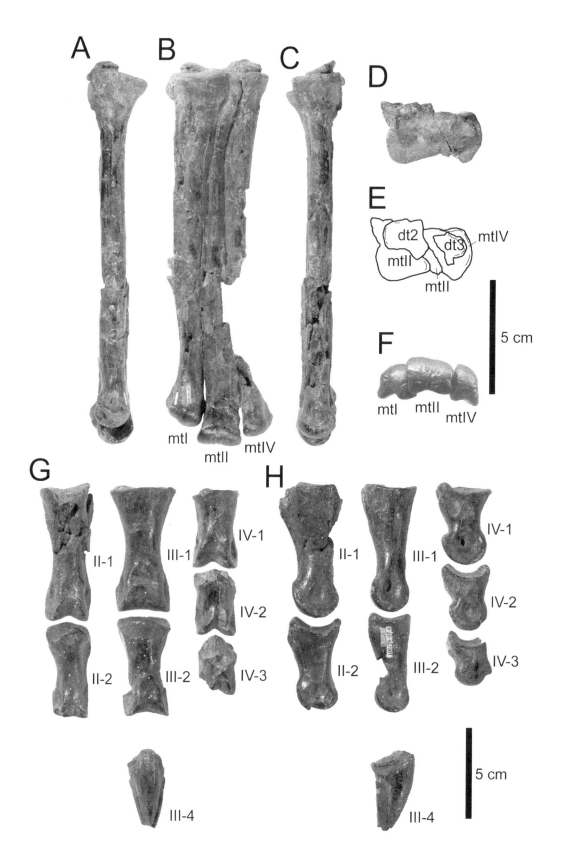

covers more than a half of the lateral surface of the tibia. The postero-lateral corner of the tibia is weakly ridged.

Astragalus. The left astragalus is missing the tip of its ascending process (Fig. 6.9C–H). In distal view, the lateral condyle is longer anteroposteriorly than the medial condyle as in other ornithomimosaurs. The lateral border of the astragalus has a notch for the calcaneum. The calcaneum participates in the lateral half of the lateral condyle, is disk-shaped, and is wider than it is tall in lateral view. The calcaneum has a prominence along the medial side, which fits into the notch of the astragalus, forming a firm contact.

Distal Tarsals. Distal tarsal III is mainly in contact with the posterior half of the proximal end of metatarsal III and partially with metatarsal II as in other ornithomimosaurs (Fig. 6.10A–E). Distal tarsal IV is incomplete and sits on metatarsal IV.

Metatarsals. Harpymimus okladnikovi preserves three metatarsals (metatarsal V is not preserved). Unlike *Garudimimus brevipes* (Barsbold 1981), it lacks digit I (although the lack might be an artifact of preservation) (Fig. 6.10A–F). Metatarsal III is the longest metatarsal, and metatarsal II is slightly shorter than metatarsal IV. Metatarsals II and IV are triangular in proximal view. The distal ends of metatarsals II and IV have a single, rounded surface for the first phalanges, and the posterior sides have lateral and medial condyles. The lateral condyle is much larger than the medial one in metatarsal II, but the condyles are equal in metatarsal IV. The distal ends of all metatarsals, except the lateral side of metatarsal IV, have deep lateral ligament fossae. The proximal end of metatarsal III is exposed on the anterior (extensor) surface, indicating the lack of arctometatarsalian condition as in *Garudimimus brevipes* (Barsbold 1981). In anterior view, metatarsal III dramatically widens distally, where there is a medial expansion. The expansion is positioned at 43 percent of metatarsal III total length from the distal end. Metatarsals II and IV partially cover the collateral ligament fossae of metatarsal III in lateral and medial views.

Pedal Phalanges. Eight phalanges are preserved from the right foot (II-1, II-2, III-1, III-2, III-4, IV-1, IV-2, and IV-3) (Fig. 6.10G,H). The most proximal phalanges (II-1, III-1, and IV-1) are similar in having undivided proximal surfaces. There are depressions on the lateral surfaces of II-1 and III-1 and on the medial surface of IV-1 near the proximal end. Phalanx III-1 is wider than it is high anteroposteriorly, differing from the other two first phalanges, and is slightly shorter than phalanx II-1. The lateral and medial condyles at the distal ends of II-1 and IV-1 are well separated by intercondylar grooves, whereas those of phalanx III-1 are weakly divided. Phalanx IV-1 is shorter than phalanges II-1 and III-1. It has unevenly sized condyles, with the medial one being larger than the lateral. Its proximal end has a ventral intercondylar process. There are depressions on the dorsal surfaces of the distal ends in phalanges II-1, III-1, III-2, and IV-1, where the dorsal intercondylar processes of the following phalanges fit. The collateral ligament fossa is deeper on the lateral surface than on the medial surface in the phalanges of digit II and is deeper on the medial surface than on the

(Opposite page)
Figure 6.10. Distal tarsals, metatarsals, and pedal phalanges of Harpymimus okladnikovi *(IGM 100/29): metatarsals with articulated distal tarsals in medial (A), anterior (B), lateral (C), proximal (D and E), and distal (F) views and pedal phalanges in dorsal (G) and medial (H) views. Abbreviations: dt—distal tarsal; mt—metatarsal.*

TABLE 6.3

Measurements of the Pelvic Girdle and Hindlimb
Elements of *Harpymimus okladnikovi*

All elements are from the left side except the ilium. The widths of all
elements are measured at the proximal ends. Asterisks indicate that an
element is incomplete (*, minimum value; **, estimated value).

Element	Length (mm)	Width (mm)
Ilium	381*	—
Metatarsal I	292	36
Metatarsal II	310	—
Metatarsal III	304**	40
Pedal phalanx II-1	72	28
Pedal phalanx II-2	51	26
Pedal phalanx III-1	67	38
Pedal phalanx III-2	54	32
Pedal phalanx III-4	44*	21
Pedal phalanx IV-1	42	25
Pedal phalanx IV-2	34	24
Pedal phalanx IV-3	33	22

lateral surface in the phalanges of digit IV. The fossae of the phalanges
of digit III are equally developed on lateral and medial surfaces.

Phalanx II-2 is the only penultimate pedal phalanx preserved in
Harpymimus okladnikovi. It is similar to phalanx II-1 in shape but
differs in lacking depressions on the lateral surface of the proximal end
and on the dorsal surface of the distal end. Phalanx II-2 is 71 percent of
the length of II-1, which is much longer than phalanx II-2 in ornithomimids (<60 percent) (Table 6.3). Phalanx III-2 is similar to III-1 but
shorter. The depression on the lateral surface near the proximal end of
the former is smaller than the depression on the latter. The distal end of
III-2 has more pronounced condyles than that of III-1, and its proximal
end has a weak dorsal intercondylar process. Phalanges IV-2 and IV-3
are similar to IV-1 in general shape but differ in having dorsal and
ventral intercondylar processes at the proximal end and ridges, which
separate the proximal surfaces into two. Phalanx IV-3 is distinguishable from IV-2 in having a ventral intercondylar process, stronger than
the dorsal one, at the proximal end. Only one pedal ungual is preserved
and is probably III-4 because it is symmetrical with respect to the
sagittal plane. It is similar to pedal phalanges in other ornithomimosaurs but has a strong flexor tubercle.

Discussion

Harpymimus okladnikovi and *Pelecanimimus polyodon* are recognized as basal ornithomimosaurs, but their interrelationship has been

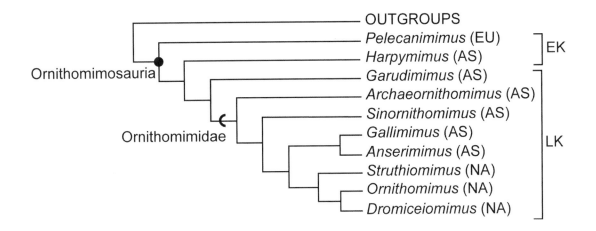

Figure 6.11. Cladogram of Ornithomimosauria by Kobayashi and Lü (2003) with temporal and geographical occurrences. Abbreviations: EU—Europe; AS—Asia; NA—North America; EK—Early Cretaceous; LK—Late Cretaceous.

problematic because of primitive characters in the skull and derived characters in the manus in *Pelecanimimus polyodon* (Pérez-Moreno et al. 1994; Pérez-Moreno and Sanz 1995). A phylogenetic analysis by Kobayashi and Lü (2003) suggests that *Harpymimus okladnikovi* is more derived than *Pelecanimimus polyodon* and is a sister taxon to the Late Cretaceous forms (Fig. 6.11). This analysis disagrees with a previous phylogenetic tree by Osmólska (1997). The monophyly of *Harpymimus okladnikovi* and the Late Cretaceous forms is supported by five unambiguous synapomorphies of the skull: the loss of premaxillary and maxillary dentition, the lack of lacrimal prominence, the ventral reflection of the anterior portion of the dentary, and the presence of two antorbital fenestrae (Kobayashi and Lü 2003).

Harpymimus okladnikovi has teeth only in the dentary, and the distribution of teeth in the anterior part of the dentary is unique (Fig. 6.1). *Pelecanimimus polyodon* is the other ornithomimosaur with dentition (Pérez-Moreno et al. 1994), but it differs in the number (75 dentary teeth and 220 teeth in total) and distribution of teeth (teeth are present in the premaxilla, maxilla, and dentary). Although the holotype of *Harpymimus okladnikovi* does not preserve teeth, they are described as subcylindrical and truncated (Barsbold and Perle 1984) and peg-like (Currie 2000), and each alveolus is small (approximately 2.5 mm in diameter) and separated from its neighbors by approximately 1 mm. The upper jaw (premaxilla and maxilla) of *Harpymimus okladnikovi* lacks teeth but might have been covered by a rhamphotheca as in *Gallimimus* and *Ornithomimus* (Norell et al. 2001a) because of the presence of neurovascular foramina in the premaxilla. *Harpymimus okladnikovi* teeth were probably used not for cutting or piercing prey, as were those of other non-avian theropods, but rather for grabbing and holding food.

As Pérez-Moreno and Sanz (1995) and Kobayashi and Lü (2003) suggested, metacarpal I of *Harpymimus okladnikovi* is proportionally shorter than in the other ornithomimosaurs (Fig. 6.12), and this is the primitive condition in ornithomimosaurs. Additionally, metacarpal I of *Harpymimus okladnikovi* has a laterally rotated (rotated counterclock-

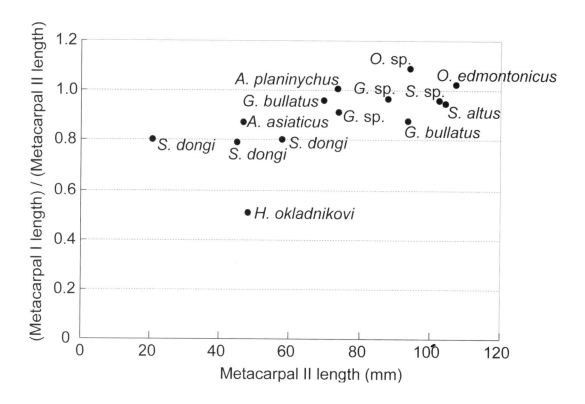

Figure 6.12. Graph of metacarpal II lengths and (metacarpal I length)/(metacarpal II length) ratios in ornithomimosaurs, showing a low ratio in Harpymimus okladnikovi.

wise for a right manus in distal view) distal end and well-developed ginglymoid metacarpal-phalangeal articulation as seen in *Allosaurus fragilis*, indicating that these features are plesiomorphic (Kobayashi and Lü 2003). Nicholls and Russell (1985) noted that medially rotated (rotated clockwise for a right manus in distal view) metacarpal I is present in *Struthiomimus altus* and *Gallimimus bullatus*. In this condition, digit I moves away from the other two digits in extension and toward the other digits in flexion. In *Gallimimus* sp. (IGM 100/14), all digits are aligned along the same plane in the flexed condition as in other ornithomimosaurs (except *Harpymimus okladnikovi*), whereas digit I of *Harpymimus okladnikovi* is more proximally placed than the other digits. In *Gallimimus* sp., the digits span a much larger area when expanded than when flexed (Fig. 6.13I,L). These features suggest a specialization for clamping and hooking in *Struthiomimus altus* and *Gallimimus* (*G. bullatus* and *G.* sp.) (Fig. 6.13G–L).

The manus condition in *Harpymimus okladnikovi* is different from that in *Struthiomimus altus* and *Gallimimus* (*G. bullatus* and *G.* sp.) but similar to that in other theropods (Carpenter 2002). During flexion and extension, digit I of *Harpymimus okladnikovi* rotates more than in *Struthiomimus altus* and *Gallimimus* sp. (the angle between flexion and extension is approximately 90° in *Harpymimus okladnikovi* and 30° in *Struthiomimus altus* and *Gallimimus* sp.), and digits diverge in flexion, which may be an adaptation for grasping and raking (Fig. 6.13A–F). In

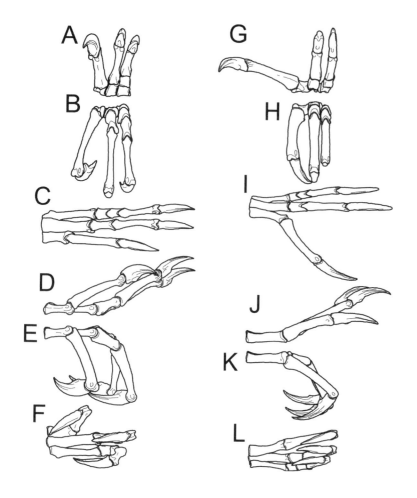

contrast, *Pelecanimimus polyodon* has subequal metacarpals and a laterally rotated distal metacarpal I, suggesting a more derived manus structure than in *Harpymimus okladnikovi.*

With respect to the relationships of *Harpymimus okladnikovi* and *Pelecanimimus polyodon* with other ornithomimosaurs, skull features suggest that *Harpymimus okladnikovi* is more derived than *Pelecanimimus polyodon.* This situation could be explained by two scenarios: convergence in *Pelecanimimus polyodon* or reversal in *Harpymimus okladnikovi.*

Elongation of caudal prezygapophyses occurs in some groups of Theropoda (e.g., *Deinonychus;* Ostrom 1969). The caudal prezygapophyses of ornithomimosaurs become elongated posterior to the mid-caudal vertebrae (seventeenth in *Harpymimus okladnikovi;* fourteenth in *Anserimimus planinychus*) (Fig. 6.14B). Most of the caudal prezygapophyses of *Harpymimus okladnikovi* do not exceed 50 percent of the centrum length, but those of ornithomimids (e.g., *Gallimimus bullatus, Anserimimus planinychus, Ornithomimus* sp.) have posterior caudal vertebrae with much longer prezygapophyses (more than 50 percent)

Figure 6.13. Comparisons of hand (left) structures of Harpymimus okladnikovi (GIN 100/29) (A–F) and Gallimimus sp. (GIN 100/14) (G–L). Extension condition in distal (A and G), anterior (C and I), and medial (D and J) views. Flexion condition in distal (B and H), medial (E and K), and posterior (F and L) views. Not to scale.

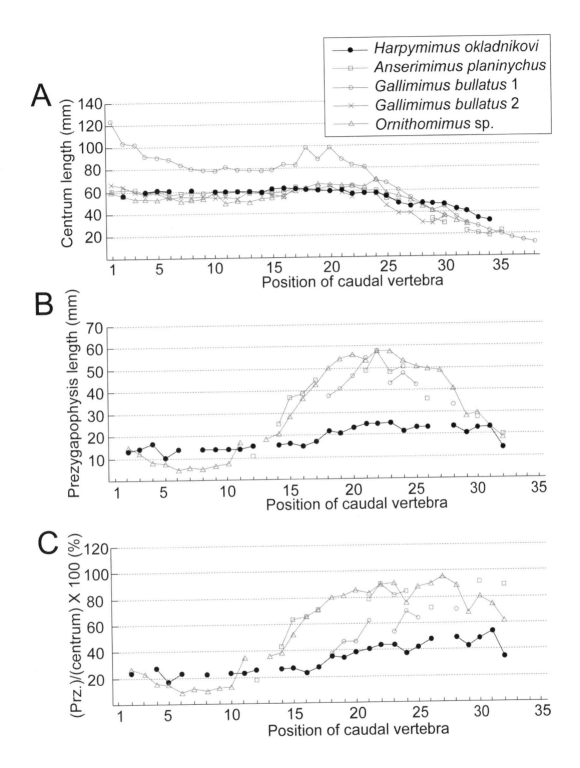

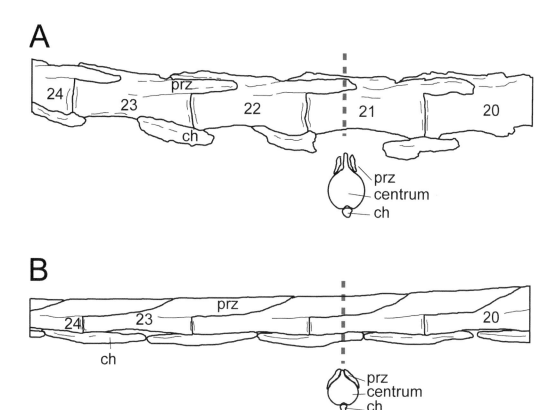

A

24 23 prz 22 21 20

ch

prz
centrum
ch

B

24 23 prz 20

ch

prz
centrum
ch

(Fig. 6.14C). The prezygapophyses of *Harpymimus okladnikovi* are low dorsoventrally and have nearly vertical articular surfaces (Fig. 6.15A). Those of *Ornithomimus* sp. are high, and the articular surfaces of the prezygapophyses face ventromedially (Fig. 6.15B). Because of these features and elongated prezygapophyses, the posterior caudal vertebrae of *Ornithomimus* sp. interlock well and have fewer degrees of freedom in dorsoventral movement. The posterior portion of the *Harpymimus okladnikovi* tail has more freedom of movement dorsoventrally, which is probably a primitive condition in ornithomimosaurs.

Khand et al. (2000) interpreted the age of the Shinekhudug Formation, yielding *Harpymimus okladnikovi*, as Hauterivian-Barremian, although it was noted as Aptian-Albian by Barsbold and Perle (1984). The tree topology of ornithomimosaur taxa by Kobayashi and Lü (2003) follows chronological appearance (*Pelecanimimus polyodon* from Barremian, Pérez-Moreno et al. 1994; *Garudimimus brevipes*, Cenomanian-Turonian, Khand et al. 2000; late Late Cretaceous for the others, Kobayashi and Lü 2003, Barsbold and Osmólska 1990, and references therein) (Fig. 6.11). The tree topology also suggests that the origin of ornithomimosaurs is in either eastern Asia or Europe prior to the Barremian. Ornithomimids dispersed to the North American continent during or prior to the Late Cretaceous.

(Opposite page)
Figure 6.14. Comparisons of changes in centrum and prezygapophysis lengths and ratios plotted against the caudal vertebra position in Harpymimus okladnikovi *(GIN 100/29),* Anserimimus planinychus *(GIN 100/300),* Gallimimus bullatus *(1 for GIN100/11 and 2 for GIN100/12), and* Ornithomimus sp. *(TMP 95.110.1).*

(Above)
Figure 6.15. Comparisons of caudal vertebrae (twentieth to twenty-fourth) of Harpymimus okladnikovi *(GIN 100/29) and* Ornithomimus sp. *(TMP 95.110.1) with schematic cross-sections the twenty-first caudal vertebra. Abbreviations: ch—chevron, prz—prezygapophysis.*

Conclusion

Reexamination of the *Harpymimus okladnikovi* skeleton revealed new important information that helps refine the diagnoses of the taxon. *Harpymimus okladnikovi* is an important taxon for understanding the early evolution of ornithomimosaurs. Detailed comparative study and phylogenetic analysis support the phylogenetic status of *Harpymimus okladnikovi* as basal to the clade of *Garudimimus brevipes* plus Ornithomimidae but more derived than *Pelecanimimus polyodon*.

The arrangement of dentary teeth in *Harpymimus okladnikovi* is unique among ornithomimosaurs. Small teeth in the lower jaw and the coverage of the upper jaw by a rhamphotheca were probably used for grabbing and holding food. The short metacarpal I of *Harpymimus okladnikovi*, with laterally rotated and ginglymoid metacarpal phalangeal articulations, suggests that digit I moved toward the other two digits (digits II and III) in extension and away in flexion, as in other theropods; this is the primitive condition in ornithomimosaurs. The prezygapophyses of the posterior caudal vertebrae in *Harpymimus okladnikovi* are shorter than those in other ornithomimosaurs. The elongated and tilted prezygapophyses of the posterior caudal vertebrae in derived ornithomimosaurs suggest less flexibility in dorsoventral movement of the tail. The tree topology of ornithomimosaur phylogeny follows chronological appearance and suggests that the origin of ornithomimosaurs is in either eastern Asia or Europe prior to the Barremian. Ornithomimids dispersed to the North American continent during or prior to the Late Cretaceous.

Acknowledgments. This chapter comprised a portion of the Ph.D. dissertation of Y. K., who would like to acknowledge his committee members, Louis L. Jacobs (Southern Methodist University), Dale A. Winkler (Southern Methodist University), Anthony R. Fiorillo (Dallas Museum of Natural History), Philip J. Currie (Royal Tyrrell Museum of Palaeontology), and the co-author (Rinchen Barsbold). We would like to thank Isao Takahashi, Kazuyuki Takahashi, and Kazuhisa Sato for helping us to access the original specimen and ornithomimosaur casts housed in Nakasato Dinosaur Center. We are grateful to Philip J. Currie and Elizabeth L. Nicholls (Royal Tyrrell Museum of Palaeontology), Mark A. Norell (American Museum of Natural History), Peter J. Makovicky (Field Museum of Natural History), Bernardino Pérez-Moreno (Universidad Autónoma de Madrid), Zhi-Ming Dong (Institute of Vertebrate Paleontology and Paleoanthropology), and Kevin Seymour (Royal Ontario Museum) for providing access to specimens. We also thank to Louis L. Jacobs, Dale A. Winkler, Philip J. Currie, Peter J. Makovicky, and Bernardino Pérez-Moreno for valuable comments on the early version of the manuscript and Kenneth Carpenter (Denver Museum of Nature and Science) and Bruce M. Rothschild (Arthritis Center of Northeast Ohio) for reviewing the manuscript. This project is supported by the Institute for the Study of Earth and Man, the Jurassic Foundation, and the Sasakawa Scientific Research Grant.

References Cited

Barsbold, R. 1976. [On the evolution and systematics of the late Mesozoic dinosaurs]. *Trudy Sovmestnaia Rossiisko-Mongol'skaia Paleontologicheskaia Ekspeditsiia* 3: 68–75. [In Russian.]

Barsbold, R. 1981. [Toothless carnivorous dinosaurs of Mongolia]. *Trudy Sovmestnaia Rossiisko-Mongol'skaia Paleontologicheskaia Ekspeditsiia* 15: 28–39. [In Russian.]

Barsbold, R. 1988. [A new Late Cretaceous ornithomimid from the Mongolia People's Republic]. *Paleontologicheskii zhurnal* 1: 122–125. [In Russian.]

Barsbold, R., and H. Osmólska. 1990. Ornithomimosauria. In D. B. Weishampel, P. Dodson, and H. Osmólska (eds.), *The Dinosauria,* pp. 225–244. Berkeley: University of California Press.

Barsbold, R., and A. Perle. 1984. [On first new find of a primitive ornithomimosaur from the Cretaceous of the MPR]. *Paleontologicheskii zhurnal* 2: 121–123. [In Russian.]

Brochu, C. A. 1996. Closure of neurocentral sutures during crocodilian ontogeny: Implication for maturity assessment in fossil archosaurs. *Journal of Vertebrate Paleontology* 16: 49–62.

Carpenter, K. 2002. Forelimb biomechanics of nonavian theropod dinosaurs in predation. *Senckenbergiana Lethaea* 82: 59–76.

Currie, P. J. 2000. Theropods from the Cretaceous of Mongolia. In M. J. Benton, M. A. Shishkin, D. M. Unwin, and E. N. Kurochkin (eds.), *The Age of Dinosaurs in Russia and Mongolia,* pp. 434–455. Cambridge: Cambridge University Press.

Hurum, J. H. 2001. Lower jaw of *Gallimimus bullatus.* In D. H. Tanke and K. Carpenter (eds.), *Mesozoic Vertebrate Life,* pp. 34–41. Bloomington: Indiana University Press.

Khand, Y., D. Badamgarav, Y. Ariunchimeg, and R. Barsbold. 2000. Cretaceous System in Mongolia and its depositional environments. In H. Okada and N. J. Mateer (eds.), *Cretaceous Environments of Asia,* pp. 49–79. Amsterdam: Elsevier Science B. V.

Kobayashi, Y., and Lü J.-C. 2003. A new ornithomimid dinosaur with gregarious habits from the Late Cretaceous of China. *Acta Palaeontologica Polonica* 48: 235–259.

Makovicky, P. J. 1995. Phylogenetic aspects of the vertebral morphology of Coelurosauria (Dinosauria: Theropoda). M.S. dissertation, Copenhagen University.

Nicholls, E. L., and A. P. Russell. 1981. A new specimen of *Struthiomimus altus* from Alberta, with comments on the classificatory characters of Upper Cretaceous ornithomimids. *Canadian Journal of Earth Sciences* 18: 518–526.

Nicholls, E. L., and A. P. Russell. 1985. Structure and function of the pectoral girdle and forelimb of *Struthiomimus altus* (Theropoda: Ornithomimidae). *Palaeontology* 28: 643–677.

Norell, M. A., P. J. Makovicky, and P. J. Currie. 2001a. The beaks of ostrich dinosaurs. *Nature* 412: 873–874.

Norell, M. A., J. M. Clark, and P. J. Makovicky. 2001b. Phylogenetic relationships among coelurosaurian theropods. In J. Gauthier and L. F. Gall (eds.), *New Perspectives on the Origin and Early Evolution of Birds: Proceedings of the International Symposium in Honor of John H. Ostrom,* pp. 29–67. New Haven, Conn.: Peabody Museum of Natural History, Yale University.

Osborn, H. F. 1916. Skeletal adaptation of *Ornitholestes, Struthiomimus, Tyrannosaurus. Bulletin of American Museum of Natural History* 35: 733–771.

Osmólska, H. 1997. Ornithomimosauria. In P. J. Currie and K. Padian (eds.), *Encyclopedia of Dinosaurs,* pp. 499–503. San Diego: Academic Press.

Osmólska, H., E. Roniewicz, and R. Barsbold. 1972. A new dinosaur, *Gallimimus bullatus* n. gen., n. sp. (Ornithomimidae) from the Upper Cretaceous of Mongolia. *Palaeontologica Polonica* 27: 103–143.

Ostrom, J. H. 1969. Osteology of *Deinonychus antirrhopus,* an unusual theropod from the Lower Cretaceous of Montana. *Bulletin of the Peabody Museum of Natural History* 30: 1–165.

Pérez-Moreno, B. P., and J. L. Sanz. 1995. The hand of *Pelecanimimus polyodon:* A preliminary report. In *Extended Abstracts: II International Symposium on Lithographic Limestones, Cuenca, Spain, July 1995,* pp. 115–117. Cuenca, Spain: Ediciones de la Universidad Autónoma de Madrid.

Pérez-Moreno, B. P., J. L. Sanz, A. D. Buscalioni, J. J. Moratalla, F. Ortega, and D. Rasskin-Gutman. 1994. A unique multitoothed ornithomimosaur dinosaur from the Lower Cretaceous of Spain. *Nature* 370: 363–367.

Russell, D. A. 1972. Ostrich dinosaurs from the Late Cretaceous of western Canada. *Canadian Journal of Earth Sciences* 9: 375–402.

Smith, D., and P. Galton. 1990. Osteology of *Archaeornithomimus asiaticus* (Upper Cretaceous, Iren Debasu Formation, People's Republic of China). *Journal of Vertebrate Paleontology* 10: 255–265.

Xu X., M. A. Norell, Wang X.-L., P. J. Makovicky, and Wu X.-C. 2002. A basal troodontid from the Early Cretaceous of China. *Nature* 415: 780–784.

7. Theropod Teeth from the Upper Cretaceous (Campanian-Maastrichtian), Big Bend National Park, Texas

Julia T. Sankey, Barbara R. Standhardt, and Judith A. Schiebout

Abstract

Big Bend National Park, Texas, has one of the southernmost terrestrial records for the Late Cretaceous in North America. Cretaceous theropod dinosaurs are not as well known from southern North America as from more northern areas. Theropod teeth were collected from microfossil sites from the Upper Cretaceous upper Aguja and lower Tornillo formations, spanning the late Campanian to late Maastrichtian (approximately 74–67 Ma). In addition to previously recognized taxa from Campanian sites, several teeth from Maastrichtian sites are unlike any previously described from Big Bend. These new morphotypes are referred to as *Saurornitholestes* n. sp.?. Theropods present in the Campanian and Maastrichtian of Big Bend include tyrannosaurids, *Saurornitholestes* cf. *S. langstoni*, *Saurornitholestes* n. sp.?, *Richardoestesia* cf. *R. gilmorei*, *R. isosceles*, and cf. *Paronychodon*. Additionally, possible bird teeth are tentatively identified in the assemblage. *Saurornitholestes* n. sp.? and cf. *Paronychodon* occur only in the Maastrichtian sites, suggesting that there were distinct Campanian and Maastrichtian theropod assemblages in Big Bend, as there were in northern areas. Absent from both the Campanian and Maastrichtian assemblages in Big Bend are *Dromaeosaurus albertensis* and *Troodon formosus,* which are common in northern areas. Also, many taxa are represented by teeth of hatchlings or juveniles, demonstrating that the animals nested in this area.

Introduction

The first thorough study of theropod teeth (Currie et al. 1990) was based on collections from the Upper Cretaceous (upper Campanian) Judith River Group of Alberta and included detailed descriptions and illustrations of tooth and denticle morphology. This collection increased in size as a result of an extensive screenwashing program (Brinkman 1990; Peng et al. 2001), which allowed Baszio (1997a,b) to document the range of variation of theropod teeth and to discern important paleoecologic patterns during the Late Cretaceous. The collection was increased to over 1,700 teeth by additional screenwashing, and Sankey et al. (2002) described and measured the collection, quantifying the range of variation in both known taxa and new morphotypes. These new morphotypes, possibly representing new taxa, are particularly significant because they document higher theropod diversity in the assemblage than was previously recognized. Bird teeth were also described. Because the collection of theropod teeth from the Judith River Group is the largest and most thoroughly studied for the late Campanian, it is frequently used in comparisons with other contemporaneous faunas, such as the theropods from Big Bend National Park, Texas.

The theropods from Big Bend are important because they are some of the southernmost records from the Late Cretaceous of North America. Big Bend (Fig. 7.1) was within the southern biogeographic province (Lehman 1997), which was characterized by the *Normapolles* palynoflora, with a warm, dry, non-seasonal climate and open canopy woodlands. Differences between the southern and northern provinces (Wyoming and north) were primarily due to differences in temperature and rainfall (Lehman 1997).

Considerably less is known about the dinosaurs in the southern province compared to the northern province, partly because there is less outcrop area and there are fewer paleontologists working in the area (Lehman 1997). However, the lack of information is also due to there being fewer and less well preserved fossils. One factor involved in this taphonomic bias in Big Bend was that uplands were relatively distant, resulting in slower sedimentation rates and condensed stratigraphic and faunal records compared to those of northern areas. Increased aridity during the Late Cretaceous, due to climate change, retreat of the Western Interior Seaway, and uplift of the western mountains, occurred earlier in the Late Cretaceous in this area than in the north. For example, the dinosaur bone beds in the Aguja Formation (upper Campanian) of Big Bend probably formed during periodic droughts that were severe enough to cause marshes to dry up (Davies and Lehman 1989).

The first screenwashing program of Cretaceous microsites in Big

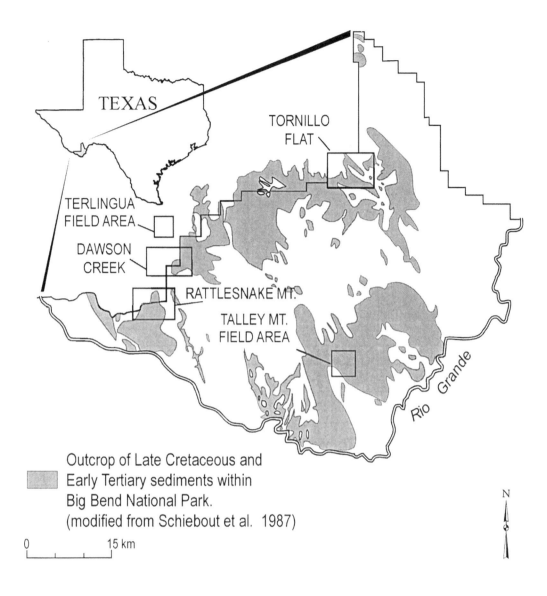

Outcrop of Late Cretaceous and
Early Tertiary sediments within
Big Bend National Park.
(modified from Schiebout et al. 1987)

0   15 km

N

*Figure 7.1. Map of Big Bend
National Park, Texas.*

Bend was developed by Judith Schiebout (Louisiana State University) and her students in the 1980s (Standhardt 1986) and continues today with Sankey (1998, 2001; Sankey and Gose 2001). These efforts have produced samples of dinosaur and other vertebrate small teeth and bones. Although the collection is considerably smaller than that from the Judith River Group because there are fewer productive microsites in Big Bend, it is important because it documents the theropods from this area. The goals of this research are to determine (1) what theropod taxa occurred in Big Bend; (2) whether they were different from northern theropods; (3) whether there were differences between the Campanian and Maastrichtian theropod assemblages; and (4) whether theropod diversity changed in Big Bend during the last 10 million years of the Cretaceous.

Maastrichtian theropods collected by Standhardt (1986) from

microsites in the uppermost Aguja (lower Maastrichtian) and lower Tornillo (Maastrichtian) are included in this paper. Within this collection there are new morphotypes of *Saurornitholestes*, which may represent one or more new species.

Rowe et al. (1992) reported late Campanian theropods collected by screening the Terlingua microsite (upper Aguja). A few of the illustrated teeth were misidentified but were corrected by Sankey (2001). Additional late Campanian theropods (and other dinosaurs) were described by Sankey (2001) from screened Talley Mountain microsites (upper Aguja), including a new species of theropod, *Richardoestesia isosceles*. Recent collections of late Campanian theropods from newly discovered microsites at Rattlesnake Mountain (upper Aguja) are reported here, further documenting Big Bend theropod variation and diversity.

Stratigraphy

Aguja Formation. The Aguja Formation (upper Campanian to lower Maastrichtian; Fig. 7.2) contains coastal and floodplain sediments deposited during the final retreat of the Western Interior Seaway from Big Bend. The Aguja is a widespread, eastward-thinning unit of 135 to 285 m of paralic and marine sandstones interbedded with shale and lignite (Lehman 1985). The Terlingua Creek sandstone member represents the last marine transgression (Regression 8 of Kauffman 1977), and the overlying upper shale member represents the last pre-Laramide tectonic sedimentation in the area (Lehman 1991). In the lower part of the upper shale member are carbonaceous mudstones, thin beds of lignite, and large siderite ironstone concretions representing distributary channels, levees, crevasse splays, and poorly drained interdistributary marshes and bays. The upper part, with variegated mudstones and sandstones containing conglomeratic lags of paleocaliche nodules, represents fluvial environments within a deltaic coastal plain and inland floodplain (Lehman 1985, 1991).

Magnetostratigraphy of the upper shale member of the Aguja Formation in the Talley Mountain area correlated the deposits to the base of Chron 32, or approximately 71–74 Ma (late Campanian–early Maastrichtian) (Sankey and Gose 2001). This correlation was constrained by the following evidence. First, the marine Terlingua Creek sandstone, which underlies the upper shale member, is middle Campanian in age (Lehman 1985; Rowe et al. 1992). Of particular importance is the presence within this unit of *Baculites maclearni* (Rowe et al. 1992), a zonal index fossil for the middle Campanian, with a duration of approximately 79.6 to 80.2 Ma (Obradovich 1993). Second, western-most outcrops of the lower portion of the upper shale member of the Aguja Formation are middle Campanian on the basis of mammals from Terlingua (Cifelli 1995; Rowe et al. 1992; Weil 1992, 1999). Third, mammals from the upper shale member of the Aguja in the Talley Mountain area include *Alphadon* cf. *A. halleyi* (LSU-6252), a cosmopolitan marsupial that is characteristic of Judithian (North American Land Mammal "Age") faunas (Lillegraven and McKenna

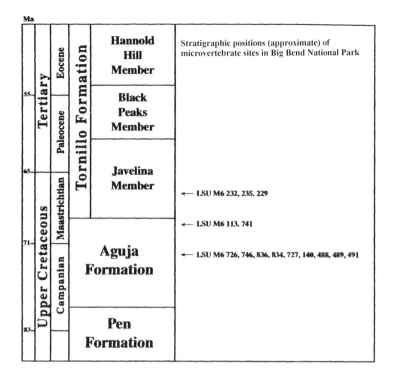

| | | | | Hannold Hill Member | Stratigraphic positions (approximate) of microvertebrate sites in Big Bend National Park |

Tertiary — Eocene / Paleocene — Tornillo Formation

Hannold Hill Member

Black Peaks Member

Javelina Member

← LSU M6 232, 235, 229

← LSU M6 113, 741

Aguja Formation

← LSU M6 726, 746, 836, 834, 727, 140, 488, 489, 491

Pen Formation

Upper Cretaceous — Maastrichtian / Campanian

1986) and is also found from Terlingua (Rowe et al. 1992). The known range of the Judithian is approximately 5 million years, from ~75 to 78 Ma (Goodwin and Deino 1989) or ~74 to 79 Ma (Lillegraven and McKenna 1986). Fourth, the uppermost part of the upper shale member is lower Maastrichtian, on the basis of vertebrates, especially mammals, and limited magnetostratigraphy (Lehman 1985, 1989, 1990; Standhardt 1986).

Tornillo Formation. Overlying the Aguja is the Tornillo Formation, which is an Upper Cretaceous through lower Tertiary deposit (stratigraphic terminology follows Schiebout et al. 1987). The Tornillo consists of mudstones and sandstones and represents fluvial floodplain deposition. The mudstones often contain prominent color banding, indicating different paleosols. Fluctuations in water table levels were due to climatic changes, sea level changes, or both, with the black and very dark red beds indicating high water tables, and the red and gray beds, containing abundant calcium carbonate nodules, indicating lower water tables (concentrating calcium carbonate in the B soil horizons; Schiebout et al. 1987).

The K/T boundary is within the Javelina member of the Tornillo Formation. At the Dawson Creek section, the boundary has been bracketed between fossil sites and is within a 35-m span of floodplain deposits. Sediments are fluvial mudstones containing strong color banding due to paleosol development. However, no physical evidence for the K/T impact event has been found in these deposits (Standhardt 1986; Schiebout et al. 1987; Lehman 1989, 1990).

Figure 7.2. Stratigraphy of Aguja and Tornillo formations, with positions of microsites (modified from Schiebout et al. 1987; Standhardt 1986).

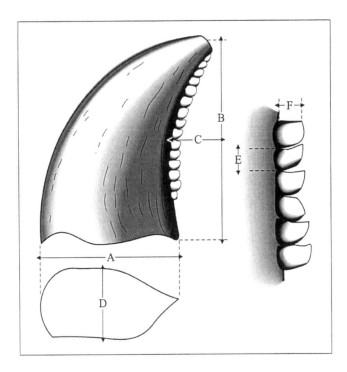

*Figure 7.3. Measurements of
teeth. (A) FABL, fore-aft basal
length, not including denticles;
(B) greatest height, from crown
tip to base (not including root, if
unshed tooth) and measured from
posterior side; (C) curvature, the
greatest distance from posterior
carina (not including the
denticles) to a perpendicular line
from tooth tip to base; (D) cross-
sectional thickness, the greatest
lateral-lingual cross-sectional
tooth thickness; (E) greatest
denticle width; (F) greatest
denticle height. Denticles/mm
(of largest denticles present).*

Materials and Methods

Fossil Sites. The specimens are from vertebrate microfossil sites from the Aguja and Tornillo formations. Fossils were collected by both surface collection and underwater screening techniques. Screened matrix was sorted with a dissecting microscope. Studies of the sedimentology, stratigraphy, palynology, and magnetostratigraphy provide a well-constrained stratigraphic framework for the sites (Lehman 1985; Standhardt 1986; Sankey 1998; Sankey and Gose 2001). All specimens are curated into the Louisiana State University Museum of Natural Science Vertebrate Paleontology Collections, where detailed locality information is on file.

When evaluating a morphologically distinct tooth, three alternatives were considered: that the tooth is (1) a variant along the tooth row; (2) a variant within the population but not expressed in each individual; (3) a distinct taxon. The third alternative is supported if the tooth morphotype has a distinct stratigraphic distribution. Although no new taxa are proposed in this study, a new and distinct morphotype of *Saurornitholestes* from a Maastrichtian microsite may represent a new species.

Measurements and Photographs. Locations for tooth measurements are illustrated in Figure 7.3. A microscope with an ocular micrometer was used by Sankey. Measurements are given in millimeters (mm) and are abbreviated as follows: FABL, fore-aft basal length; ht, height; curv, curvature; CST, cross-sectional thickness; CSS, cross-sectional shape; and dent/mm, denticles/mm. Measurement resolution is 0.5 mm for denticles/mm and 0.1 mm for denticle width and height. Measurements for specimens are available from Sankey.

TABLE 7.1

Tyrannosaurid Measurements (in mm)

—Approximate measurements, due to broken tooth, represented by [ ]. If measurement not possible, represented by —.
—Denticle measurements were made on largest denticles present. Abbreviations: curvature (curv); fore-aft basal length (FABL); height (ht); width (wd); cross-sectional thickness (CST); cross-sectional shape (CSS); denticle (dent); denticles/mm (dent/mm); round (rd); fragment (frag); ant. (anterior); post. (posterior); flat (flattened).

LSUMG Spec. no.	LSUM. Local no.	Geo. fm.	Age	Curv	FABL	Ht	CST	CSS	Ant. dents present	Dents/mm post.	Dent.wd post.	Dent. ht. post.
6201	726	Aguja	late Camp	—	—	—	—	—	—	2	0.4	0.5
6209	726	Aguja	late Camp	—	—	—	—	—	yes	2.5	0.3	[0.2]
6218	726	Aguja	late Camp	no	8.3	[15]	5.8	rd/oval	yes	5	0.2	0.2
6219	726	Aguja	late Camp	—	—	—	—	—	—	—	—	—
6274	746	Aguja	late Camp	no	8.5	[20]	—	oval	yes	2.5	0.3	0.4
6227	746	Aguja	late Camp	—	—	—	—	—	—	2	0.5	0.4
6282	747	Aguja	late Camp	—	—	—	—	—	yes	—	—	—
6272	746	Aguja	late Camp	—	—	—	—	—	—	—	—	—
6262	746	Aguja	late Camp	—	[8.5]	15	5	oval	—	—	—	—
8199	836	Aguja	late Camp	no	[6.0]	—	5.5	rd	no	minute	—	—
8211	834	Aguja	late Camp	0.5	7	[10]	3.5	oval	yes	3.5	0.3	3
5914	727	Aguja	late Camp.	—	14	—	9	oval	yes	2	0.5	0.5
6042	727	Aguja	late Camp.	—	—	—	—	—	yes	4	0.2	—
6039	726	Aguja	late Camp.	—	—	—	—	—	—	—	—	—
6186	140	Aguja	late Camp	—	2.8	—	1.4	oval	yes	6	0.2	0.1
6236	489	Aguja	late Camp	—	3.5	—	2.6	rd	no	5.5	0.2	0.17
6239	489	Aguja	late Camp	—	4.8	—	[3.5]	oval/rd	no	5.5	0.2	0.3
5580	489	Aguja	late Camp	—	—	—	—	—	—	—	—	—
5483	488	Aguja	late Camp	—	8.7	—	5.7	oval/rd	yes	4	0.3	0.3
5987	741	Aguja	Camp/Maast	—	16	—	9	oval	yes	2.5	0.4	0.3

TABLE 7.1 (cont.)

Tyrannosaurid Measurements (in mm)

LSUMG Spec. no.	LSUM. Local no.	Geo. fm.	Age	Curv	FABL	Ht	CST	CSS	Ant. dents present	Dents/mm post.	Dent.wd post.	Dent. ht. post.
6024	741	Aguja	Camp/Maast	—	1.2	—	0.9	oval	no	7	0.2	0.2
5921	741	Aguja	Camp/Maast	—	—	—	—	—	—	—	—	—
5920	741	Aguja	Camp/Maast	—	[13]	—	[0.8]	oval	yes	2	0.4	0.4
5919	741	Aguja	Camp/Maast	—	12	[16]	7	oval	yes	2.5	0.4	0.3
6018	741	Aguja	Camp/Maast	—	[3.0]	—	1.9	oval	no	4	0.2	0.2
6015	741	Aguja	Camp/Maast	—	—	—	—	—	—	—	—	—
5922	741	Aguja	Camp/Maast	—	[14]	—	8	oval	yes	2.5	0.3	0.3
6017	741	Aguja	Camp/Maast	—	—	—	—	—	—	—	—	—
6020	741	Aguja	Camp/Maast	—	—	—	—	—	—	—	—	—
6016	741	Aguja	Camp/Maast	—	—	—	—	—	yes	4	0.3	0.2
6027	741	Aguja	Camp/Maast	—	—	—	—	—	—	—	—	—
6036	741	Aguja	Camp/Maast	—	—	—	—	—	—	—	—	—
6023	741	Aguja	Camp/Maast	—	[2.0]	3	1.2	oval/rd	no	4.5	0.2	0.2
6021	741	Aguja	Camp/Maast	—	—	—	—	—	yes	5.5	0.2	0.2
6028	741	Aguja	Camp/Maast	—	—	—	—	—	—	—	—	—
6035	741	Aguja	Camp/Maast	—	—	—	—	—	—	—	—	—
6025	741	Aguja	Camp/Maast	—	—	—	—	—	—	—	—	—
6033	741	Aguja	Camp/Maast	—	—	—	—	—	yes	5	0.2	0.2
6031	741	Aguja	Camp/Maast	—	—	—	—	—	—	5	0.2	0.23
6029	741	Aguja	Camp/Maast	—	—	—	—	—	—	4	0.3	[0.2]
6032	741	Aguja	Camp/Maast	—	—	—	—	—	yes	6.5	0.1	[0.1]
5034	741	Aguja	Camp/Maast	—	—	—	—	—	—	—	—	—
6045	741	Aguja	Camp/Maast	—	—	—	—	—	yes	—	0.2	—
6046	741	Aguja	Camp/Maast	—	—	—	—	—	—	—	—	—

TABLE 7.1 (*cont.*)

Tyrannosaurid Measurements (in mm)

LSUMG Spec. no.	LSUM. Local no.	Geo. fm.	Age	Curv	FABL	Ht	CST	CSS	Ant. dents present	Dents/mm post.	Dent.wd post.	Dent. ht. post.
5926	741	Aguja	Camp/Maast	—	—	—	—	—	—	—	—	—
6030	741	Aguja	Camp/Maast	—	—	—	—	—	—	—	—	—
1375	113	Aguja	Ea. Maast	[0.5]	3.6	[5.5]	3.2	rd	no	6.5	0.1	0.2
1313	113	Aguja	Ea. Maast.	—	—	—	—	—	yes	2.5	0.4	0.3
1312	113	Aguja	Ea. Maast.	—	—	—	—	—	—	—	—	—
5994	113	Aguja	Ea. Maast.	—	—	—	—	—	—	—	—	—
6003	113	Aguja	Ea. Maast.	—	—	—	—	—	—	—	—	—
6009	232	Tornillo	Maast	—	10.8	[20]	8	oval	yes	2.5	0.3	0.3
6011	232	Tornillo	Maast	none	15	[30]	19	oval	yes	2	0.4	—
6001	232	Tornillo	Maast	—	—	—	—	—	—	2	0.5	0.4
6012	232	Tornillo	Maast	—	—	—	—	—	—	2	0.5	[0.4]
6013	232	Tornillo	Maast	—	—	—	—	—	—	3.5	0.3	0.5
6008	232	Tornillo	Maast	—	—	—	—	—	—	—	—	—
6007	232	Tornillo	Maast	—	—	—	—	—	—	2.5	0.5	0.3
6006	232	Tornillo	Maast	—	14	—	8	oval	yes	3	0.3	—
5985	229	Tornillo	Maast	—	—	—	—	—	—	3	0.3	0.5
5957	229	Tornillo	Maast	—	—	—	—	—	—	4	0.3	0.3
5991	229	Tornillo	Maast	—	—	—	—	—	—	4	0.3	0.4
5992	229	Tornillo	Maast	—	—	—	—	—	—	—	—	—
5988	229	Tornillo	Maast	—	—	—	—	—	—	—	—	—
5944	229	Tornillo	Maast	—	—	—	—	—	—	—	—	—
1088	229	Tornillo	Maast	2	12	24	8	rd	yes	2.5	0.4	0.4
3059	235	Tornillo	Maast	—	—	—	—	—	—	—	—	—
1930	235	Tornillo	Maast	2	12	21	8	rd	yes	2.5	0.4	0.37
6005	235	Tornillo	Maast	—	—	—	—	—	—	2	0.4	0.3

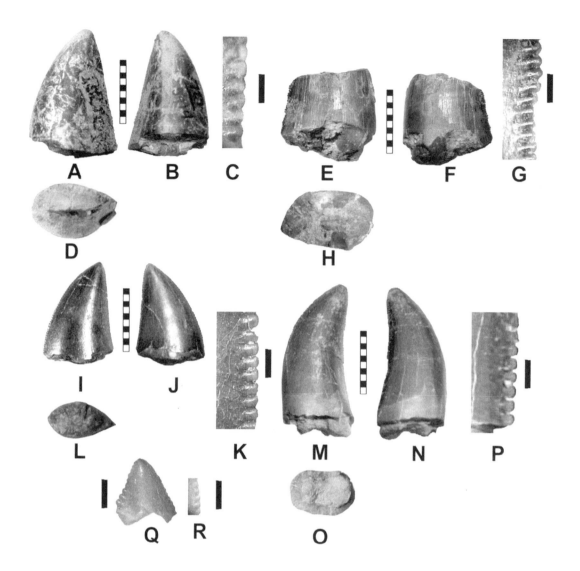

Figure 7.4. Tyrannosaurid teeth from the Aguja and Tornillo formations, Big Bend National Park, Texas, with views of lingual and labial sides, cross-section through base of tooth, and close-up of denticles. (A–D) LSU 727:5914; (E–H) LSU 741:5987; (I–L) LSU 741:5919; (M–O) LSU 229:1088; (Q–R) LSU 741:5926. Alternating black and white scale = 1 cm; solid black scale = 1 mm.

Institutional Abbreviations. LSU, Louisiana State University Museum of Natural Science, Baton Rouge; RTMP, Royal Tyrrell Museum of Palaeontology, Drumheller, Alberta; UALVP, University of Alberta, Lab for Vertebrate Paleontology, Edmonton; UCMP, University of California Museum of Paleontology.

Selected Systematic Paleontology
Order Saurischia Seeley 1888
Family Tyrannosauridae Osborn 1905
Tyrannosauridae Indeterminate
Table 7.1, Figure 7.4

Description. Teeth vary in size, ranging from large (Fig. 7.4A) to small (Fig. 7.4Q); and tooth shape cross-section varies from oval (Fig.

7.4L) to D-shaped (Fig. 7.4P). The degree of recurvature ranges from slightly recurved to none. Denticles are chisel-shaped with pointed tips. Denticles are present on both the anterior and the posterior carinae and are approximately equal in size on both carinae, although slightly larger on the posterior. Denticle size is approximately uniform from tooth base to tooth tip. Denticles/mm on posterior carinae range from 2 to 6/mm (x = 3.6, n = 41). Denticle width ranges from 0.2 to 0.5 mm (x = 0.3, n = 42). Denticle height ranges from 0.1 to 0.5 mm (x = 0.3, n = 28). (All denticle measurements were made on the largest denticles present on the posterior carinae; Fig. 7.3.)

Premaxillary teeth (LSU 489:6236, 6239; 113:1375) are round in cross-sectional shape, have little recurvature, and have no denticles on the anterior carinae.

Discussion. Tyrannosaurid teeth are more round in cross-sectional shape than other theropod teeth (Abler 1997). Denticles occur on both the anterior and the posterior carinae, denticles are approximately equivalent in size on both carinae, and there is less variation in denticle sizes between tooth base and tooth tip. Denticles are typically large and chisel-shaped, denticle width (labial-lingual) is greater than height (proximodistal), and denticles usually occur 3/mm (Currie et al. 1990). Tyrannosaurid denticles are typically the largest in labial-lingual width and have the largest and deepest interdenticle spaces of any theropod (Sankey et al. 2002). Abler (1997) and Baszio (1997b) discuss tyrannosaurid teeth in more detail.

Juvenile tyrannosaurid teeth (including fragments) are often found with small theropod teeth in upper Cretaceous vertebrate microfossil sites (Currie et al. 1990). For example, in one Big Bend microfossil site from Rattlesnake Mountain (uppermost Aguja; LSU-741; Figs. 7.1 and 7.4), juvenile tyrannosaurid teeth are the most abundant dinosaur specimens present. However, in other Big Bend microsites, tyrannosaurid teeth (juvenile and adult) are rare. For example, only one specimen was found out of 3,347 specimens from the Talley Mountain microsites (Sankey 2001). The Talley Mountain sites are from more coastal environments, and the Rattlesnake Mountain microsite is from a more inland environment, suggesting that tyrannosaurids (both juvenile and adult) lived in more inland areas in Big Bend. Different tyrannosaurids occurred in the Campanian and Maastrichtian. However, because distinguishing tyrannosaurid taxa with teeth is difficult, no differences could be determined between the Big Bend Campanian and Maastrichtian tyrannosaurids.

Family Dromaeosauridae Mathew and Brown 1922
Genus *Saurornitholestes* Sues 1978
Saurornitholestes cf. *S. langstoni* Sues 1978
Table 7.2, Figure 7.5

Description. Teeth are small, recurved, and flattened (labial-lingually). Denticles are slender and pointed.

Referred Specimens. LSU 726:5923, 5924, 5928, 6204; 741:5927, 5936, 5930, 5935, 5932, 5931, 5929; 796:6270, 6280; 140:6139,

TABLE 7.2

Sauromitholestes spp. Measurements (in mm)

Approximate measurements, due to broken tooth, represented by [ ]. If measurement not possible, represented by —.
Denticle measurements were made on largest denticles present. Abbreviations: curvature (curv.); fore-aft basal length (FABL); height (ht); cross-sectional thickness (CST); cross-sectional shape (CSS); denticles/mm (dent/mm); rd (round); dent (denticles); ht (height); wd (width); frag (fragment); ant. (anterior); post. (posterior); flat (flattened).

LSUMG Spec. No.	LSUMG Local. No	Geo. Fm.	Age	Curv	FABL	HT	CST	CSS	Ant. dents present	Dents/mm post.	Dent. wd. post.	Dent. ht. post.
5923	726	Aguja	late Camp	0.4	4.0	5.3	2.0	oval	minute	4.5	0.2	0.3
6204	726	Aguja	late Camp	0.5	3.5	6	—	oval	minute	5	0.2	0.2
6270	746	Aguja	late Camp	—	4	—	1.8	oval	minute	6.5	0.1	0.1
6281	746	Aguja	late Camp	0.5	4.8	7.2	2.5	oval/rd	minute	6.5	0.2	0.2
6280	746	Aguja	late Camp	[0.5]	4	[7.5]	—	oval	minute	4	0.3	0.2
5924	726	Aguja	late Camp	0.4	4.0	—	—	oval	?	4.5	0.2	0.3
5928	726	Aguja	late Camp	[0.5]	[3.2]	—	—	oval	no	4.5	0.2	0.3
6139	140	Aguja	late Camp	0.2	3.7	[5.2]	1.6	oval	minute	5	0.2	worn
6185	140	Aguja	late Camp	—	3.3	—	1.5	oval	no	5	0.2	0.2
6183	140	Aguja	late Camp	—	—	—	—	—	—	—	—	—
6132	140	Aguja	late Camp	—	—	—	—	—	—	—	—	—
6184	140	Aguja	late Camp	—	—	—	—	—	yes	—	0.1	—
6229	488	Aguja	late Camp	—	—	—	—	—	—	5	0.2	0.3
5659	489	Aguja	late Camp	[0.3]	2.3	2.8	1.4	oval	no	5	0.2	0.3
6234	489	Aguja	late Camp	—	—	—	—	—	yes	7	0.1	0.1
5950	491	Aguja	late Camp	—	2	[2.2]	1.1	oval	no	6	0.1	0.2
5980	491	Aguja	late Camp	—	—	—	—	—	—	—	—	—
5927	741	Aguja	Camp/Maas	—	—	—	—	—	?	4.5	0.2	0.2
5936	741	Aguja	Camp/Maas	—	—	—	—	—	?	6	0.1	0.2
5930	741	Aguja	Camp/Maas	—	3	—	1.3	oval	?	—	—	—

TABLE 7.2 (cont.)

Sauromitholestes spp. Measurements (in mm)

LSUMG Spec. No.	LSUMG Local. No	Geo. Fm.	Age	Curv	FABL	HT	CST	CSS	Ant. dents present	Dents/mm post.	Dent. wd. post.	Dent. ht. post.
5935	741	Aguja	Camp/Maas	—	—	—	1.5	oval	no	[5]	0.2	0.3
5932	741	Aguja	Camp/Maas	—	—	—	—	oval	no	4	0.2	0.3
5931	741	Aguja	Camp/Maas	—	[2.5]	—	1.3	oval	—	6	0.1	0.2
5929	741	Aguja	Camp/Maas	—	[3.8]	—	1.8	oval	—	4	0.2	0.3
5938	741	Aguja	Camp/Maas	—	2	—	0.8	oval	yes	7	0.1	0.1
5937	741	Aguja	Camp/Maas	—	[2.5]	—	1	oval	—	5	0.1	0.1
5109	113	Aguja	Ea. Maast.	0.2	2.8	[4.0]	1.2	oval	—	6	0.2	0.2
5940	113	Aguja	Ea. Maast.	—	—	—	—	—	—	—	0.2	0.2
1355	113	Aguja	Ea. Maast.	—	—	—	—	—	—	—	0.2	0.2
5942	113	Aguja	Ea. Maast.	—	—	—	—	—	—	—	0.1	0.2
1309	113	Aguja	Ea. Maast.	[0.2]	1.8	[2.2]	0.8	oval	no	7	0.2	worn
1307	113	Aguja	Ea. Maast.	0.5	4	6.2	1.8	oval	minute	9	0.2	worn
5156	113	Aguja	Ea. Maast.	—	[2.5]	[4.0]	—	—	yes	5	0.2	0.2
1308	113	Aguja	Ea. Maast.	0.5	4	[5.5]	1.9	oval	no	5	0.2	0.2
5943	113	Aguja	Ea. Maast.	—	—	—	1	oval, v. flattened	no	6	0.2	0.2
5953	229	Tornillo	Maast.	—	3.3	[4.0]	—	oval	no	5	0.2	0.3
5963	229	Tornillo	Maast.	0.5	3.6	4.5	1.8	oval	minute	4.5	0.2	0.3
1089	229	Tornillo	Maast.	0.3	4.8	7.5	—	—	?	5	0.2	0.3
3128	229	Tornillo	Maast.	0.4	[4.0]	[6.0]	1.7	oval	yes	5	0.2	0.2
5987	229	Tornillo	Maast.	—	[3.5]	—	—	oval	minute	4	0.2	0.2
3069	235	Tornillo	Maast.	—	—	—	—	oval	yes	—	—	0.2
1931	235	Tornillo	Maast.	0.5	3	3.5	1.3	oval	?	7	0.1	0.1
6000	232	Tornillo	Maast.	—	3.3	—	1.8	oval/round	no	6	0.1	0.2

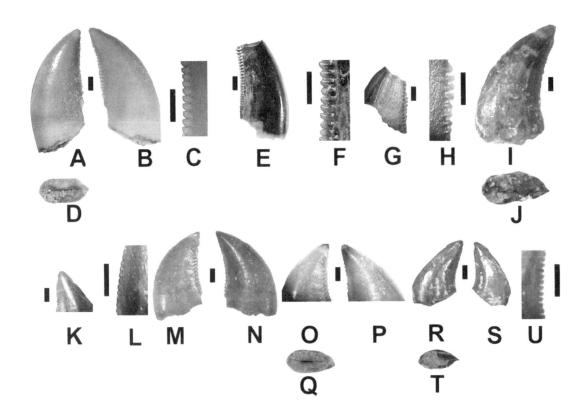

Figure 7.5. Saurornitholestes teeth from the Aguja and Tornillo formations, Big Bend National Park, Texas, with views of lingual and labial sides, cross-section through base of tooth, and close-up of denticles. Saurornitholestes cf. S. langstoni: (A–D) LSU 726:5923; (E–F) LSU 726:5924; (G–H) LSU 741:5927; (I–L) LSU 113:1307. Saurornitholestes n. spA (M–Q) LSU 229:3128. Saurornitholestes n. sp. B (R–U) LSU 235:1931. Scale = 1 mm.

6185; 489:5659; 491:5950; 488:6229; 229:5953, 5963, 1089; 113: 1308, 5943.

Description. Specimen LSU 726:5923 (Fig. 7.5A–D) is a typical example of this group. The tooth is small, recurved, sharply pointed, laterally compressed (labial-lingually), and oval in cross-sectional shape. Denticles on the anterior carinae, if present, are minute. Denticles on the posterior carinae are higher than they are wide, are sharply pointed toward the tip of the tooth, and vary in size from the tooth base to tooth tip. There are deep, narrow interdenticle spaces. The distinctive characteristics are (1) tooth recurved and flattened (labial-lingually); (2) absent or minute denticles on anterior carinae; and (3) long and pointed denticles on posterior carinae.

Denticles/mm on the posterior carinae range from 4 to 6/mm (x = 5.2, n = 24). Denticle width ranges from 0.1 to 0.2 mm (x = 0.2, n = 24). Denticle height ranges from 0.2 to 0.3 mm (x = 0.2, n = 24).

Discussion. These teeth are similar in both tooth and denticle shape and size to those identified and described as *Saurornitholestes* cf. *S. langstoni* from the Judith River Group of Alberta (Currie et al. 1990; Baszio 1997b; Sankey et al. 2002). In particular, these teeth closely match the specimen that is illustrated in Sankey et al. 2002 (figs. 4.14 –4.17) and referred to Morphotype D (Sankey et al. 2002, Appendix 1.3) except that in the Big Bend specimens, denticles on the anterior carinae are either absent or minute. These teeth also closely resemble specimens illustrated in Currie et al. 1990 (fig. 8.2T–W) and in Baszio 1997b (plate I, 19–22, Dinosaur Park Group). The measurements of these samples are similar to measurements from this sample (Table 7.2). Sankey et al. (2002) measured ninety-nine *Saurornitholestes* cf. *S. langstoni* teeth, which had an average denticle height of 0.23 mm and an average denticle width of 0.18 mm. Also, Sankey et al. (2002) reported a wide range of variation in denticle size within the large sample of teeth studied, reflecting the presence of two or three distinct morphotypes in the sample. Most specimens are from sites within the upper Aguja Formation (upper Campanian to Campanian-Maastrichtian). However, three specimens (LSU 5953, 5963, 1089) are also from the lower Tornillo Formation (Maastrichtian). Additional specimens from these sites will test this pattern of relative abundance.

cf. *Saurornitholestes* n. sp.? A

Referred Specimens. 113:1307, 5156, 5109; 229:5987, 3128; 746: 6281.

Description. Specimen LSU 229:3128 (Fig. 7.5N–R) is a typical example of this morphotype. The tooth is small, recurved, flattened (labial-lingually), and oval in cross-sectional shape. Denticles on anterior carinae are present, but they are small, considerably smaller than posterior denticles. Denticles on posterior carinae are higher than they are wide and are slightly pointed. Denticles extend along the length of the carinae and are slightly smaller at the tooth tip. These teeth are similar to those of *S.* cf. *S. langstoni*, except in having anterior denticles and less-pointed posterior denticles. Denticles/mm on posterior carinae range from 4 to 9/mm (x = 5.8, n = 5). Denticle width ranges from 0.17 to 0.23 mm (x = 0.79, n = 5). Denticle height ranges from 0.17 to 0.23 mm (x = 0.2, n = 5).

Discussion. The presence of denticles on the anterior carinae is not size dependant, because both LSU 3128 and 5923 (*S.* cf. *S. langstoni*) are similar in size to teeth referred to cf. *S.* n. sp.? As discussed for *S.* cf. *S. langstoni* teeth, these teeth are also similar to those identified as *Saurornitholestes* cf. *S. langstoni* (Currie et al. 1990; Baszio 1997b; Sankey et al. 2002). They most closely match specimens illustrated in Baszio 1997b (plate II, 23–26, Horseshoe Canyon Formation). These teeth have been found only in Maastrichtian sites, and only from Big Bend. LSU site 113 is within the uppermost Aguja (lower Maastrichtian), and LSU site 229 is in the lower Tornillo (Maastrichtian; Standhardt 1986). Additional specimens of this morphotype would confirm this as a distinct taxon, especially if its stratigraphic distribution is restricted to Maastrichtian.

cf. *Saurornitholestes* n. sp.? B

Referred Specimen. 235:1931.

Description. Specimen LSU 1931 (Fig. 7.5S–U) is small, flattened (labial-lingually), oval in cross-sectional shape, and strongly recurved. Posterior denticles are unique. Denticle tips are not pointed, but are rounded in outline (the tooth is not worn). Interdenticle spaces are present. Anterior denticles are also present. Although the anterior carinae are missing, holes for denticles are present at the tooth base. Posterior denticles occur 7/mm and are 0.13 mm wide and 0.13 mm high (i.e., equally wide as they are high).

Discussion. LSU 1931 is unique in the following characteristics: (1) extremely recurved; (2) denticles present on both the anterior and posterior carinae; (3) denticles are small, approximately uniform in size from base to tip of tooth, and not pointed. The tooth is unlike any theropod tooth examined in the Tyrrell's or UCMP's collections. In tooth shape, the specimen resembles *Saurornitholestes* cf. *S. langstoni*, such as in Baszio 1997b (plate II, 28–Scollard Formation). However, in denticle shape and size, it resembles *Richardoestesia gilmorei* (Currie et al. 1990; Baszio 1997b; Sankey et al. 2002). The specimen is from a site (LSU 235) in the lower Tornillo Formation and is Maastrichtian in age. More specimens of this morphotype are needed to clarify whether this morphotype is restricted to the Maastrichtian in Big Bend.

cf. *Saurornitholestes* n. sp.? C

Referred Specimen. 232:6000.

Description. Specimen LSU 6000 is small, flattened (labial-lingually), and recurved. Because the cross-sectional shape is a rounded oval and because there are no denticles on the anterior carinae, this may be a premaxillary tooth, as discussed for *Saurornitholestes* in Sankey et al. 2002. Posterior denticles are long and slender. Denticles are 6/mm and are 0.13 mm wide and 0.2 mm high. Denticle tips are not pointed, but are rounded in outline. Distinct interdenticle spaces are present.

Discussion. Specimen LSU 6000 is a probable premaxillary tooth. It does not closely match any theropod teeth examined from the Judith River Group (Sankey et al. 2002). It is similar to *Dromaeosaurus* in the Scollard and Lance formations (Baszio 1997b, plates I, 9, and 13). This specimen is from a site, LSU-232, within the lower Tornillo Formation and is Maastrichtian in age.

Infraorder Maniraptora Gauthier 1986
Family Unknown
Genus *Richardoestesia* Currie, Rigby, and Sloan 1990
Richardoestesia isosceles Sankey 2001
Table 7.3

Referred Specimens. See Table 7.3.

Description. Teeth are small, with no or little recurvature, and are oval in cross-sectional shape. Denticles are present on both the anterior and the posterior carinae. Posterior denticles are minute. They occur 8–12/mm and are 0.1 mm wide and 0.1 mm high (i.e., equally high as they are wide). Denticles are not pointed.

Diagnosis. Because the holotype for this species is from Big Bend (Aguja; late Campanian; Sankey 2001) and the species was further

TABLE 7.3.

Richardoestesia Measurements (in mm)

Approximate measurements, due to broken tooth, represented by [ ]. If measurement not possible, represented by —. Denticle measurements were made on largest denticles present. Abbreviations: curvature (curv.); fore-aft basal length (FABL); height (ht); cross-sectional thickness (CST); cross-sectional shape (CSS); denticles/mm (dent/mm); rd (round); dent (denticles); ht (height); wd (width); frag (fragment); ant. (anterior); post. (posterior); flat (flattened).

LSUMG Spec. No.	LSUM. Local No.	Geo. Fm.	Age	Curv	FABL	Ht	CST	CSS	Ant. dents present	Dents/mm post.	Dent.wd post.	Dent. ht. post.
6140	140	Aguja	late Camp	—	2.2	—	1	oval	yes	minute	—	—
6237	489	Aguja	late Camp	—	2	[2.5]	0.9	oval	—	10	0.1	0.1
6238	489	Aguja	late Camp	straight	—	—	—	—	yes	7.5	0.1	0.1
6235	489	Aguja	late Camp	straight	1.7	[3.0]	0.7	oval	yes	11	0.1	0.1
6233	489	Aguja	late Camp	straight	—	—	—	round	yes	8	0.1	0.1
6264	492	Aguja	late Camp	straight	—	—	—	—	—	4?	0.3	0.2
6050	489	Aguja	late Camp	straight	1.7	[3.5]	1	oval	yes	10	0.1	0.1
6051	140	Aguja	late Camp	—	—	—	—	—	yes	9	0.1	0.1
5933	741	Aguja	Camp/Maas	—	[1.8]	[2.3]	0.9	oval	no	9	0.1	0.1
5934	741	Aguja	Camp/Maas	—	—	—	1	oval	?	8	0.1	0.1
5939	113	Aguja	Ea. Maast.	straight	[2.0]	[4.0]	1	oval	yes	12	0.1	0.1

described from specimens from the Judith River Group, Alberta (Sankey et al. 2002), descriptions from those papers are included here.

"Teeth straight; narrow; shaped like an isosceles triangle in lateral view (as mentioned for *Richardoestesia* sp. in Currie et al. 1990 and in Baszio 1997b). Shape of tooth in basal cross-section is labiolingually flattened oval. Denticles minute (0.1 mm in height and in anteroposterior width); square; uniformly-sized from base to tip of tooth; extend length of carinae. Anterior denticles, if present, often considerably smaller than posterior denticles. Interdenticle spaces usually minute and barely visible; denticles closely spaced. Denticle tips straight or faintly rounded, but not pointed. 7–11 denticles/mm" (Sankey 2001, p. 213).

Discussion. Sankey (2001) listed eight specimens from the Talley Mountain microsites (See Table 7.3). Three additional fragmentary specimens are reported here, all from the uppermost Aguja (Campanian-Maastrichtian). One fragmentary specimen of *R. gilmorei* was reported from Big Bend in Sankey 2001. However, no additional specimens have been recovered.

Straight teeth of *Richardoestesia* from the Judith River Group of Alberta were first recognized as a taxon distinct from *R. gilmorei* by Currie et al. (1990) and were referred to as *Richardoestesia* sp. Additional specimens from this area were described by Baszio (1997b) and Peng (1997; Peng et al. 2001). Baszio (1997a) made the important observation that teeth from *R. gilmorei* and *Richardoestesia* sp. have different relative abundance patterns in the Late Cretaceous of Alberta, reflecting their different paleoecologies and further supporting the idea that they represent distinct taxa. Sankey (2001) formally named *Richardoestesia* sp. on the basis of specimens from Big Bend. Sankey et al. (2002) described and measured further specimens from the Judith River Group, documenting qualitative and quantitative differences from *R. gilmorei*. *R. isosceles* is included in the genus *Richardoestesia* because of the presence of small denticles. Differences include the following: in *R. gilmorei*, teeth are usually shorter and more recurved, denticles are pointed, and small interdenticle spaces are present; whereas in *R. isosceles*, denticles are not pointed, and denticles are present on both anterior and posterior edges (Sankey et al. 2002).

Family Unknown
Genus *Paronychodon* Cope 1876
Paronychodon cf. *P. lacustris* Cope 1876

Referred Specimens. LSU 113:1310, 5107, 1311, 5993, 5996.

Description. Teeth are small, slightly recurved, flattened (labial-lingually), with one flattened side, and with distinct longitudinal ridges on both sides of tooth. No denticles are present on the anterior carinae. On the posterior carinae, denticles are absent, except in one specimen (LSU 5939) where they are minute.

Discussion. Teeth closely match specimens referred to *P. lacustris* from the Judith River Group, Alberta, in Sankey et al. 2002. Teeth from this poorly understood taxon were described, illustrated, and mea-

TABLE 7.4.

Paronychodon Measurements (in mm)

Approximate measurements, due to broken tooth, represented by [ ]. If measurement not possible, represented by —. Denticle measurements were made on largest denticles present. Abbreviations: curvature (curv.); fore-aft basal length (FABL); height (Ht); cross-sectional thickness (CST); cross-sectional shape (CSS); denticles/mm (dent/mm); rd (round); dent (denticles); ht (height); wd (width); frag (fragment); ant. (anterior); post. (posterior); flat (flattened).

LSUMG Spec. No.	LSUM. Local No.	Geo. Fm.	Age	Curv	FABL	Ht	CST	CSS	Ant. dents present	Dents/mm post.
1310	113	Aguja	Ea. Maast.	[0.2]	3.6	[5.5]	2	oval, 1 flat side	no	none
5107	113	Aguja	Ea. Maast.	[0.2]	1.6	[3.0]	1.1	round, 1 flat side	no	none
1311	113	Aguja	Ea. Maast.	—	2.2	—	1.3	flat side	no	none
5993	113	Aguja	Ea. Maast.	—	1.7	[2.5]	0.9	oval	no	none
5996	113	Agua	Ea. Maast.	—	—	—	—	1 flat side	—	—

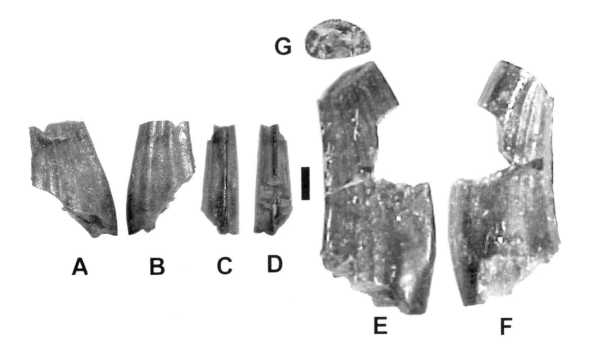

Figure 7.6. Paronychodon *teeth from the Aguja and Tornillo formations, Big Bend National Park, Texas, with views of lingual and labial sides, cross-section through base of tooth, and close-up of denticles. (A–D) LSU 113:5939, (E–G) LSU 113–1310. Scale = 1 mm.*

sured, and the features that distinguish them from similar taxa were pointed out (Sankey et al. 2002). Teeth have one flattened surface, numerous and well-developed longitudinal ridges, and no denticles. Two morphotypes were recognized. Morphotype A is larger, straighter, and not constricted at the base. Morphotype B is smaller and more recurved and has a distinct constriction below the crown. Although the Big Bend specimens are fragmentary, they more closely resemble Morphotype A because they are straight or only slightly recurved and have no constriction below the crown.

All Big Bend specimens reported here are from one lower Maastrichtian site (LSU 113) from the uppermost Aguja in the Dawson Creek area. They most closely resemble *Paronychodon lacustris* Morphotype B (Sankey et al. 2002, figs. 5.25–5.27). *Paronychodon* teeth are uncommon in most Late Cretaceous assemblages. However, they are more common in Maastrichtian sites, and it is important to note that this same pattern occurs in Big Bend.

cf. Class Aves Linnaeus, 1758
Table 7.5

Referred Specimens. LSU 113:5995; 232:5999.

Description. Teeth are small, round to oval in cross-sectional shape, and with little recurvature. No denticles are present on either the posterior or the anterior carinae.

Discussion. Teeth resemble those referred to birds in Sankey et al. 2002 in the following characteristics: teeth are small, round to oval in cross-sectional shape, slightly laterally compressed, straight, indented

TABLE 7.5.

cf. Aves Measurements (in mm)

Approximate measurements, due to broken tooth, represented by [ ]. If measurement not possible, represented by —.
Denticle measurements were made on largest denticles present. Abbreviations: curvature (curv.); fore-aft basal length (FABL);
height (Ht); cross-sectional thickness (CST); cross-sectional shape (CSS); denticles/mm (dent/mm); rd (round); dent (denticles);
ht (height); wd (width); frag (fragment); ant. (anterior); post. (posterior); flat (flattened).

LSUMG Spec. No	LSUMG ocal.	Geo. Fm.	Age	Curv.	FABL	HT	CST	CSS	A.Dents. present.	Dents/mm Post.	Dent. Wd.Post.	Dent. Ht. Post.
5995	113	Aguja	Ea. Maast.	[0.1]	1.2	[2.0]	0.7	ova/rd	none	none	none	none
5999	232	Tornillo	Maast	0.1	1	1.5	—	—	oval	none	none	none

at their base, with absent or minute denticles, some with a thin carina on both the anterior and the posterior edges, and oval to oval-flattened in cross section.

The two Big Bend specimens tentatively referred to bird are both from Maastrichtian sites. LSU 5995 is from the uppermost Aguja (LSU site 113), and LSU 5999 is from the lower Tornillo (LSU site 232). They are the first possible bird teeth reported from Big Bend.

Distinguishing bird teeth from small teeth of *Richardoestesia gilmorei* and *R. isosceles* can be difficult. However, *R. isosceles* teeth are not indented at the base; and *R. gilmorei* are more recurved, have larger denticles, and lack the distinctive hourglass of the base of the bird teeth (Sankey et al. 2002). Although Sankey et al. (2002) reported that the bird teeth from the Judith River Group of Alberta are similar to those from *Hesperornis*, they are also similar to those from some small non-avian dinosaurs (e.g., Xu et al. 2000). Therefore, referral of the Big Bend specimens to bird is tentative.

Discussion

Baszio (1997a) documented theropod relative abundance patterns from the mid-Campanian to upper Maastrichtian of Alberta based on large collections of theropod teeth from the RTMP and UALVP. Using cluster analysis of relative abundance of theropod taxa he found two groups. Assemblage B is the Dinosaur Park–Horseshoe Canyon cluster, characterized by abundant *Troodon* and few *Richardoestesia* sp. Assemblage A is the Milk River–upper Maastrichtian cluster, characterized by a relatively high abundance of *Richardoestesia* sp. and *Paronychodon*. Baszio (1997a) suggested that these two distinct assemblages occupied different geographic areas, and that the assemblages shifted north-south according to regional climatic changes or shifted coastal-inland according to transgressions and regressions. Thus, Baszio demonstrated that there were different theropod assemblages characteristic of certain paleoenvironmental and paleoclimatic conditions, and that the geographic areas these assemblages occupied shifted through time.

Within Big Bend, there also appear to be two distinct theropod assemblages, one late Campanian–early Maastrichtian and one Maastrichtian. The Campanian–early Maastrichtian theropod assemblage is from the upper Aguja, which is more coastal; this assemblage is characterized by a higher abundance of *Saurornitholestes* cf. *S. langstoni* and *Richardoestesia isosceles*. The Maastrichtian theropod assemblage is from the uppermost Aguja and lower Tornillo, which is more inland; this assemblage is characterized by a higher abundance of *Saurornitholestes* n. sp.? and *Paronychodon*.

The Judith River Group of Alberta is often considered the standard for late Campanian theropod diversity in North America. Sankey et al. (2002) recognized several new morphotypes that are possible new taxa, making theropod diversity in the Judith River Group even higher than previously recognized (Currie et al. 1990; Baszio 1997b). However, compared to the Judith River Group, Big Bend theropod diversity appears to be significantly lower. For example, *Troodon* and *Dromaeo-saurus*, present in the Judith River Group, are absent from Big Bend. The reasons for Big Bend's lower diversity are not known. Additionally, valid comparisons between Big Bend and Alberta theropod diversity are difficult to make, owing to differences in paleogeography, depositional setting, outcrop area, and amount of collection effort.

Among the Big Bend theropods, there are some interesting patterns. For example, distinct morphotypes of *Saurornitholestes* may represent distinct species. This is supported by morphologic as well as stratigraphic information: teeth of *Saurornitholestes*. n. sp.? have been found only in the early Maastrichtian sites.

This sample of theropod teeth from Big Bend demonstrates that the theropod assemblages in this area were distinct from northern areas. More samples are needed to better characterize the differences in these taxa. The sample also demonstrates that there were distinct theropod assemblages in the Campanian and Maastrichtian of Big Bend. However, considerably more information is needed on the Maastrichtian theropods from Big Bend in order to document theropod diversity patterns during the last 10 million years of the Cretaceous of both Big Bend and across a wider geographic gradient in North America.

Conclusions

The theropods present in the Big Bend area are identified and compared to contemporaneous assemblages such as the Judith River Group of Alberta. This was accomplished by describing, illustrating, and measuring theropod teeth collected both by surface collecting and by screening of microsites in the upper Aguja (upper Campanian–lower Maastrichtian) and the lower Tornillo (Maastrichtian) formations of Big Bend. Variation within the theropods is documented, in particular with respect to *Saurornitholestes;* and the first possible *Paronychodon* and bird specimens are recorded from Big Bend. The presence of two theropod assemblages, one from the late Campanian–early Maastrichtian and one from the Maastrichtian, is also documented. Further work in Big Bend is needed in order to (1) test whether *Saurornitholestes* n. sp.? is a distinct taxon; (2) test whether *Troodon* and *Dromaeosaurus* were really absent; (3) determine more differences between the Campanian and Maastrichtian theropod assemblages; and (4) increase the number of samples of the new morphotypes reported here. This work will improve our knowledge of theropod assemblages from this southern biogeographic province and document how the assemblages changed during the last 10 million years of the Cretaceous, leading up to the K/T extinctions.

Acknowledgments. Sankey appreciates funding for her Big Bend work from the Dinosaur Society, Jurassic Foundation, Earthwatch Durfee Student Challenge Award, and South Dakota School of Mines and Technology's Haslem and Nelson Funds, and for her work in Alberta from the United States–Canada Fulbright Program. Funding to Schiebout supporting Standhardt's Big Bend research was obtained from the National Science Foundation. Standhardt appreciates field and lab help by LSU students during the 1980s. Sankey appreciates help in field and lab from Jean Sankey, students from LSU (1995–98), students from Earthwatch's Durfee Student Challenge Awards Program (2000), and students from SDSM&T (2002). Sankey's Big Bend National Park fieldwork was done under BBNP permit numbers BIBE-2000–046, 2001–SCI-0010, and 2002–SCI-0001. We thank Vidal Davilla, Don Corrick, and Rito Rivero for their assistance. Donald Brinkman and Philip Currie (Royal Tyrrell Museum of Palaeontology) taught Sankey about theropod teeth, and she appreciates their help and interest in this work. Illustrations were prepared by Lisa Pond (LSU).

References Cited

Abler, W. L. 1997. Tooth serrations in carnivorous dinosaurs. In P. J. Currie and K. Padian (eds.), *Encyclopedia of Dinosaurs,* pp. 740–743. San Diego: Academic Press.

Baszio, S. 1997a. Investigations on Canadian dinosaurs: Palaeoecology of dinosaur assemblages throughout the Late Cretaceous of southern Alberta, Canada. *Courier Forschungsinstitut Senckenberg* 196: 1–31.

Baszio, S. 1997b. Investigations on Canadian dinosaurs: Systematic palaeontology of isolated dinosaur teeth from the Latest Cretaceous of south Alberta, Canada. *Courier Forschungsinstitut Senckenberg* 196: 33–77.

Brinkman, D. B. 1990. Paleoecology of the Judith River Formation (Campanian) of Dinosaur Provincial Park, Alberta, Canada: Evidence from vertebrate microfossil localities. *Palaeogeography, Palaeoclimatology, Palaeoecology* 78: 37–54.

Cifelli, R. L. 1995. Therian mammals of the Terlingua Local Fauna (Judithian), Aguja Formation, Big Bend of the Rio Grande, Texas. *Contributions to Geology, University of Wyoming* 30: 117–136.

Cope, E. D. 1876. Descriptions of some vertebrate remains from the Fort Union beds of Montana. *Proceedings of the Academy of Natural Sciences of Philadelphia* 28: 248–261.

Currie, P. J., J. K. Rigby Jr., and R. E. Sloan. 1990. Theropod teeth from the Judith River Formation of southern Alberta, Canada. In K. Carpenter and P. J. Currie (eds.), *Dinosaur Systematics: Approaches and Perspectives,* pp. 107–125. Cambridge: Cambridge University Press.

Davies, K., and T. M. Lehman. 1989. The WPA Quarries. In A. B. Busbey III and T. M. Lehman (eds.), *Vertebrate Paleontology, Biostratigraphy, and Depositional Environments, Latest Cretaceous and Tertiary, Big Bend Area, Texas.* Guidebook, Field Trip nos. 1a, b, c, pp. 32–42. Society of Vertebrate Paleontology 49th Annual Meeting, Austin, Texas.

Gauthier, J. 1986. Saurischian monophyly and the origin of birds. In K. Padian (ed.), *The Origin of Birds and the Evolution of Flight,* pp. 1–55. Memoirs of California Academy of Science, no. 8. San Francisco: California Academy of Sciences.

Goodwin, M. B., and A. L. Deino. 1989. The first radiometric ages from the Judith River Formation (Upper Cretaceous), Hill County, Montana. *Canadian Journal of Earth Sciences* 26: 1384–1391.

Kauffman, E. G. 1977. Geological and biological overview: Western Interior Cretaceous Basin. *The Mountain Geologist* 14: 75–99.

Lehman, T. M. 1985. Stratigraphy, sedimentology, and paleontology of

Upper Cretaceous (Campanian-Maastrichtian) sedimentary rocks in Trans-Pecos, Texas. Ph.D. dissertation, University of Texas at Austin.

Lehman, T. M. 1989. *Chasmosaurus mariscalensis*, sp. nov., a new ceratopsian dinosaur from Texas. *Journal of Vertebrate Paleontology* 9: 137–162.

Lehman, T. M. 1990. Paleosols and the Cretaceous/Tertiary transition in the Big Bend region of Texas: *Geology* 18: 362–364.

Lehman, T. M. 1991. Sedimentation and tectonism in the Laramide Tornillo Basin of west Texas. *Sedimentary Geology* 75: 9–28.

Lehman, T. M. 1997. Late Campanian dinosaur biogeography in the Western Interior of North America. In D. A. Wolberg and E. Stump (eds.), *Dinofest International: Proceedings of a Symposium Sponsored by Arizona State University*, pp. 223–240. Philadelphia: Academy of Natural Sciences.

Lillegraven, J. A., and M. C. McKenna. 1986, Fossil mammals from the "Mesaverde" Formation (Late Cretaceous, Judithian) of the Bighorn and Wind River basins, Wyoming, with definitions of the Late Cretaceous North American land mammal "ages." *American Museum Novitates*, no. 2840: 1–68.

Linnaeus, C. 1758. *Systema naturae per regna tria naturae*. 10th ed., rev. 2 vols. Holmiae: L. Salmii.

Matthew, W. D., and B. Brown. 1922. The family Deinodontidae, with notice of a new genus from the Cretaceous of Alberta. *Bulletin of the American Museum of Natural History* 46: 367–385.

Obradovich, J. D. 1993. A Cretaceous time scale. In W. G. E. Caldwell and E. G. Kauffman (eds.), *Evolution of the Western Interior Basin*, pp. 379–396. Geological Association of Canada, Special Paper, no. 39. St. John's, Newfoundland: Geological Association of Canada.

Osborn, H. F. 1905. *Tyrannosaurus* and other Cretaceous carnivorous dinosaurs. *Bulletin of the American Museum of Natural History* 21: 259–265.

Peng J.-H. 1997. Palaeoecology of vertebrate assemblages from the upper Cretaceous Judith River Group (Campanian) of southeastern Alberta, Canada. Ph.D. dissertation, University of Calgary.

Peng J.-H., A. P. Russell, and D. B. Brinkman. 2001. *Vertebrate Microsite Assemblages (Exclusive of Mammals) from the Foremost and Oldman Formations of the Judith River Group (Campanian) of Southeastern Alberta: An Illustrated Guide*. Provincial Museum of Alberta, Natural History Occasional Paper, no. 25. Edmonton: Curatorial Section, Provincial Museum of Alberta.

Rowe, T., R. L. Cifelli, T. M. Lehman, and A. Weil. 1992. The Campanian Terlingua local fauna, with a summary of other vertebrates from the Aguja Formation, Trans-Pecos, Texas. *Journal of Vertebrate Paleontology* 12: 472–493.

Sankey, J. T. 1998. Vertebrate Paleontology and Magnetostratigraphy of the Upper Aguja Formation (Late Campanian), Talley Mountain Area, Big Bend National Park, Texas. Ph.D. dissertation, Louisiana State University.

Sankey, J. T. 2001. Late Campanian Southern Dinosaurs, Aguja Formation, Big Bend, Texas. *Journal of Paleontology* 75: 208–215.

Sankey, J. T., and W. A. Gose. 2001. *Late Cretaceous Mammals and Magnetostratigraphy, Big Bend, Texas*. Occasional Papers of the Museum of Natural Science, no. 77. Baton Rouge: Museum of Natural Science, Louisiana State University.

Sankey, J. T., D. B. Brinkman, M. Guenther, and P. J. Currie. 2002. Small theropod and bird teeth from the Judith River Group (Late Campanian), Alberta. *Journal of Paleontology* 76: 751–763.

Schiebout, J. A., C. A. Rigsby, S. D. Rapp, J. A. Hartnell, and B. R. Standhardt. 1987. Stratigraphy of the Cretaceous-Tertiary and Paleocene-Eocene transition rocks of Big Bend National Park, Texas. *Journal of Geology* 95 (3): 359–375.

Seeley, H. G. 1888. The classification of the Dinosauria. *Report British Association for the Advancement of Science* 1887: 698–699.

Standhardt, B. R. 1986. Vertebrate paleontology of the Cretaceous/Tertiary transition of Big Bend National Park, Texas. Ph.D. dissertation, Louisiana State University.

Sues, H.-D. 1978. A new small theropod dinosaur from the Judith River Formation (Campanian) of Alberta Canada. *Zoological Journal of the Linnean Society* 62: 381–400.

Weil, A. 1992. The Terlingua local fauna: Stratigraphy, paleontology, and multituberculate systematics. M.A. thesis, University of Texas at Austin.

Weil, A. 1999. Multituberculate phylogeny and mammalian biogeography in the Late Cretaceous and earliest Paleocene Western Interior of North America. Ph.D. dissertation, University of California at Berkeley.

Xu X., Zhou Z.-H., and Wang X.-L. 2000. The smallest known non-avian theropod dinosaur. *Nature* 408: 705–708.

8. Last Patagonian Non-Avian Theropods

RODOLFO A. CORIA AND
LEONARDO SALGADO

Abstract

Isolated theropod remains are reported from uppermost terrestrial Cretaceous sediments (Allen Formation) from Salitral Moreno (northern Patagonia, Río Negro Province). The assemblage is composed of many disarticulated, weathered, and isolated cranial and postcranial elements assigned to the Theropoda. The bones include a fragmentary braincase, caudal vertebrae, and limb elements. No apomorphical features have been identified to link these elements with the typical South American abelisaurs and carcharodontosaurs. In contrast, the low and slender caudal centra, dorsomedially oriented femoral head, and laterally compressed proximal metatarsal III suggest higher tetanuran affinities. The stratigraphic provenance of all the material corresponds with a faunal association with both endemic South American and immigrant North American dinosaurs.

Introduction

South American theropods include several taxa recorded from the Late Triassic (Reig 1963; Colbert 1970; Sereno et al. 1993; Arcucci and Coria 2003) and Middle Jurassic (Bonaparte 1986) and a high concentration of discoveries from the second half of the Cretaceous (Novas 1997a). Among the diverse and numerous South American geographical regions, Patagonia is one of the richest dinosaur fossil–bearing areas. Located in the southernmost part of the continent, it is the source of much of our knowledge of dinosaur evolution from the Cenomanian to the Campanian (Coria and Salgado 1995, 1998; Novas 1997b, 1998). Unlike the dinosaurian record from the Northern Hemisphere, the record of the theropod dinosaurs that inhabited southern South America after the Campanian is very poor. One of the reasons is likely the scarcity of latest Cretaceous outcrops in Patagonia.

However, an important Late Cretaceous dinosaur association has been identified in geological formations (i.e., Colitoro, Allen, and Loncoche formations) regarded as Campanian to Maastrichtian. This was the time when South and North America were connected by a land bridge, allowing a faunal interchange. The herbivorous components of this dinosaur assemblage included hadrosaurs, titanosaurs, and ankylosaurs (Bonaparte et al. 1984; Powell 1987; Salgado and Coria 1993, 1996; Coria and Salgado 2001). In contrast, only one theropod species has been described from a very incomplete and fragmentary specimen (Coria 2001).

The isolated elements assigned to Theropoda that are described here were located in the Vertebrate Paleontology Collection of the Carlos Ameghino Provincial Museum, Cipolletti, Río Negro Province. Over the years, many persons and expeditions collected this material, and because all the specimens are isolated elements, their exact taxonomic identifications remain obscure. Nevertheless, the specimens fill a gap of information related to the last non-avian meat-eating dinosaurs that lived in Patagonia at the end of the Cretaceous.

Institutional Abbreviations. MPCA-PV, Museo Provincial Carlos Ameghino, Paleontología de Vertebrados, Cipolletti, Río Negro Province, Argentina.

Description

MPCA-PV-80. Ventroposterior fragment of braincase. The fragment of a theropod braincase (Fig. 8.1) is composed of a partial basioccipital that includes the occipital condyle, which is weathered, and the vertical region ventral to the condyle that articulated with the opisthotic dorsally and the basisphenoid anteriorly. The condyle has a well-marked neck with a sharp edge, which projects strongly forward, at the perimeter (Fig. 8.1B,D). The condyle is rather flat dorsally along the ventral surface of the neural channel (Fig. 8.1A), and it projects straight backward (Fig. 8.1D). In ventral view, the basisphenoidal recess forms a single, transversely wide cavity, unlike that in most theropods, where the cavity is divided by a septum (Fig. 8.1B). This cavity is slightly asymmetrical and is broader on the left side. In dorsal view, the floor of the neural channel shows several pneumatic pores that connect with the bone's highly pneumatized interior (Fig. 8.1A). Only a small portion of the basisphenoid is preserved, solidly fused to the anterior surface of the basioccipital. Nevertheless, the suture shows along the metotic strut, indicating that the bones of the braincase were not completely co-ossified, which suggests that the bone likely belongs to an immature individual.

In posterior view, the basioccipital extension below the condyle does not seem to have either the medial depressed area or the well-

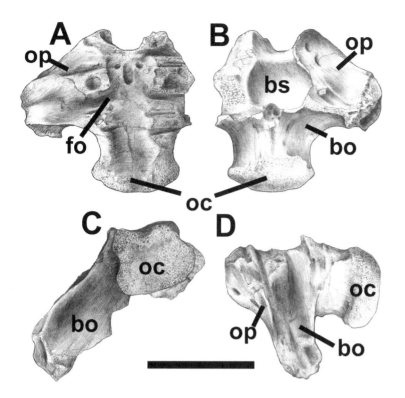

Figure 8.1. MPCA-PV-80: ventroposterior fragment of braincase in (A) dorsal, (B) ventral, (C) occipital, and (D) lateral views. Abbreviations: bo— basioccipital; bs—basisphenoidal recess; fo—fenestra ovalis; oc— occipital condyle; op—opisthotic. Scale bar = 5 cm.

defined basal tubera present in most theropods (Fig. 8.1C). This condition in specimen MPCA-PV-80 resembles that described for abelisaurs (Bonaparte et al. 1990; Bonaparte 1991). The occipital condyle and the ventral part of the basioccipital are almost perpendicular as in abelisaurids (Bonaparte 1991), *Acrocanthosaurus* (Currie and Carpenter 2001), *Allosaurus* (Madsen 1976), *Dromaeosaurus* (Currie 1995), *Herrerasaurus* (Sereno and Novas 1993), *Piatnitzkysaurus* (Bonaparte 1986), and *Troodon* (Currie 1985). MPCA-PV-80 does not have the sharp angle described for *Giganotosaurus* (Coria and Currie 2002) and *Sinraptor* (Currie and Zhao 1993).

In dorsal view, the fenestra ovalis is level with the floor of the foramen magnum, and the vestibules of the middle ear are directed ventrally and slightly posteriorly (Fig. 8.1A).

MPCA-PV-81. Caudal centrum. This specimen is a distal caudal vertebra (Fig. 8.2). Only the base of the neural arch is preserved, and vertebra lacks the neural spine, pre- and postzygapophyses, and the anterior part of the centrum (Fig. 8.2A,B). The transverse processes are reduced to a pair of distally thickened, longitudinal ridges on both sides of the centrum (Fig. 8.2A–C). In ventral view, there is a low sagittal keel, and there are two parasagittal swellings proximal to the midpoint of the vertebra (Fig. 8.2D). This is a derived combination of features for the caudal vertebrae. The centrum, although incomplete, seems to be slightly amphycoelic.

MPCA-PV-83. Caudal centrum. This specimen is almost two-thirds of a distal caudal centrum (Fig. 8.3), because transverse pro-

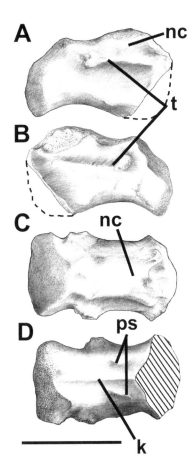

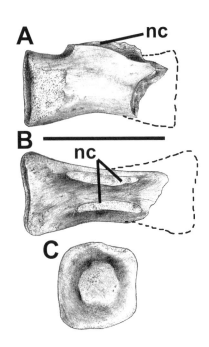

Figure 8.2. (above left)
MPCA-PV-81: caudal centrum in
(A) right lateral, (B) left lateral,
(C) dorsal, and (D) ventral views.
Abbreviations: k—ventral keel;
nc—neural canal base; ps—
parasagittal swellings; t—
transverse process. Dotted lines,
reconstructed bone; hatched area,
broken bone. Scale bar = 5 cm.

Figure 8.3. (above right)
MPCA-PV-83: distal caudal
centrum in (A) lateral, (B) dorsal,
and (C) distal views.
Abbreviations and patterns as in
Figure 8.2. Scale bar = 5 cm.

cesses are absent (Fig. 8.3A). The preserved portion suggests that the base of the neural arch is placed in the mid-part of the centrum, as in most theropods (Fig. 8.3A,B). Two interesting characters of this specimen are the square outline of the articular surface (Fig. 8.3C) and the flat ventral surface.

MPCA-PV-84. Proximal femur. The specimen consists of the proximal end of a left femur (Fig. 8.4) including most of the head. The head projects medially and slightly upward, and expands slightly distally (Fig. 8.4A). In proximal view, the femoral head curves slightly caudally, producing a concave posterior margin and a convex anterior margin as in most theropods (Fig. 8.4C). The head is apomorphically wider at its base, becoming narrower toward the distal end, but it lacks a distinct neck. The dorsal surface is smooth and craniocaudally convex. The greater trochanter is flat laterally and expands craniocaudally (Fig. 8.4B,C). Although very weathered, it is clearly separated from the femoral head by a shallow groove (Fig. 8.4A,C). There are no remains of the anterior trochanter. Very little of the hollow shaft is preserved.

MPCA-PV-86. Metatarsal III. The distal half of an elongated metatarsal III is strongly compressed at its mid-shaft (Fig. 8.5A,C). The shaft is roughly triangular in cross-section, suggesting that the bone was pinched between metatarsals II and IV, in a manner similar to that seen

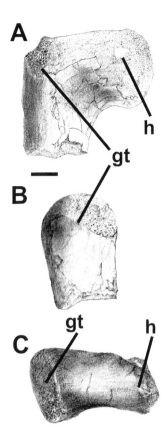

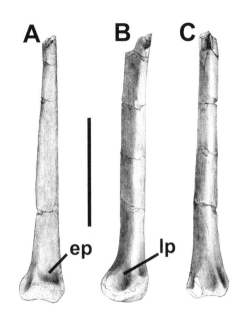

Figure 8.4. (above left)
MPCA-PV-84: proximal femur in
(A) caudal, (B) lateral, and
(C) proximal views.
Abbreviations: gt—greater
trochanter; h—femoral head.
Scale bar = 1 cm.

in arctometatarsalian theropods, although not to the degree seen in tyrannosaurids. The transversely expanded distal condyle is almost symmetrical with a deep anterior pit for the extensor tendon (Fig. 8.5A), and there are two deep collateral ligament pits on both sides of the distal end (Fig. 8.5B). The bone clearly belonged to a slender, long-legged animal unlike any previously reported theropod from any Patagonian locality.

MPCA-PV-100. *Quilmesaurus curriei* Coria, holotype (Fig. 8.6). *Quilmesaurus curriei* (Coria 2001) is a theropod species with a highly specialized knee join. The scientific name was dedicated to the extinct Argentinean native people, the Quilmes, and to Dr. Philip J. Currie. The specimen consists of the distal part of a right femur (Fig. 8.6A–E) and a right tibia (Fig. 8.6F–K). *Quilmesaurus* is the only named theropod recognized from this assemblage and is the youngest record of a non-avian dinosaur from Patagonia. The femur is characterized by a strong, well-developed mediodistal crest (Fig. 8.6A–D), and the tibia by a hooked cnemial crest (Fig. 8.6F–J). The asymmetrical distal end of the tibia has a lateral malleolus that is twice the size of medial (Fig. 8.6F,G) (Coria 2001). Owing to the lack of fusion of the proximal tarsals, it was not recognized as a ceratosaur and is considered related to basal tetanurans because of the presence of a notch on the distal articular surface of the tibia (Fig. 8.6K), a character shared with *Giganotosaurus* and *Sinraptor* (Coria 2001).

Figure 8.5. (above right)
MPCA-PV-86: metatarsal III in
(A) cranial, (B) lateral, and
(C) caudal views. Abbreviations:
ep—extensor pit; lp—lateral pit.
Scale bar = 5 cm.

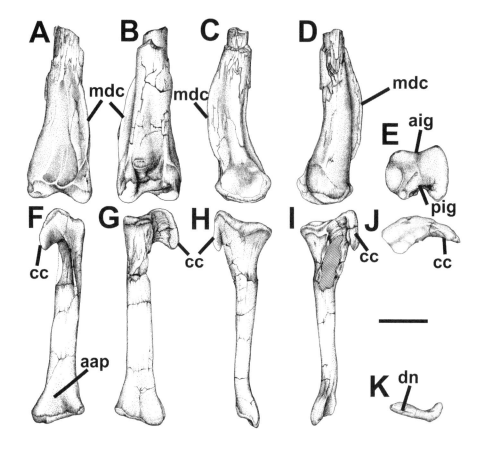

Figure 8.6. MPCA-PV-100: Quilmesaurus curriei, holotype. Right femur in (A) cranial, (B) caudal, (C) medial, (D) lateral, and (E) distal views; and right tibia in (F) cranial, (G) caudal, (H) medial, (I) lateral, (J) proximal, and (K) distal views. Abbreviations: aap—articulation for ascending process of astragalus; aig— anterior intercondylar groove; cc—cnemial crest; dn—distal notch; mdc; mediodistal crest; pig—posterior intercondylar groove. Scale bar = 10 cm. (Modified from Coria 2001.)

Discussion

Interestingly, none of the elements described above can be related to either of the two typical Cretaceous South American theropod taxa, the abelisaurs and the carcharodontosaurs. Those have occipital condyles formed two-thirds by the basioccipital, a subdivided basisphenoidal recess, caudal vertebrae that lack ventral keels, and a robust third metatarsal (Bonaparte et al. 1990; Coria and Currie 2002; Coria et al. 2002).

Carcharodontosaurids bear dorsomedially directed femoral heads, like the head of MPCA-PV-84. However, the completely formed articular surface of the specimen indicates not a juvenile but instead a sub-adult individual of a small meat eater. All known carcharodontosaurids represent large animals, over 10 m long at maturity. On the other hand, the femoral head of this fragmentary Patagonian tapers medially apomorphically, which is the opposite condition for any known theropod.

Provisionally, we assign the specimens described in this paper to the Tetanurae, mainly because they share derived features (dorsomedially directed femoral head; presence of notch in distal end of tibia; slender, proximally compressed metatarsal III; and ventrally keeled caudal vertebrae) that are absent in abelisaurs. We realize, nevertheless, that the assignment is based on isolated elements. The specimens probably represent more than one species. Consequently other than *Quilmesaurus,* we do not assign any of these elements to specific taxa, even though some of the bones have clearly apomorphic conditions (undivided basisphenoidal recess, medially tapering femoral head, and parasagittal ventral keels in caudal centra).

The local fauna of Salitral Moreno is, to date, one of the most diverse assemblages of dinosaurs in Patagonia. In a relatively small area, at least four species of sauropods are present (García and Salgado 2002), two hadrosaur species are recognized (Bonaparte et al. 1984; Powell 1987), and the ankylosaurs (Coria and Salgado 2001) complete the list of herbivorous dinosaurs. Apparently, the small to medium-size tetanurans occupied the role of meat eaters.

Acknowledgments. We thank Mr. Carlos Muñoz for access to material in his care and P. J. Currie for valuable comments on the manuscript. Reviews by Drs. R. Molnar and K. Carpenter significantly improved the content of the paper. Illustrations were prepared by Alfredo Jeréz. Research was supported by Municipalidad de Plaza Huincul.

References Cited

Arcucci, A. B., and R. A. Coria. 2003. A new Triassic carnivorous dinosaur from Argentina. *Ameghiniana* 40: 217–228.

Bonaparte, J. F. 1986. Les dinosaures (carnosaures, allosauridés, sauropodes, cétiosauridés) du Jurassique moyen de Cerro Cóndor (Chubut, Argentina). *Annales de Paléontologie (Vertébrés-Invertébrés)* 72: 247–289.

Bonaparte, J. F. 1991. The Gondwanian theropod families Abelisauridae and Noasauridae. *Historical Biology* 5: 1–25.

Bonaparte, J. F., M. R. Franchi, J. E. Powell, and E. C. Sepulveda. 1984. La Formación Los Alamitos (Campaniano-Maastrichtiano) del sudeste de Río Negro, con descripción de *Kritosaurus australis* nov. sp. (Hadrosauridae). Significación paleobiogeográfica de los vertebrados. *Revista de la Asociación Geológica Argentina* 39: 284–299.

Bonaparte, J. F., F. E. Novas, and R. A. Coria. 1990. *Carnotaurus sastrei Bonaparte, the Horned, Lightly Built Carnosaur from the Middle Cretaceous of Patagonia.* Contributions in Science, no. 416. Los Angeles: Natural History Museum of Los Angeles County.

Colbert, E. H. 1970. A saurischian dinosaur from the Triassic of Brazil. *American Museum Novitates,* no. 2405: 1–39.

Coria, R. A. 2001. A new theropod from the Late Cretaceous of Patagonia. In D. H. Tanke and K. Carpenter (eds.), *Mesozoic Vertebrate Life,* pp. 3–9. Bloomington: Indiana University Press.

Coria, R. A., and P. J. Currie. 2002. The braincase of *Giganotosaurus carolinii. Journal of Vertebrate Paleontology* 22: 802–811.

Coria, R. A., and L. Salgado. 1995. A new giant carnivorous dinosaur from the Cretaceous of Patagonia. *Nature* 377: 224–226.

Coria, R. A., and L. Salgado. 1998. A basal Abelisauria Novas, 1992 (Theropoda-Ceratosauria) from the Cretaceous of Patagonia, Argentina. B. P. Pérez-Moreno, T. Holtz Jr., J. L. Sanz, and J. Moratalla (eds.), *Gaia: Aspects of Theropod Paleobiology,* vol. 15, pp. 89–102. Lisbon: Museu Nacional de História Natural.

Coria, R. A., and L. Salgado. 2001. South American ankylosaurs. In K.

Carpenter (ed.), *The Armored Dinosaurs,* pp. 159–168. Bloomington: Indiana University Press.

Coria, R. A., L. M. Chiappe, and L. Dingus. 2002. A new close relative of *Carnotaurus sastrei* Bonaparte (Abelisauridae: Theropoda) from the Late Cretaceous of Patagonia. *Journal of Vertebrate Paleontology* 22: 460–465.

Currie, P. J. 1985. Cranial anatomy of *Stenonychosaurus inequalis* (Saurischia, Theropoda) and its bearing on the origin of birds. *Canadian Journal of Earth Sciences* 22: 1643–1658.

Currie, P. J. 1995. New information on the anatomy and relationships of *Dromaeosaurus albertensis* (Dinosauria: Theropoda). *Journal of Vertebrate Paleontology* 15: 576–591.

Currie, P. J., and K. Carpenter. 2000. A new specimen of *Acrocanthosaurus atokensis* (Theropoda, Dinosauria) from the Lower Cretaceous Antlers Formation (Lower Cretaceous, Aptian) of Oklahoma, USA. *Geodiversitas* 22: 207–246.

Currie, P. J., and Zhao X.-J. 1993. A new carnosaur (Dinosauria, Theropoda) from the Jurassic of Xinjiang, People's Republic of China. *Canadian Journal of Earth Sciences* 30: 2037–2081.

García, R., and L. Salgado. 2002. La fauna de saurópodos de la localidad de Salitral Moreno (Cretácico superior, Provincia de Río Negro). Symposium of dinosaurs and Mesozoic reptiles. Abstracts from VIII *Congreso Argentino de Paleontología y Bioestratigrafía,* pp. 62. Corrientes.

Madsen, J. H., Jr. 1976. Allosaurus fragilis: *A Revised Osteology.* Utah Geological and Mineral Survey Bulletin, no. 109. Salt Lake City: Utah Geological and Mineral Survey, Utah Department of Natural Resources.

Novas, F. E. 1997a. South American dinosaurs. In P. J. Currie and K. Padian (eds.), *Encyclopedia of Dinosaurs,* pp. 678–689. New York: Academic Press.

Novas, F. E. 1997b. Anatomy of *Patagonykus puertai* (Theropoda, Avialae, Alvarezsauridae), from the Late Cretaceous of Patagonia. *Journal of Vertebrate Paleontology* 17: 137–166.

Novas, F. E. 1998. *Megaraptor namunhuaiquii,* gen. et sp. nov., a large-clawed, Late Cretaceous theropod from Patagonia. *Journal of Vertebrate Paleontology* 18: 4–9.

Powell, J. E. 1987. Hallazgo de un dinosaurio hadrosáurido (Ornithischia, Ornithopoda) en la Formación Allen (Cretácico superior) de Salitral Moreno, Provincia de Río Negro, Argentina. *X Congreso Geológico Argentino, Actas* 3: 149–152. Tucumán, Argentina.

Reig, O. A. 1963. La presencia de dinosaurios saurisquios en los "Estratos de Ischigualasto" (Mesotriásico superior) de las provincias de San Juan y La Rioja (República Argentina). *Ameghiniana* 3: 3–20.

Salgado, L., and R. A. Coria. 1993. Un nuevo titanosaurino (Sauropoda-Titanosauridae) de la Fm. Allen (Campaniano-Maastrichtiano) de la Provincia de Río Negro, Argentina. *Ameghiniana* 30: 119–128.

Salgado, L., and R. A. Coria. 1996. First evidence of an ankylosaur (Dinosauria. Ornithischia) in South America. *Ameghiniana* 33: 367–371.

Sereno, P. C., and F. E. Novas. 1993. The skull and neck of the basal theropod *Herrerasaurus ischigualastensis. Journal of Vertebrate Paleontology* 13: 451–476.

Sereno, P. C., C. A. Forster, R. R. Rogers, and A. M. Monetta. 1993. Primitive dinosaur skeleton from Argentina and the early evolution of Dinosauria. *Nature* 361: 64–66.

II. Theropod Working Parts

9. Enamel Microstructure Variation within the Theropoda

KATHY STOKOSA

Abstract

Isolated teeth from various members of the Theropoda were compared by means of scanning electron microscopy, and the results indicate that enamel structures vary within this suborder. The teeth examined belong to *Troodon, Paronychodon, Richardoestesia,* and the Tyrannosauridae. All are from the Upper Cretaceous Fox Hills (Meade County, western South Dakota), Hell Creek (Harding County, northwestern South Dakota), and Lance (Niobrara County, eastern Wyoming) formations.

Small theropods, such as *Troodon* and *Richardoestesia,* have teeth composed of "simple" parallel enamel. In contrast, larger theropods, such as *Tyrannosaurus* and *Albertosaurus,* have teeth composed of columnar enamel. Columnar enamel contains crystallites that are more structurally complex and have a higher level of organization than crystallites in parallel enamel. The existence of these two distinct enamel types may reflect biomechanical adaptations to different feeding habits.

The columnar enamels present in various members of the Tyrannosauridae were compared. *Tyrannosaurus* teeth are composed of poorly developed columnar enamel. The crystallites lack complex structural organization, making it difficult to discern columnar unit boundaries. In contrast, *Albertosaurus* and Genus "A" (?*Albertosaurus*) have well-developed columnar enamel. The crystallites are highly organized with defined columnar unit boundaries. The columnar enamel crystallite differences found within the Tyrannosauridae are distinct and appear to be generically specific. Enamel comparisons, supported by morphological comparison (i.e., denticle shape), suggest that at least two tyrannosaur species were present during the deposition of the Hell Creek and Lance formations.

Introduction

Theropod teeth are recurved, serrated, and laterally compressed. This design, often termed ziphodont, can be traced back to some of the earliest members of the Dinosauria. The presence of ziphodont teeth in ancestral forms implies that the theropod condition is plesiomorphic. In general, theropod teeth have thin enamel, especially relative to the large tooth size some taxa attained (e.g., *Tyrannosaurus rex*). The enamel of small theropods, such as *Troodon* and *Richardoestesia*, is structurally distinct from that of larger theropods such as the tyrannosaurids. This difference in enamel structure may be a result of dietary constraints on tooth function.

Theropod teeth are composed of an outer layer of enamel and an inner core of dentine. Enamel is composed of the phosphatic mineral hydroxyapatite. The hydroxyapatite is in the form of small, micrometer-size crystallites. In the specimens analyzed, these crystallites were arranged in two distinct ways, forming either parallel enamel or columnar enamel. (The terms *parallel* and *columnar* are based on the terms defined in Sander 1999 for reptile enamel.)

Enamel is highly mineralized and is therefore difficult to modify after it has been deposited by the ameloblasts (Sander 1999). Enamel deposition occurs at the enamel-dentine junction (EDJ) and extends outward. The enamel layers protect the underlying dentine and aid in maintaining the shape of the tooth. Normally, the enamel surface is smooth to reduce the friction between the tooth and the food item and to allow maximum penetration of the tooth into the food. The enamel surface morphology forms as enamel layers are deposited (Sander 1999).

Previous Studies

Unlike the enamel of mammals, which has been extensively studied, the enamel of reptiles is much less well known, because of the structure of reptile teeth. Mammalian enamel is composed of prisms that can be viewed under polarized light (Sander 1999), but reptile enamel consists of crystallites that cannot be adequately studied under polarized light. Instead, they must be viewed under a scanning electron microscope (SEM) in order to view the crystallite orientations. Early studies, such as that of Erler (1935), attempted to understand reptile enamel structures using polarized light. Erler noted that crocodile teeth had structures such as enamel voids, tubules, and columns of enamel that consisted of alternating dark and bright bands. Yet because he was unable to obtain a three-dimensional view of these structures, he was not able to understand the full complexity of enamel surface morphology. These same alternating bands of dark and light enamel were noted by Schmidt (1947) in a study dealing with aquatic and marine reptiles. Schmidt termed these bands *Saulengliederung.*

Poole (1956) discussed in great detail the structures formed by the orientation and arrangement of enamel crystallites in synapsids. Poole noted enamel structures that appeared to be quite complex; however, his observations were limited because he used polarized light.

With the advent of the SEM, complex structures formed by crystallites became easier to observe. Cooper and Poole (1973) found that the teeth of *Uromastyx* possessed prismatic enamel, making it the only reptile known to have this type of enamel. Their work led to the prevailing view that reptile enamel was very uniform, consisting of parallel crystallites that, at most, made simple, cylindrical structures (Sander 1999). Carlson and Bartels (1986) stated that the enamel of crocodiles and lepidosaurs was more complex and diverse than once thought, but, except for some photomicrographs published in Carlson (1990), the results of this study were not published in detail.

Unfortunately, the enamel terminology developed for mammal teeth has resulted in confusion when applied to reptilian teeth. As a result, many of the past reptile enamel studies were disregarded, leading to the false belief that reptiles possessed only prismatic enamel. One of the most recent and extensive studies of reptilian enamel has been by Sander's (1999) investigation of fifteen orders and thirty-one families of recent and fossil reptiles. Sander established a new enamel terminology for the Reptilia, using concise, descriptive terms in order to convey the structural complexity and diversity that exist in reptile enamel. Sander's terminology is used below on the dinosaur teeth.

Past studies (e.g., Russell 1970; Paul 1988) have focused on describing theropod tooth morphology for systematic accounts of theropods (Farlow et al. 1991). With the aid of the SEM, research has expanded from analyzing the outer surface tooth morphology to exploring and understanding changes within the internal enamel morphology as well. Enamel microstructure, coupled with outer tooth morphology analysis, may provide the information necessary to identify isolated theropod teeth at the familial, generic, or even specific level with a confidence not previously attained.

Stokosa (2000) combined outer tooth morphology characters with enamel characters to identify isolated theropod teeth from the Fox Hills, Lance, and Hell Creek formations. The enamel of the smaller theropods, such as *Troodon formosus, Richardoestesia gilmorei,* and *Richardoestesia* sp., was composed of parallel crystallites. These taxa could not be separated from one another on the basis of enamel. However, Stokosa found that within the Tyrannosauridae, enamel orientations for *Tyrannosaurus rex* and *Albertosaurus* appeared to be genera specific, as described below.

Materials and Methods

The theropod teeth used in this study are archived at the Museum of Geology at the South Dakota School of Mines and Technology. Thin sections of the specimens were prepared for SEM analysis. Before being

Enamel Types

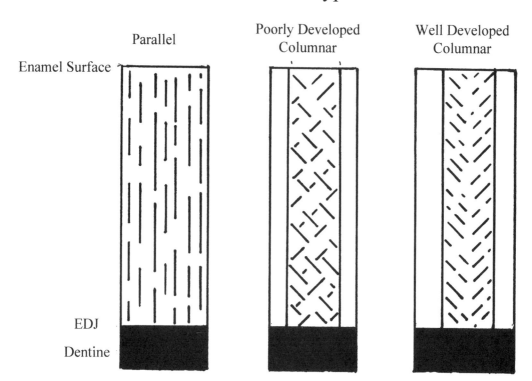

Figure 9.1. Schematic representation of the three major enamel types.

cut, the teeth were embedded in resin. The samples were cut in three planes: a tangential view, parallel to the enamel outer surface; a cross-sectional view, through the enamel layer at a 90° angle to the enamel-dentine junction in a frontal plane of the animal; and a longitudinal view, in a parasagittal plane. Thin sections were ground and then polished. After being polished, the samples were etched for 10 seconds in a 2 N HCl solution, rinsed, and dried. Specimens were set on microscope mounts and analyzed with a JEOL 840A scanning electron microscope at the South Dakota School of Mines and Technology.

Institutional Abbreviations. SDSM, South Dakota School of Mines and Technology, Rapid City, S.D.; TMP, Royal Tyrrell Museum of Palaeontology, Drumheller, Alberta, Canada.

Enamel Descriptions

Within the Theropoda, two distinct types of enamel microstructure organization were found: parallel and columnar (Fig. 9.1). These two enamel types differ in the orientation of the hydroxyapatite crystallites with respect to the enamel-dentine junction. In parallel enamel, the crystallites are deposited parallel to one another and normal to the plane of the enamel-dentine junction. The crystallites are closely packed against one another, reducing the amount of open space between crys-

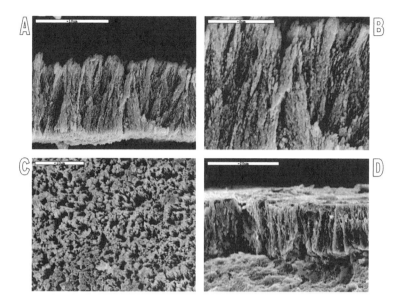

tallites. Columnar enamel, in contrast, contains crystallites that are more organized (Fig. 9.1). Columnar enamel crystallites make up units or bundles. These units are parallel to one another and normal to the enamel-dentine junction. Within the bundles, the crystallites exist in a V-shaped orientation; however; the ends of the crystallites do not touch. These V-shaped crystallites diverge toward the enamel-dentine junction. Well-developed columnar enamel has bundles that are easy to define. Conversely, poorly developed columnar enamel has bundles that are hard to define (Fig. 9.1). These differences in columnar enamel are apparent within the Tyrannosauridae.

Enamel Analysis

Troodon sp. cf. *Troodon formosus*
Figure 9.2

Specimen Analyzed. SDSM 14518, one tooth from the Fox Hills Formation of western South Dakota. SDSM 14518 was identified as a dentary tooth on the basis of published figures (Carpenter 1982) and descriptions (Currie 1987).

Enamel Thickness. The enamel thickness ranges from 10 microns at the center of the serration row to 15 microns at the tooth tip.

Enamel Type. In this specimen, parallel crystallite enamel (cf. Sander 1999, fig. 10e) makes up the *schmelzmuster,* a term defined in Sander 1999 as denoting the spatial arrangement of the enamel types and major discontinuities within a single tooth. The presence of parallel crystallite enamel corresponds with thin enamel. Sander (1999) noted poorly developed columnar units in his *Troodon* samples as well. Columnar units were not observed in SDSM 14518.

Figure 9.2. Troodon *sp. cf.* Troodon formosus *(SDSM 14518). (A) Longitudinal section, view of parallel crystallite enamel. EDJ at the bottom of page; scale = 10 μm. (B) Longitudinal section, close-up of A; scale = 5 μm. Note the presence of gold aggregate on the enamel. (C) Tangential section, view showing no evidence of structural orientation of the parallel crystallites; scale = 10 μm. (D) Longitudinal section, slight oblique view of parallel crystallites; scale = 20 μm. The presence of gold aggregate is not so obvious.*

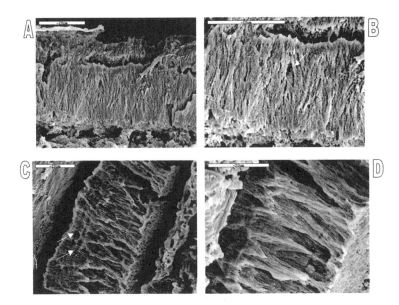

Figure 9.3. Dromaeosaurus *sp. cf.* Dromaeosaurus albertensis *(SDSM 14566). (A) Longitudinal section, view of parallel crystallite enamel. EDJ at the bottom of page. (B) Longitudinal section, close-up of parallel enamel. Note the enamel void near the center of the picture. (C) Cross-section, poorly developed columnar enamel with incremental lines present near the top of the enamel. EDJ on the right side of the picture. (D) Cross-section, close-up of columnar enamel. EDJ to right of the picture. Scales = 20 μm.*

Dromaeosaurus sp. cf. *D. albertensis*
Figure 9.3

Specimen Analyzed. SDSM 14566, a tooth fragment from the Fox Hills Formation of western South Dakota. Owing to the incompleteness of the specimen, dental location could not be determined. This tooth was identified as belonging to *Dromaeosaurus,* owing to the presence of the twisting anterior carinae typical of this taxon.

Enamel Thickness. The enamel thickness range is 40–45 microns along the serration row.

Enamel Type. Parallel crystallites (Fig. 9.3A,B) are present near the surface of the tooth; however, in cross-sectional views (Fig. 9.3C,D), columnar enamel is visible. The crystallites appear to be structured and slightly divergent. Incremental lines are present in certain areas corresponding with the presence of columnar enamel. The tooth appears to be made up of columnar enamel (from the enamel-dentine junction) but changes to parallel crystallite enamel approximately 10 microns from the outer surface.

Tyrannosauridae
?*Albertosaurus* Genus "A" Indeterminate

Specimens Analyzed. SDSM 12737 and SDSM 15143 from the Lance Formation of eastern Wyoming; SDSM 64351 from the Hell Creek Formation of northwestern South Dakota. SDSM 15143 and SDSM 64351 appear to be either posterior dentary or posterior maxillary teeth, owing to the lack of twist in the anterior carinae. SDSM 12737 is a tooth tip, preventing position determination.

Enamel Thickness. Thickness ranges from 40 to 55 microns on SDSM 12737, from 180to 200 microns on SDSM 64351, and from 95

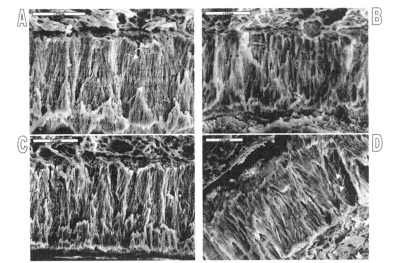

*Figure 9.4. SEM micrograph of enamel, genus "A" indeterminate (?Albertosaurus).
(A) Longitudinal section, view of parallel crystallite enamel with several incremental lines present. EDJ at the bottom of the picture.
(B) Longitudinal section, another view of the parallel crystallite enamel. (C) Longitudinal section, parallel crystallite enamel on opposite edge of tooth. EDJ at the bottom of the picture.
(D) Cross-section, well-developed columnar enamel with faint incremental lines present. EDJ at the bottom of the picture. Scales = 20 μm.*

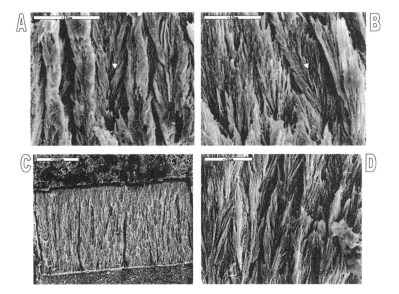

*Figure 9.5. Genus "A" indeterminate (?Albertosaurus). (A) Longitudinal section, well-developed columnar enamel with a close-up of the crystallites; scale = 10 μm. Note the well-defined zone of crystallite divergence in the center of the picture. EDJ at the bottom of the picture. (B) Longitudinal section, close-up of divergent crystallites; scale = 10 μm. EDJ to the right of the picture. (C) Cross-section, schmelzmuster composed of highly organized columnar enamel; scale = 100 μm. EDJ at the bottom of the picture.
(D) Cross-section, close-up of divergent crystallites; scale = 10 μm. Again note the central line of crystallite divergence. EDJ at the bottom of the picture.*

to 100 microns on SDSM 15143. The enamel is thickest near the tooth tip on all three specimens.

Enamel Type. Parallel crystallites are present near the tooth's surface (Fig. 9.4A,C). This parallel enamel layer is approximately 20 microns thick. Beneath this layer, the crystallites begin to diverge and form well-defined columnar enamel. Incremental lines are present (Figure 9.4A) within the columnar enamel. In cross-section, SDSM 12737 has crystallites showing well-developed crystallite divergence (Fig. 9.4D). SDSM 64351 shows well-defined zones of divergent crystallites (Figs. 9.5A,B,D) in longitudinal views.

Figure 9.6. Cf. Tyrannosaurus rex *(SDSM 15135). (A) Longitudinal section, poorly developed columnar enamel with a close-up of the crystallites; scale = 10 μm. EDJ at the bottom of the picture. (B) Tangential section, slight oblique view of enamel; scale = 5 μm. The large holes appear to be enamel tubules. EDJ to the right of the picture. (C) Cross-section, view of poorly developed columnar enamel with at least two pronounced incremental lines present; scale = 20 μm. EDJ is to the right of the picture. (D) Cross-section, close-up of columnar enamel in C; scale = 20 μm.*

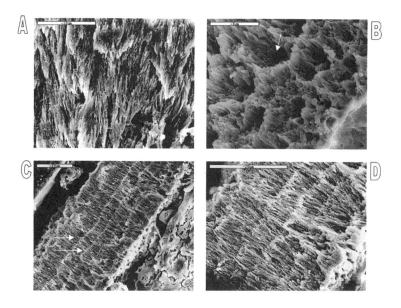

Figure 9.7. Cf. Tyrannosaurus rex *(SDSM 64287). (A) Cross-section, view of poorly developed columnar enamel; scale = 20 μm. EDJ on right side of picture. (B) Cross-section, another view of poorly developed columnar enamel; scale = 20 μm. Note the large enamel void which appears to be the result of a crack in the enamel. EDJ on right side of the picture. (C) Tangential section, view showing the absence of any crystallite orientation; scale = 10 μm.*

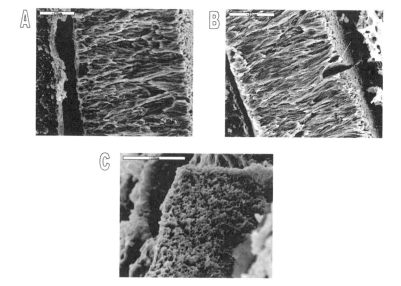

Tyrannosaurus rex and cf. *Tyrannosaurus rex**
Figures 9.6–9.8

Specimens Analyzed. *SDSM 15135 (tooth tip), SDSM 15115 (tooth fragment), and *SDSM 64287 (whole tooth). SDSM 15135 and SDSM 15115 are from the Lance Formation of eastern Wyoming. SDSM 64351 is from the Hell Creek Formation of northwestern South Dakota. SDSM 15115 is from the base of a posterior premaxillary tooth. SDSM 64287 is either a posterior dentary tooth or a posterior maxillary tooth, as indicated by the lack of twist in the anterior carinae. The positioning of SDSM 15135 could not be determined.

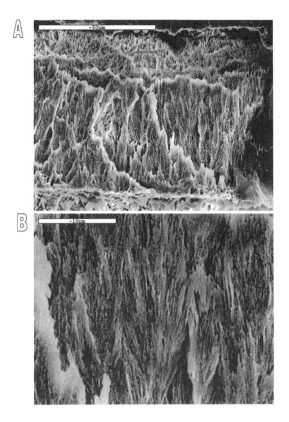

Figure 9.8. Cf. Tyrannosaurus rex *(SDSM 15115). (A) Longitudinal section,* schmelzmuster *composed of poorly developed columnar enamel; scale = 50 μm. EDJ at the bottom of the picture. (B) Longitudinal section, close-up of divergent crystallites; scale = 10 μm. Note the close packing of the crystallites making it difficult to discern columnar unit boundaries. EDJ is at the bottom of the picture.*

Enamel Thickness. Thickness ranges from 45 to 50 microns on SDSM 15135 (Fig. 9.6), from 60 to 75 microns on SDSM 64287 (Fig. 9.7), and from 80 to 90 microns on SDSM 15115 (Fig. 9.8).

Enamel Type. SDSM 15135 (Fig. 9.6) has poorly developed columnar enamel that lacks the central line of divergence seen in Genus "A" (Fig. 9.5A). The columnar unit boundaries are often difficult to define. Parallel enamel is present near the surface of the tooth and is only about 15 microns thick. Incremental lines are quite distinct in cross-section on SDSM 15135 (Fig. 9.6C,D). Figure 9.6B,C shows the enamel in tangential view. The crystallites appear to lack any geometric arrangement such as the polygonal columnar units that Sander (1999) noted in some specimens; however, enamel tubules are present (Fig. 9.6B).

Albertosaurus sp. Indeterminate
Figure 9.9

Specimen Analyzed. TMP 2001.12.01, tooth from the Dinosaur Park Formation of the Judith River Group, Alberta, Canada. This tooth may be either a posterior dentary tooth or a posterior maxillary tooth.

Enamel Thickness. Thickness range is 100–120 microns.

Enamel Type. Parallel enamel is present near the surface of the tooth (approximately 15 microns thick), with columnar enamel making up the rest of the enamel thickness. The columnar unit boundaries

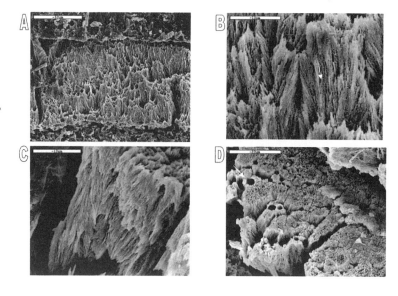

Figure 9.9. Albertosaurus. *(A) Longitudinal section,* schmelzmuster *composed of moderately to well-developed columnar enamel; scale = 50 μm. EDJ at the bottom of the picture. (B) Longitudinal section, close-up of divergent crystallites; scale = 10 μm. Note some of the columnar units have a zone of crystallite divergence. EDJ at the bottom of the picture. (C) Cross-section, close-up of divergent crystallites; scale = 10 μm. EDJ is on the top, right of the picture. (D) Tangential section, view of enamel surface showing enamel tubules and the polygonal cross-section shape of the columnar units; scale = 20 μm.*

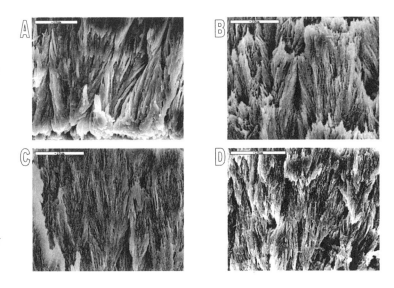

Figure 9.10. Comparison of tyrannosaurid enamel crystallites. (A) Longitudinal section, close-up of crystallites from Genus "A." Note the well developed columnar unit boundaries as well as the zones of crystallite divergence. EDJ at the bottom of the picture. (B) Longitudinal section, close-up of crystallites from Albertosaurus. *The columnar unit boundaries are defined and the crystallites have a zone of crystallite divergence. This zone is not as defined as in Genus "A." EDJ at the bottom of the picture. (C) Longitudinal section, close-up of crystallites from* Tyrannosaurus rex. *Note columnar unit boundaries are difficult to discern. The crystallites are divergent however they lack a distinct zone of divergence. EDJ at the bottom of the picture. (D) Longitudinal section, close-up of crystallites of cf.* Tyrannosaurus rex. *EDJ at the bottom of the picture. Scales = 10 μm.*

are more defined than those of *Tyrannosaurus* and closely resemble those observed in Genus "A." The crystallites are organized in such a way that a zone of crystallite divergence can be observed (Fig. 9.9B,C). This zone is similar that observed in Genus "A." Enamel tubules are present (Fig. 9.9D) in tangential view with columnar units that appear to be polygonal in cross-section.

Comparison of Tyrannosaurid Enamel Crystallites. Specimens compared include SDSM 15143 (Genus "A"), SDSM 15115 (*Tyrannosaurus*), SDSM 64287 (cf. *Tyrannosaurus*), and TMP 2001.12.01 (*Albertosaurus*). Crystallite orientations vary greatly even though all these teeth

belong to the Tyrannosauridae (Fig. 9.10). On the basis of enamel structure, Genus "A" shares features with *Albertosaurus*, suggesting that Genus "A" may in fact be *Albertosaurus*. Genus "A" and *Albertosaurus* both have defined columnar unit boundaries, and both teeth have crystallites that demonstrate a zone of crystallite divergence. These features contrast with the poorly developed columnar enamel of *Tyrannosaurus*, in which columnar unit boundaries can be difficult to discern owing to the close packing of the crystallites. The crystallites in *Tyrannosaurus* are divergent; however, they lack the distinct zone of divergence observed in Genus "A" and *Albertosaurus*.

Family Indeterminate
Richardoestesia sp. cf. *R. gilmorei*
Figure 9.11

Specimen Analyzed. SDSM 64372b from the Hell Creek Formation of northwestern South Dakota. Currie et al. (1990) noted that anterior dentary teeth are distinct from posterior, with the latter having more curvature. This specimen appears to have been an anterior dentary tooth, on the basis of illustrations in Currie et al. (1990). However, upper jaws have not been described for this species, making it difficult to determine the precise dental position of this specimen.

Enamel Thickness. Thickness is 10 microns.

Enamel Type. The individual crystallites are difficult to differentiate on this specimen; however, the orientation appears to be simple parallel.

Richardoestesia sp. Indeterminate
Figure 9.12

Specimen Analyzed. SDSM 14516, a worn tooth from the Fox Hills Formation of western South Dakota. The dental position of this

Figure 9.11. Richardoestesia cf. Richardoestesia gilmorei. *(A) Longitudinal section, view of parallel crystallites; scale = 10 μm. EDJ on right side of picture. (B) Longitudinal section, three-dimensional picture of enamel that has pulled away from the dentine; scale = 10 μm. The dentine is at the bottom of the picture and cannot be seen. This picture illustrates the importance of getting a three dimensional picture of the enamel in order to fully understand its structure. (C) Tangential section, view of parallel crystallites; scale = 5 μm. (D) Cross-section, view of parallel crystallites; scale = 10 μm.*

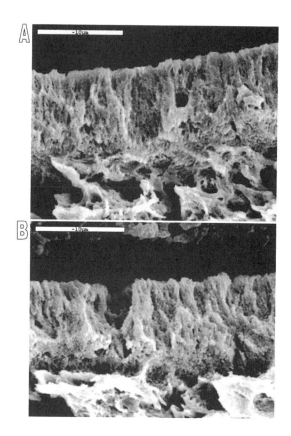

Figure 9.12. Richardoestesia *sp.* indeterminate. *(A) Longitudinal section, view of parallel crystallite. Note the presence of gold aggregate on the crystallites, making it difficult to clearly see the enamel. EDJ on bottom of the picture. (B) Longitudinal section, another view of the parallel crystallites. EDJ on bottom of page. Scales = 10 μm.*

tooth could not be determined, owing to the absence of cranial material for this genus.

Enamel Thickness. Thickness range is 10–15 microns.

Enamel Type. This specimen has parallel crystallite enamel.

Paronychodon sp. cf. *P. lacustris*
Figure 9.13

Specimen Analyzed. SDSM 15101b from the Lance Formation of eastern Wyoming. This tooth is identified as *Paronychodon*, on the basis of the lack of serrations and the flat lingual surface. The position of this tooth could not be determined, owing to the absence of cranial material for this genus.

Enamel Thickness. Thickness range is 0–15 microns.

Enamel Type. SDSM 15101b has parallel crystallites (Fig. 9.13A,B).

Biomechanical Differences between Parallel and Columnar Enamel

From a biomechanical perspective, columnar enamel is more resistant to deformation caused by mediolateral bending stress than is parallel enamel (Sander 1999). Teeth belonging to such forms as

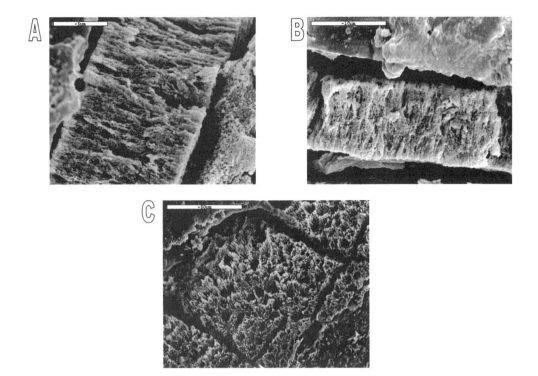

Troodon, Richardoestesia, and *Paronychodon* have enamel layers that are rigid owing to the parallel orientation of the crystallites. In contrast, the numerous columnar unit boundaries present in columnar enamel allow minute movement or "flex" to occur, preventing enamel damage. Parallel enamel lacks the ability to flex and therefore cannot withstand the same bending stresses as columnar enamel. The inability to withstand bending stress may weaken the integrity of the enamel permanently, resulting in the flaking off of enamel layers. Although parallel enamel lacks the flexibility of columnar enamel, it may be more resistant to wear than columnar enamel (Sander 1999). *Dromaeosaurus, Tyrannosaur,* and *Albertosaurus* teeth use a combination of these two enamel types. Parallel enamel makes up the outer tooth surface, and the inner enamel is columnar. The resulting tooth is resistant to surface wear and able to withstand mediolateral bending stresses more effectively than a tooth composed of only parallel enamel.

Figure 9.13. Paronychodon. (A) Longitudinal section, view of parallel crystallites; scale = 5 μm. Note the presence of gold aggregate on the structures. EDJ on the right side of the picture. (B) Longitudinal section, another view of the parallel crystallites; scale = 10 μm. EDJ on the bottom of picture. (C) Tangential section, view of crystallites; scale = 50 μm. Note the presence of larger holes that appear to be enamel tubules, indicating some sort of crystallite orientation may be present.

Discussion of Tyrannosaurid Enamel

The columnar enamel of different tyrannosaurids was found to exhibit structural differences involving the degree of structural orientation of the crystallites relative to the EDJ (see Figures 9.5 and 9.6). Three reasons are postulated to explain why two members of the same family with similar tooth morphologies have such a difference in crystallite orientation: (1) tooth location, (2) specimen maturity, and (3) distinct genera.

Tooth Location. Dental position morphological variation is associated with heterodont reptiles (Sander 1999). *Troodon* is the only theropod in this study that appears to have heterodont dentition. Bakker et al. (1988) noted structural differences between the anterior and posterior maxillary teeth of *Tyrannosaurus:* the more posterior teeth are blunt and stout, whereas the anterior are taller and more lateromedially compressed. Bakker et al. (1988, p. 24) stated, "Clearly the function of these small [posterior] teeth was different from that of the tall, anterior teeth—perhaps the posterior teeth were used to crack bones." Tyrannosaurids may have used posterior teeth differently than teeth more anteriorly positioned; however, this does not necessarily imply heterodonty. Therefore, dental position should not effect enamel crystal orientation within the Tyrannosauridae.

Specimen Maturity. Sander (1999) noted that a mature tooth of *Omphalosaurus* (an ichthyosaur) was composed of compound microunit enamel. The immature, replacement teeth found within the same jaw were composed of columnar enamel. Sander (1999) suggested that an extracellular proteinaceous enamel matrix guides crystal growth in some way. Amelogenesis is poorly understood within modern reptiles and is even less understood in extinct forms. Enamel differences between mature and immature teeth within the same jaw do not appear to be prevalent in nature (Sander 1999).

Distinct Genera. The simplest explanation for the difference in crystallite orientation is that these unidentified tyrannosaur teeth represent two distinct taxa. Sander (1999) also noted enamel variation within the Tyrannosauridae. Morphological analysis of tyrannosaur teeth used in this study, from the Fox Hills, Hell Creek, and Lance formations, has led to the differentiation of two morphologically distinct tyrannosaurids: *Tyrannosaurus rex* and *Albertosaurus.* The morphological differences are reflected in the enamel structure as well as noted above.

Conclusions

The differences between parallel and columnar enamel may reflect the role of diet in the development of teeth. The presence of columnar enamel may indicate that forms like *Dromaeosaurus* and some tyrannosaurs ingested bone as well as soft tissue. A diet that included bone would require a tooth that could withstand the bending force encountered when crushing bone and yet still maintain structural integrity. The short, stout denticles on dromaeosaur and tyrannosaur teeth suggest that these forms used their teeth to crush bone (Currie et al. 1990; Farlow et al. 1991; Abler 1992). The percentage of bone in the diet of a taxon may be reflected in the amount of divergent crystallite organization and denticle morphology.

In contrast to specimens having columnar enamel, *Troodon, Richardoestesia,* and *Paronychodon* have parallel enamel, which may indicate that bone was not ingested. The denticles of *Troodon* are so coarse that they appear to be very susceptible to breakage and bending stress

(Farlow et al. 1991). On the basis of the denticle morphology, it is not surprising that *Troodon* teeth are composed of parallel enamel. The denticles of *Richardoestesia gilmorei* are very generalized, providing little information regarding diet (Currie et al. 1990); however, the presence of parallel enamel suggests that the teeth were unable to withstand large bending stresses. Bone does not appear to be part of the diet of *Richardoestesia gilmorei*. *Richardoestesia* sp. teeth have been morphologically compared to teeth of recent and fossil piscivores (Baszio 1997), although the presence of parallel enamel cannot confirm such a diet. Because *Paronychodon* teeth lack serrations, it is difficult to infer much about the function of these teeth; however, the presence of parallel enamel suggests that bone was not ingested.

Analysis of nine theropod taxa has resulted in the following conclusions: The Theropoda exhibit two enamel types: "simple" parallel enamel and the more complex columnar enamel. Dietary preferences may be reflected in enamel structure. Columnar enamel may have imparted a biomechanical advantage for breaking bone. Parallel enamel is more rigid than columnar, yet may be more resistant to wear.

Crystallite variation is evident in the columnar enamel of certain taxa within the Tyrannosauridae. Forms such as *Tyrannosaurus rex* have enamel crystallites that lack complex structural organization. Forms such as Genus "A" and *Albertosaurus* have enamel crystallites that are more structurally complex and organized than in *Tyrannosaurus rex*. This variation appears to be distinct enough to suggest that identification within the Tyrannosauridae may be determined to a generic level on the basis of enamel.

Acknowledgments. I thank Dr. Cynthia Marshall for her reading of this manuscript in its early stages and Dr. Kenneth Carpenter and Yoshitsugu Kobayashi for their critical reading in its later stages. I thank my advisor Dr. Jack Horner and Dr. James Martin, Dr. James Fox, and Dr. Edward Duke (South Dakota School of Mines and Technology) for SEM funding. Despite Dr. Duke's claim of not knowing much about enamel microstructures, his comments regarding my specimens were very helpful. I would also like to thank Mark Gabel for access to the SEM at Black Hills State. I thank Pete and Neal Larson at the Black Hills Institute for allowing me to look at their theropod material, as well as their time and assistance in identifying my specimens. Sara Black translated a much needed article for my research. C. B. Wood and Dr. Martin Sander generously sent reference articles to me, along with advice involving enamel. I especially thank Dr. Sander for his comments regarding my results. Thomas Carr made helpful comments regarding the Tyrannosauridae. Dr. Don Brinkman from the Royal Tyrrell Museum of Palaeontology in Alberta, Canada, sent me an *Albertosaurus* tooth and allowed me to thin section it for my enamel research. I acknowledge Mike Greenwald, Tammie Bouchard, Dr. Phil Bjork, and Dr. James Martin for collecting the material I used in this study. Finally, I thank my parents, Joe and Pam Stokosa, and last, but not least, Josh Edwards, who provided me with lots of help and encouragement.

References Cited

Abler, W. L. 1992. The serrated teeth of tyrannosaurid dinosaurs, and biting structures in other animals. *Paleobiology* 18: 161–183.

Bakker, R. T., M. Williams, and P. J. Currie. 1988. *Nanotyrannus,* a new genus of pygmy tyrannosaur, from the latest Cretaceous of Montana. *Hunteria* 1 (5): 1–30.

Baszio, S. 1997. Investigations on Canadian dinosaurs: Systematic palaeontology of isolated dinosaur teeth from the Latest Cretaceous of southern Alberta, Canada. *Courier Forschungsinstitut Senckenberg* 196: 33–77.

Carlson, S. J. 1990. Vertebrate dental structures. In J. G Carter (ed.), *Skeletal Biomineralization: Patterns, Processes and Evolutionary Trends,* vol. 1, pp. 531–556. New York: Van Nostrand Reinhold.

Carlson, S. J., and W. S. Bartels. 1986. Ultrastructural complexity in reptilian tooth enamel. *Geological Society of America Abstracts with Programs* 18: 558.

Carpenter, K. 1982. Baby dinosaurs from the Late Cretaceous Lance and Hell Creek Formations and a description of a new species of theropod. *Contributions to Geology, University of Wyoming* 20: 123–134.

Cooper, J. S., and D. F. G. Poole. 1973. The dentition and dental tissues of the agamid lizard, *Uromastyx. Journal of Zoology, London* 169: 85–100.

Currie, P. J. 1987. Theropods of the Judith River Formation of Dinosaur Provincial Park, Alberta, Canada. In P. J. Currie and E. H. Koster (eds.), *Fourth Symposium on Mesozoic Terrestrial Ecosystems, Short Papers,* pp. 52–60. Occasional Papers of the Royal Tyrrell Museum of Palaeontology, no. 3. Drumheller, Alberta: Royal Tyrrell Museum of Palaeontology.

Currie, P. J., J. K. Rigby Jr., and R. E. Sloan. 1990. Theropod teeth from the Judith River Formation of Southern Alberta, Canada. In K. Carpenter and P. J. Currie (eds.), *Dinosaur Systematics: Approaches and Perspectives,* pp. 107–125. Cambridge: Cambridge University Press.

Erler, G. 1935. Uber den Zahnschmelz der Krokodile. *Zeitschrift für Zellforschung und Milroskopische Anatomine* 23: 589–606.

Farlow, J. O., D. L. Brinkman, W. L. Abler, and P. J. Currie. 1991. Size, shape, and serration density of theropod dinosaur lateral teeth. *Modern Geology* 16: 161–198.

Paul, G. S. 1988. The small predatory dinosaurs of the mid-Mesozoic: The horned theropods of the Morrison and Great Oolite—*Ornitholestes* and *Proceratosaurus*—and the sickle claw theropods of the Cloverly, Djadokhta and Judith River—*Deinonychus, Velociraptor* and *Saurornitholestes. Hunteria* 2: 1–9.

Poole, D. F. 1956. The structure of the teeth of some mammal-like reptiles. *Quarterly Journal Microscopical Science* 97: 303–312.

Russell, D. A. 1970. *Tyrannosaurs from the Late Cretaceous of Western Canada.* National Museums of Canada, National Museum of Natural Sciences, Publications in Palaeontology, no. 1. Ottawa.

Sander, M. P. 1999. The microstructure of reptilian tooth enamel: Terminology, function, and phylogeny. *Münchner Geowissenschaftliche Abhandlungen, Reihe* A, 38: 1–102.

Schmidt, W. J. 1947. Uber den Zahnschmelz fossiler Crocodilier. *Zeitschrift für Zellforschung* 34: 55–77.

Stokosa, K. 2000. Morphology and enamel microstructure of isolated theropod teeth from the late Cretaceous Fox Hills, Hell Creek, and Lance formations. Master's thesis, South Dakota School of Mines and Technology, Rapid City.

10. Bite Me

Biomechanical Models of Theropod Mandibles and Implications for Feeding Behavior

François Therrien, Donald M. Henderson, and Christopher B. Ruff

Abstract

A biomechanical approach is used to study feeding behavior in non-avian theropods. Mandibles can be modeled as beams undergoing bending loads during food ingestion. The bite force applied at any given point along the mandible should be proportional to the external dimensions of the mandibular ramus at that location. Thus, patterns of variation in these dimensions reflect the adaptation of the jaw to specific loads, which are related to the method of killing prey. These beam models were compared to those of the extant varanids *Varanus komodoensis* (Komodo dragon, an ambush predator with a slashing bite) and *Varanus niloticus* (Nile monitor, molluscivorous) to gain insight into the feeding behavior of theropods.

On the basis of our results, we identified five distinct theropod feeding categories: (1) the allosauroid "*Antrodemus valens*" and the abelisaurids *Majungatholus atopus* and *Carnotaurus sastrei* share the mandibular properties of the Komodo dragon, and these shared properties, combined with the similarity between the craniodental morphology of "*Antrodemus*" and *Majungatholus* and that of the varanid, suggest that they were probably large-prey hunters delivering slashing bites; (2) dromaeosaurids have mandibular properties reminiscent of those of *V. komodoensis* for slashing bites, but differences can be identified between *Dromaeosaurus* and velociraptorines, the former having a stronger bite than the latter, suggesting that it possibly relied more on its jaws to capture and kill prey; (3) *Suchomimus tenerensis* and *Dilophosaurus wetherilli* exhibit mandibular adaptations related to the capture of prey smaller than themselves, the former probably practicing a bite-and-hold strategy and the latter finishing its prey with slashing bites; (4) *Ceratosaurus nasicornis, Allosaurus fragilis, Acrocanthosaurus atokensis,* and *Giganotosaurus carolinii* demonstrate adaptations of the anterior extremity of the mandible for capturing prey and delivering powerful bites to bring down prey or deliver the final blow; and (5) tyrannosaurids, unlike any other theropods, exhibit mandibular adaptations to resist high torsional stresses at the anterior of the mandible induced during prey capture, bone crushing, or both.

The mandibular models were also used to infer relative bite force in theropods. Velociraptorines appear to have had a maximum bite force similar to that of *V. komodoensis*, whereas that of *Dromaeosaurus* was three times as great. *Suchomimus, Allosaurus,* "*Antrodemus,*" and *Ceratosaurus* were capable of exerting maximum bite forces as great as *A. mississippiensis*, and those of abelisaurids and *Albertosaurus* were twice as powerful. Among the largest theropods, *Acrocanthosaurus* and *Giganotosaurus* were surpassed by *Daspletosaurus* and *Tyrannosaurus*. The high estimates obtained for tyrannosaurids are consistent with previously published values that suggest bone-cracking abilities.

Growth series for *Allosaurus fragilis, Albertosaurus sarcophagus, Gorgosaurus libratus,* and *Tyrannosaurus rex* were also studied in order to determine whether mandibular properties changed during ontogeny. Significant changes were observed in *Allosaurus,* especially in bending rigidity, indicating that juveniles did not feed the same way as adults; juveniles probably delivered simple slashing bites. Unfortunately, the question of parental care in *Allosaurus* cannot be resolved in light of our results. The mandibular properties of tyrannosaurids were not found to vary significantly during ontogeny, other than in terms of bite force. This finding strongly suggests that juveniles were apt predators, capable of subduing their own prey rather than relying on carrion or parental care to survive.

Introduction

Determining the feeding behavior of non-avian theropods has always been a nebulous topic of dinosaur paleoecology. A few specimens, owing to exceptional preservation, give insight into predator-prey interactions, such as the "fighting dinosaurs" *Velociraptor* and *Protoceratops* (for a review, see Carpenter 1998) and stomach contents for *Compsognathus* (Ostrom 1978), *Sinosauropteryx* (Chen et al. 1998), *Coelophysis* (Colbert 1989), *Baryonyx* (Charig and Milner 1997), and *Daspletosaurus* (Varricchio 2001). The potential prey of other theropods has generally been inferred on the basis of contemporaneous faunal assemblages. Previous attempts to reconstruct theropod feeding behavior were based principally on features such as skull and tooth morphology, dental wear patterns, tooth-marked bones, and bone fragments in coprolites (e.g., Farlow 1976; Paul 1987, 1988; Farlow et al. 1991; Fiorillo 1991; Abler 1992; Erickson and Olson 1996; Erickson et al. 1996; Bakker 1998; Carpenter 1998; Chin et al. 1998; Chure et al. 1998; Henderson 1998; Jacobsen 1998, 2001; Rogers et al. 2003) or on comparisons with modern mammalian and reptilian predators, particularly the Komodo dragon (Auffenberg 1981; Paul 1987, 1988; Molnar and Farlow 1990). However, until recently, little attention had been paid to the diverse mandibular adaptations seen in theropods and how these could reflect feeding behavior (Van Valkenburgh and Molnar 2002).

Although biomechanical models have been constructed for extant and extinct synapsids (e.g., Van Valkenburgh and Ruff 1987; Biknevicius and Ruff 1992a; Crompton 1995; Greaves 1995; Biknevicius and Van Valkenburgh 1996; and references therein), crocodilians (e.g., Molnar 1969; Busbey 1989, 1995; Drongelen and Dullemeijer 1982; Sinclair and Alexander 1987; Therrien and Ruff, in prep.), some ornithischian dinosaurs (Ostrom 1961, 1964; Weishampel 1983, 1984; Norman 1984), and the gigantic bird *Diatryma* (Witmer and Rose 1991), only recently have theropod jaw biomechanics been investigated in detail (e.g., Mazzetta et al. 1998; Molnar 1998; Rayfield et al. 2001). We expand upon this work, taking a biomechanical approach to analyze feeding behavior in different groups of non-avian theropods.

Beam theory can be used to evaluate mandibles in terms of their ability to resist bending loads generated during biting. Modern mammals with distinct killing techniques and feeding habits are known to differ in the bending force profiles along the mandibular ramus, and these differences reflect the specialization of the jaw to resist forces associated with specific feeding behaviors (Biknevicius and Ruff 1992a; Therrien, in prep.). The construction of beam models for theropod mandibles, followed by their comparison with similar biomechanical models for extant varanids (*Varanus komodoensis* and *Varanus niloticus*) and crocodilians (Therrien and Ruff, in prep.) of known feeding behaviors, can give insight into the hunting techniques adopted by these animals. The comparison reveals that, although some theropods share similarities with *V. komodoensis,* a great diversity of mandibular adaptation exists among them. These differences in mandibular biomechanical properties can be attributed to possible differences in feeding behavior and, tentatively, ecological niches.

Institutional Abbreviations. AMNH, American Museum of Natural History, New York; BHI, Black Hills Institute of Geological Research, Inc., Hill City, S.D.; BYU, Brigham Young University, Provo, Utah; CM, Carnegie Museum of Natural History, Pittsburgh, Penn.; CMNH, Cleveland Museum of Natural History, Cleveland, Ohio; FMNH, Field Museum of Natural History, Chicago; LACM, Los Angeles County Museum, Los Angeles; MACN-CH, Museo Argentino de Ciencias Naturales "Bernardino Rivadavia," Buenos Aires; MNN, Musée National de Niger, Niamey, Republic of Niger; MOR, Museum of the Rockies, Bozeman, Mont.; MUCPv-CH, Museo de la Universidad Nacional del Comahue, El Chocón collection, Neuquén, Argentina; NCSM, North Carolina State Museum of Natural Sciences, Raleigh; NMC, Canadian Museum of Nature, Ottawa; ROM, Royal Ontario Museum, Toronto; TMP, Royal Tyrrell Museum of Palaeontology, Drumheller, Alberta; UCMP, University of California Museum of Paleontology, Berkeley; UMNH, Utah Museum of Natural History, Salt Lake City; USNM, United States National Museum of Natural History, Smithsonian Institution, Washington, D.C.; UUVP, University of Utah, Vertebrate Paleontology collection, Salt Lake City; YPM, Yale Peabody Museum of Natural History, New Haven, Conn.; and YPM-PU, Yale Peabody Museum—Princeton University Collection.

Beam Theory and Analytical Assumptions

Beam Theory. The external dimensions of the mandible can be used to model it as a simplified solid elliptical beam undergoing bending loads during food ingestion (Biknevicius and Ruff 1992a). The occurrence of teeth within alveoli is ignored because mandibles are better approximated by closed-section models than by open-section models, that is, models incorporating gaps for teeth (Daegling et al. 1992; Daegling and Hylander 1998). The second moment of area, I, is a measure of the distribution of bone around a given axis and can be expressed as

$$I = \int y^2 dA, \quad (1)$$

where dA is an elemental strip of area and y is the distance of the elemental strip from the centroid, or neutral axis, of the mandible (here assumed to be through the center of the mandible; Biknevicius and Ruff 1992a). For a solid ellipse, the formula can be simplified to

$Ix = \pi ba^3/4$, distribution of bone about the labiolingual axis (in cm^4);
$Iy = \pi ab^3/4$, distribution of bone about the dorsoventral axis (in cm^4);

where a represents the dorsoventral radius and b represents the labiolingual radius (Biewener 1992, fig. 1).

The section modulus, Z, a measure of strength in bending, is determined by the cortical bone cross-sectional area and the distribution of bone around the neutral axis of the mandible. It can be expressed as

$$Z = I/y \quad (2)$$

where I is the second moment of area and y the distance from the centroid (or neutral axis) to the outer edge of the bone in the plane of bending or radius of the corpus (Biknevicius and Ruff 1992a). Consequently, the section modulus will differ according to the orientation being considered (Fig. 10.1), and two formulae are thus obtained for a solid ellipse model:

$Zx = Ix/a$, bending strength in the dorsoventral plane (or about the labiolingual axis, in cm^3);

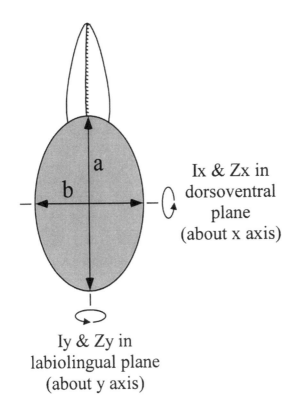

a

b

Ix & Zx in
dorsoventral
plane
(about x axis)

Iy & Zy in
labiolingual plane
(about y axis)

$Zy = Iy/b$, bending strength in the labiolingual plane (or about the dorsoventral axis, in cm³).

Figure 10.1. Cross-section of a hypothetical mandible demonstrating dimensions measured and orientation of cross-sectional properties evaluated.

The maximum bending stress of a structure (σ_B), ignoring differences in material properties, can be expressed as

$$\sigma_B = My/I, (3)$$

where M is the bending moment, y is the radius orthogonal to the axis investigated, and I is the second moment of area about the axis being investigated. Since the bending stress is a material property (of bone, in this case) and compact bone material properties have been shown to be relatively constant in vertebrates (McAllister and Moyle 1983; Cowin 1989), if we assume a constancy of safety factor in the vertebrate mandible (i.e., that the normal bending strain that mandibles undergo during feeding is a constant fraction of the maximum bending strain they can resist before failing), we can assume a constant value of σ_B for comparing the beam models. Consequently, the following equation is obtained:

$$M = I/y = Z. (4)$$

The flexure formula (Timoshenko and Gere 1972) states that the bending moment is equal to the product of the applied force, F, and the moment arm length, L:

$$M = FL (5)$$

Since $M = Z$ (see equation 4), the following transformations can be made:

$$Z = FL \text{(6)}$$
$$F = Z/L \text{ (in cm}^2\text{)(7)}$$

Thus, the maximum force applied at any given point along the mandible should be proportional to the ratio of the section modulus of the mandibular corpus to the distance from the articular fossa to that point on the mandible.

For animals with generally similar jaw muscle architectures (i.e., carnivorous predators), the maximum force is, of course, influenced by the body weight of the animal (the larger the animal, the more powerful its bite; see Meers 2002), but the force also reflects the adaptation of the jaw to a specific load, which is related to the method of killing prey. Therefore, a comparison of Z/L values between taxa will reveal variations in the magnitude of applied force at different locations on the mandible, and the Zx/Zy ratio will determine the relative mandibular strength in different planes, or the overall mandibular shape. Hylander (1979) has shown that corpora that are deeper than they are wide are adapted to resist significant sagittal bending, that corpora that are deep and wide are adapted to sustain important sagittal bending and torsional loads, and that corpora that are wider than they are deep are adapted to resist important torsional stresses. In other words, since Zx/Zy is proportional to the ratio of the dorsoventral and mediolateral radii (a/b in Fig. 10.1), a ratio greater than 1 will indicate an adaptation toward dorsoventral loads (mandibular ramus deeper than it is wide), a ratio lower than 1 an adaptation toward labiolingual loads (mandibular ramus wider than it is deep), and a ratio of 1 an equal adaptation toward dorsoventral and labiolingual loads (deep and wide mandibular ramus). Dorsoventral loads along the mandible are primarily related to the bite force exerted upon the prey in the sagittal plane. In contrast, the labiolingual force is interpreted as an adaptation against stresses induced while biting prey, either due to struggling motions or to other transverse or torsional loads produced by objects encountered during jaw closing (e.g., bone, shell), since labiolingual loads related to muscular activity, mediolateral jaw movements, or both are minimal in non-mammals (e.g., Drongelen and Dullemeijer 1982; Sinclair and Alexander 1987; Busbey 1995; Cleuren et al. 1995). Comparison of these mandibular biomechanical properties has the potential to reveal similarities and differences in feeding behaviors between predatory animals (Biknevicius and Ruff 1992a).

For the beam models to be valid, three assumptions are made: (1) the mandible is principally loaded in bending; (2) the mandible functions as a single unit of homogeneous composition without dissipation of stress, that is, along suture lines and at the intramandibular joint; and (3) the mandible consists of a perfect ellipse of solid bone. First, the mandibular ramus is known to behave as a straight beam loaded in bending; however, the symphyseal region is more complex. Hylander (1981, 1984, 1985) has shown that, in primates, dorsoventral shear, bending, and torsional stresses occur in the mandibular symphysis, although bending is dominant. The contraction of the adductor muscu-lature, which attaches on the lateral side of the mandible in mammals, tends to twist the hemi-mandibles about their axes, creating tensile stresses on the inferior border of the symphysis and compressive stresses on the superior border. However, diapsids differ in having their ad-ductor musculature inserting on both the medial and the lateral side of the mandible, although the bulk of it inserts on the lateral side in squamates and on the medial side in non-avian archosaurs (e.g., Fraz-zetta 1962; Iordansky 1964, 1970; Haas 1973; Schumacher 1973; Busbey 1989). In theory, the non-avian archosaur muscle architecture with a predominant medial pull should result mostly in bending stresses at the symphysis of opposite polarity to the stresses observed in pri-mates and squamates (tension on the superior border and compression on the inferior border), but this cannot be confirmed, owing to the dearth of strain gauge analysis for crocodilian mandibular symphyses. Therefore, we assume that bending stresses remain the significant load-ing pattern in the diapsid symphysis.

The second assumption is one of a theoretical nature. Varanids and non-avian theropods reportedly have an intramandibular joint (IMJ) between the dentary-splenial and postdentary bones that allows an increase in gape (e.g., Frazzetta 1962; Paul 1987, 1988; Bakker et al. 1988; Rieppel and Zaher 2000; Bluhm 2002), although primitive theropods and tyrannosaurids may lack it (Sereno et al. 1993; Hurum and Currie 2000; Bluhm 2002). In varanids, the IMJ is a passive joint that can move any time the animal bites on a hard object, but the range of motion is controlled or limited by the action of the intramandibu-laris anterior muscle (Frazzetta 1962; Bluhm 2002). The action of this muscle would prevent involuntary flexion of the IMJ while biting prey, which would result in a loss of bite force. Consequently, the varanid mandible can, theoretically, function as a solid unit without dissipa-tion of stress during biting if the intramandibularis anterior muscle prevents movement at the IMJ; we will assume that a similar stabiliz-ing muscular activity was present in theropods. Varanids and theropods also possess a loose mandibular symphysis that allows, to a certain degree, independent movement of one hemi-mandible relative to the other, the reptilian equivalent of the Class I symphysis of Scapino (1981). However, in carnivorans with unfused mandibles, the efficient transmission of vertical bite force between hemi-mandibles with mini-mal force dissipation has been demonstrated (Dessem 1985). There-fore, the second assumption for the beam models appears to be at least theoretically valid.

The third assumption has its origin in the methods used. By using external measurements to model a mandible as a solid elliptical beam, we ignore both the variation in the internal structure of the mandible (i.e., thickness of cortical bone, hollow mandible in the postdentary region versus solid mandible in the dentary region; see Molnar 1998; Brochu and Ketcham 2002) and its deviation from a perfect elliptical shape. Fortunately, comparison of results obtained from computed tomography (CT) scans and external dimensions in *Alligator mississip-piensis* (Therrien and Ruff, in prep.) has revealed that a simplified beam model based on external measurements can be used to study accurately the *relative* biomechanical changes along the mandible, although the model somewhat overestimates their *absolute* values when the mandible is not truly solid (also see Biknevicius and Ruff 1992b). However, since the main interest here lies in making interspecific comparisons of changes in dorsoventral, labiolingual, and relative mandibular force along the mandible rather than providing bite force estimates in absolute values (in Newtons), the assumption of a solid mandible will not invalidate the beam models presented here. Furthermore, because the mandibles of theropods can presumably be considered to be truly solid (Molnar 1998; Van Valkenburgh and Molnar 2002), the cross-sectional properties determined from the simplified beam models should be close to the real values. More-accurate biomechanical properties could be determined by conducting CT scans of mandibles, but, especially in the case of large or mounted specimens, specimens filled with matrix, or specimens located at institutions without CT facilities, this approach is problematic, time consuming, and costly. As scanning techniques improve and CT scans of specimens become more common, future work will allow more accurate determination of the biomechanical properties of theropod mandibles leading to a better understanding of their feeding adaptations.

Materials and Methods

Samples. External mandibular dimensions (width and depth) of fourteen theropod taxa and two varanid taxa were determined using digital calipers and elephant calipers (data available from Therrien). Among extant taxa, only wild-caught individuals were measured (except for a single zoo specimen of *V. komodoensis*), and the largest specimens of each taxon were preferentially selected to minimize the allometric difference between theropods and extant animals. Among extinct taxa, the most complete and least deformed specimens were selected to insure accurate representation of the original mandibular dimensions. When original specimens were unavailable, museum-quality casts were measured. More details on specimen selection and preservation condition, as well as taxonomic allocations, are given below with the results for each group.

The Komodo dragon, or ora (*Varanus komodoensis*), being the largest terrestrial predatory diapsid alive today, has often been used as a modern analogue to reconstruct theropod feeding behavior (see Auffenberg 1981; Molnar and Farlow 1990). Despite the fact that the varanid mandibular architecture is quite distinct from that of archosaurs, similar feeding behaviors should impose similar loads and stresses along the mandible, stresses to which the mandible has to respond by depositing or resorbing bone, thus affecting its cross-sectional properties. Consequently, if theropods fed in a manner akin to *V. komodoensis*, the mandibular models should show similar biomechanical adaptations. A single zoo specimen, typically fed with small (rats, mice) or dead prey, was also measured in order to compare the effect of diet differences on biomechanical properties of the mandible.

Recently, arguments have been made, on the basis of dental structure and coprolites, that tyrannosaurids may have been able to process bone prior to ingestion (Bakker et al. 1988; Farlow et al. 1991; Erickson and Olson 1996; Erickson et al. 1996; Chin et al. 1998). This bone-crushing ability would be a derived characteristic of tyrannosaurids; other theropods typically avoid bones while feeding (Fiorillo 1991; Chure et al. 1998; Jacobsen 1998). If bone crushing occurred, the tyrannosaurid models should exhibit unique biomechanical properties, unlike those seen in other theropods, which could be interpreted as adaptations to sustain the high stresses induced during bone crushing. But how does one recognize crushing abilities in a beam model?

The adult Nile monitor (*Varanus niloticus*) is a molluscivorous lizard with dental and mandibular adaptations to crush snails, mussels, and crabs (Rieppel and Labhardt 1979; Losos and Greene 1988). Although smaller than *V. komodoensis,* the general mandibular architecture of *V. niloticus* is similar and, consequently, provides an excellent opportunity to observe variations in biomechanical properties of the mandible related to differences in feeding behavior. In turn, these variations should give insight into typical mandibular adaptations related to durophagy and provide a model against which tyrannosaurid models can be compared to test earlier claims of bone-crushing abilities.

Methods. Mandibles were oriented in an approximate *in vivo* position by aligning the symphyseal surface along the vertical plane prior to measurement (Fig. 10.2). Mandibular depth is defined as the dorsoventral dimension of the mandible and was determined with the caliper blades oriented mediolaterally. Mandibular width is defined as the labiolingual dimension of the mandible and was determined with the caliper blades oriented dorsoventrally. Owing to the curved mandibular architecture at the symphysis, the mandibular depth and width at the second tooth alveolus were not measured in the same plane (see Fig. 10.2A); the mandibular depth determined in the mediolateral plane is generally less than the depth determined in the labiolingual plane (as the former does not include the entire vertical dimension of the symphysis). Consequently, the biomechanical properties calculated for this landmark will be lower than their theoretical values, but measurement consistency among theropods should prevent any problems.

In order to construct and compare beam models for the different taxa, measurements were made at analogous and easily identifiable landmarks along the mandible: (1) at the second tooth alveolus; (2) at middentary (halfway between the anterior extremity of the mandibular symphysis and the superior dentary suture with the surangular); (3) at the upper dentary suture with the surangular; (4) at the lower dentary suture with the angular; and (5) at the location of maximum mandibular depth (Fig. 10.2). However, the postdentary portion of the mandible proved to be biomechanically problematic to interpret for two reasons: (1) since it is the attachment site of powerful muscles, this portion of the mandible will reflect the localized stresses induced during muscle contraction; and (2) at least in archosaurs, the postdentary region is a hollow ellipse with incomplete walls due to local openings (e.g., mandibular fenestra, intermandibularis foramina, mandibular adductor

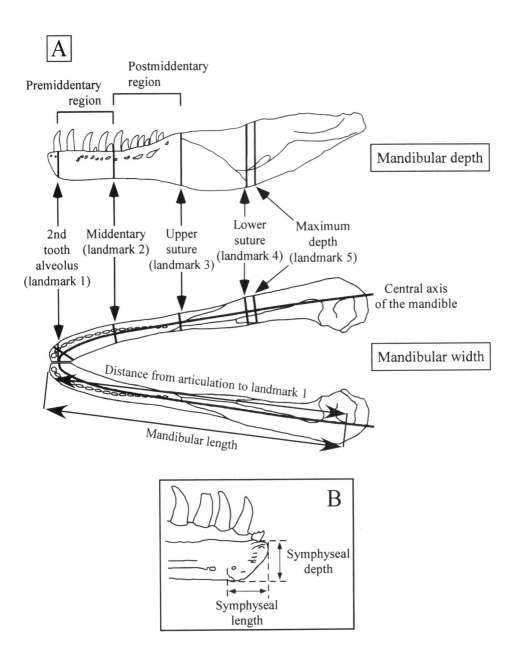

A

Premiddentary region

Postmiddentary region

Mandibular depth

2nd tooth alveolus (landmark 1)

Middentary (landmark 2)

Upper suture (landmark 3)

Lower suture (landmark 4)

Maximum depth (landmark 5)

Central axis of the mandible

Mandibular width

Distance from articulation to landmark 1

Mandibular length

B

Symphyseal depth

Symphyseal length

fossa), which assuredly affect stress distribution. For these two reasons, a solid beam model cannot predict accurately the force patterns occurring in the postdentary portion of the mandible. Additionally, the relative positions of the fourth and fifth landmarks vary among theropods, thus diminishing their value for comparative purposes. Nevertheless, since the biomechanical profiles of the postdentary region can contribute information about the state of preservation of the specimen (see *Dilophosaurus* below), these results are also presented.

In addition to the external mandibular dimensions, the distance from each landmark to the mandibular articulation (middle of articular fossa) was also determined in order to calculate the force applied along the mandible (moment arm length, see equation 7). Mandibular length, here defined as the distance separating the articular fossa and the anterior extremity of the mandible (excluding any retroarticular element), was also determined for each specimen.

The tooth-bearing portion of the mandible is the main mandibular segment involved in feeding. It has been subdivided into two sections: (1) a *pre-middentary* section, located between the second tooth alveolus (landmark 1) and the middentary (landmark 2); and (2) a *post-middentary* section, located between the middentary (landmark 2) and the superior dentary suture with the surangular (landmark 3) (Fig. 10.2). By dividing the tooth row into anterior and posterior halves and studying the force variation between landmarks, it will be possible to determine the relative changes in biomechanical properties along the tooth row, which should reflect mandibular adaptations toward specific loads interpretable in terms of feeding behavior.

Results and Discussion

Comparison of Mandibular Models

Measurements were averaged for each taxon prior to biomechanical calculation, and results are presented, where relevant, as semi-log plots to reduce the effect of allometry (see Smith 1984, 1993). Furthermore, each graph has been standardized for a unitary jaw length (the abscissa representing the relative position of a landmark along the jaw) to facilitate comparisons between homologous locations on the mandible.

Varanids. The mandible of *Varanus komodoensis* is relatively shallow, gradually decreases in depth anterior to the intramandibular joint, and is slightly curved upward at the anterior extremity. Because the hemi-mandibles rotate at the symphysis during jaw opening (Frazzetta 1962), this curvature allows the teeth to be oriented vertically at wide gape. In *Varanus niloticus,* the mandible has undergone a few modifications, such as a shortening of the tooth row and a deepening of the posterior portion of the dentary, to gain a biomechanical advantage for crushing hard shells (Rieppel and Labhardt 1979). The contact between the two hemi-mandibles is relatively equidimensional (dorsoventral dimensions nearly equal to anteroposterior dimensions) in *V. niloticus* (Fig. 10.3), whereas it appears slightly deeper than it is long in *V. komodoensis* (Therrien, pers. obs. based on a photograph).

Mandibular Force Profiles. In both varanid species, the maximum applied force in the dorsoventral plane (Zx/L) decreases anteriorly at a constant rate along the mandible, the pre-middentary slope being nearly equal to the post-middentary slope (Fig. 10.4). As such, the varanid mandible behaves as a simple lever, regardless of feeding behavior. Differences in the mandibular force profiles of *V. komodoensis* and *V. niloticus* are observed in the absolute force values that their respective mandibles can withstand: the Komodo dragon has a more powerful

Figure 10.2. (opposite pageg) (A) Landmarks on a theropod jaw. Measurements were taken in the dorsoventral and labiolingual planes and perpendicular to the central axis of the mandible, except at the second alveolus (see explanations in text). The dentary has been divided in two segments around landmark 2: a pre-middentary region and a post-middentary region. The properties of the tooth-bearing dentary should reflect adaptations related to the feeding behavior of the animal, whereas the postdentary section should reflect adaptations for the jaw musculature. "Mandibular length" represents the distance between the articular fossa and the anterior extremity of the mandible, excluding any retroarticular element. (B) Symphyseal dimensions. "Symphyseal length" refers to the dimension of the symphyseal surface parallel to the long axis of the mandible, and "symphyseal depth" refers to the dimension perpendicular to the long axis of the mandible. Note that these dimensions do not correspond to the longest and shortest diameters of the mandibular symphysis.

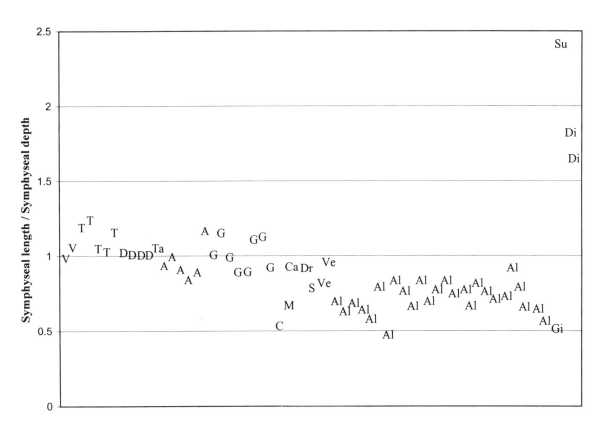

Figure 10.3. Symphyseal morphology in theropods. The Nile monitor (V), tyrannosaurids (Tyrannosaurus rex, T; Albertosaurus sarcophagus, A; Gorgosaurus libratus, G; Tarbosaurus bataar, Ta), dromaeosaurids (Dromaeosaurus albertensis, Dr; Saurornitholestes langstoni, S; Velociraptor mongoliensis, Ve), and Carnotaurus sastrei (Ca) have nearly equidimensional symphyses; allosauroids (Allosaurus fragilis, Al; Giganotosaurus carolinii, Gi), Ceratosaurus dentisulcatus (C), and Majungatholus atopus (M) have symphyses that are deeper than they are long; Suchomimus tenerensis (Su) and Dilophosaurus wetherilli (Di) have symphyses that are much longer than they are deep.

bite than the Nile monitor (greater Zx/L values). Notice that the force profiles for the zoo and wild-caught specimens of *V. komodoensis* are nearly identical.

In the labiolingual plane, the maximum applied force (Zy/L) is, once again, decreasing at a nearly constant rate anteriorly along the mandible. In *V. komodoensis*, the pre-middentary slope is slightly less negative than the post-middentary, whereas the opposite is true for *V. niloticus* (Fig. 10.4). The mandible of wild-caught Komodo dragons is slightly stronger than that of both the zoo specimen and *V. niloticus*.

It is in terms of relative mandibular strength (or overall mandibular shape, Zx/Zy) that *V. komodoensis* and *V. niloticus* differ the most (Fig. 10.4). For the wild-caught Komodo dragon, the Zx/Zy ratio produces a convex curve with a maximum value at the middentary. The Zx/Zy values are higher than 2.00 in the post-middentary section of the mandible and decrease to 1.63 at the second tooth alveolus. In comparison, the Zx/Zy values for the zoo specimen of *V. komodoensis* are higher than those of the wild-caught specimens, being around 2.60 in the posterior half of the dentary and reaching a value of 2.00 at the second alveolus. For the Nile monitor, the Zx/Zy ratio produces a concave curve of low amplitude with a minimum at the middentary. The Zx/Zy values of *V. niloticus* are relatively low compared to those of *V. komodoensis*, ranging between 0.97 at the second tooth alveolus and 1.58 at the upper suture of the dentary.

Interpretation. The mandibular force profiles reveal that the rela-

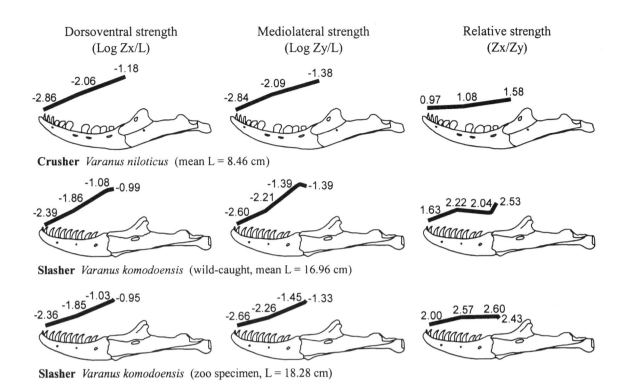

Dorsoventral strength (Log Zx/L)	Mediolateral strength (Log Zy/L)	Relative strength (Zx/Zy)

Crusher *Varanus niloticus* (mean L = 8.46 cm)

Slasher *Varanus komodoensis* (wild-caught, mean L = 16.96 cm)

Slasher *Varanus komodoensis* (zoo specimen, L = 18.28 cm)

tive maximum force distribution along the mandible is similar in the two varanid species. The fact that the dorsoventral and labiolingual force profiles decrease at a nearly linear rate along the tooth row, the pre- and post-middentary slopes being almost equal, indicates that the varanid mandible behaves as a simple lever. This is consistent with animals whose jaws are primarily used not as a means of prehension to capture prey but rather for delivering rapid bites and ingesting food.

The distinct feeding behaviors of the two varanid species can be recognized in their relative mandibular strength profiles (Zx/Zy). *V. komodoensis* has post-middentary values greater than 2.00, indicating that the mandible in this region is at least twice as deep as it is wide and, thus, that dorsoventral loads are twice as important as the labiolingual loads, presumably in the form of sagittal bending. The mandible becomes slightly rounder at the second alveolus, where the Zx/Zy value decreases to 1.63, suggesting that labiolingual loads are more important toward the mandibular symphysis than they are in the posterior portion of the tooth row. In light of these facts, we interpret the post-middentary region of the mandible of *V. komodoensis* as being adapted for slicing and the pre-middentary region as slightly better adapted toward labiolingual loads and torsion that may occur during biting, prey handling, and food acquisition. The dentition supports this interpretation—the anterior-most teeth (usually first through second or third on the dentary and first through fifth on the premaxillary) are conical, whereas the others are laterally compressed and serrated (Therrien, pers. obs.).

Figure 10.4. Mandibular properties of varanids. From left to right: landmark 1 (second tooth alveolus), landmark 2 (mid-dentary), and landmark 3 (upper dentary suture); if present, the next point represents landmark 5 (maximum depth). Mandibular length L represents the distance between the articular fossa and the anterior extremity of the mandible.

These patterns are consistent with the ambush-style feeding behavior described for the Komodo dragon (Auffenberg 1981). Specialized for hunting large mammalian prey (Rieppel 1979; Auffenberg 1981; Losos and Greene 1988), *V. komodoensis* generally delivers fast slashing bites to wound the prey and then waits for the animal to die from the septic wound, although Auffenberg (1981, p. 245–249) relates instances where individuals occasionally bit and held onto prey relatively smaller than themselves in order to subdue it. The giant varanid then dismembers its prey by biting and forcefully moving its head fore and aft, using its laterally compressed, serrated teeth to cut large chunks of flesh (Auffenberg 1981). This behavior is quite different from the feeding behavior reported in some conical-toothed crocodilians, which grab a fold of flesh with the anterior extremity of their jaws and either forcefully pull or roll on themselves in order to tear the flesh away (e.g., Cott 1961; Meyer 1984; Taylor 1987).

Interestingly, the differences in mandibular force profiles between the wild-caught and zoo specimens are statistically different only in the Zx/Zy profile at the second alveolus (*t*-test, $p < 0.036$), where it reaches a value of 2.00 in the zoo specimen and 1.63 in the wild-caught specimens. The results mean that mandibular adaptations for food processing (along the tooth row) do not vary between wild individuals and those kept in captivity, but that the anterior of the mandible does differ between individuals that have to catch and subdue live prey (wild animals) and those being fed small or dead prey. Animals feeding on relatively smaller or dead prey will have significantly higher Zx/Zy values at the second alveolus, indicating a better dorsoventrally buttressed mandible because it is not required to be adapted for the (transverse or torsional) loads exerted by struggling prey and needs only to be adapted for food processing. In contrast, *V. niloticus* has Zx/Zy values ranging from 1.58 to 1.08 in the post-middentary region and a Zx/Zy value of 0.97 at the second alveolus, indicating that the mandible is rounder because loads exerted on the mandible are not as well constrained in the dorsoventral plane as in *V. komodoensis*.

In his work on primates, Hylander (1979) concluded that mandibular corpora that are as deep as they are thick ($Zx/Zy \sim 1.00$) were adapted to resist relatively large dorsoventral bending loads as well as relatively large twisting moments around their long axes induced by powerful unilateral chewing. Furthermore, Hylander (1981) described how torsion can be induced in the mandible if the bite point is not situated exactly over the long axis of the mandibular ramus and how the twisting moment increases as the bite point migrates posteriorly along the tooth row. Accordingly, torsional moments have been reported to occur in the mandibular symphysis of primates during chewing (Hylander 1981, 1984, 1985). Given that *V. niloticus* has a non-occluding dentition and crushes shells with its posterior teeth, it is not surprising that the mandible, in addition to the other adaptations toward a molluscivorous diet described by Rieppel and Labhardt (1979), adopts a shape to resist the high torsional moment associated with durophagy. In light of this adaptation, the Zx/Zy value close to 1.00 at the second tooth alveolus in *V. niloticus* can be interpreted as an adaptation to resisting torsion while crushing hard shells with the posterior teeth rather than to resisting struggling prey (snails and mussels should offer very little resistance to capture!).

In summary, the beam models allow us to distinguish different feeding behaviors in extant varanids. *Varanus komodoensis,* an ambush predator that uses its jaws to deliver rapid, slashing bites has a mandible that behaves as a simple lever in terms of dorsoventral and labiolingual force profiles. The mandible is also strongly dorsoventrally buttressed along the tooth row but is slightly better adapted, near the mandibular symphysis, to sustain labiolingual loads that may be exerted by struggling prey or during food ingestion. In contrast, *Varanus niloticus,* an animal that traded a rapid slashing bite for a slower but more powerful crushing bite, also preserves a simple-lever mandible as in *V. komodoensis;* however, the mandible is nearly as strong in the labiolingual as in the dorsoventral plane, especially at the second alveolus, and that strength provides great resistance against torsional moments generated while crushing hard shells. Thus, mandibular properties, especially at the anterior extremity of the mandible, do reflect changes in behavior or in diet, a fact that will prove useful in interpreting theropod feeding behavior below.

Theropods

Several non-avian theropod taxa were studied to gain insight into the diversity of feeding behaviors among Mesozoic predators (Fig. 10.5). Although they are generally similar, the mandibular strength profiles of theropods exhibit subtle differences between and within taxonomic groups, suggesting differences in feeding behaviors. These groups will be addressed in phylogenetic sequence.

Ceratosauria and Dilophosaurus. The ceratosaur taxa studied here consist of *Ceratosaurus nasicornis* (USNM 4735) and the abelisaurids *Carnotaurus sastrei* (cast of MACN-CH 894) and *Majungatholus atopus* (cast of FMNH PR 2100). Although its exact phylogenetic position is still subject to debate, the Early Jurassic *Dilophosaurus wetherilli* (UCMP 77270) is also discussed in this section.

The specimens of *Dilophosaurus* and *Ceratosaurus* are not perfectly preserved: the mandible of *Dilophosaurus* is mediolaterally compressed, most noticeably in the postdentary region, and that of *Ceratosaurus* is slightly deformed at the intramandibular joint (at the upper and lower suture of the dentary with the postdentary bones). Although the original, undeformed dimensions of *Ceratosaurus* could be estimated, those of *Dilophosaurus* could not. This implies that the labiolingual values might be underestimated in the respective deformed areas and, consequently, that the ratio of Zx/Zy might be overestimated.

The mandible of *Dilophosaurus* is relatively shallow, and its dorsal margin is concave, reaching a minimum depth at the seventh tooth (equivalent to middentary; see Welles 1984, p. 157, fig. 7). This concavity appears to correspond with the convex margin and enlarged teeth of the maxilla (Welles 1984), a condition somewhat reminiscent of that observed in nonpiscivorous crocodilians. The mandible becomes deeper

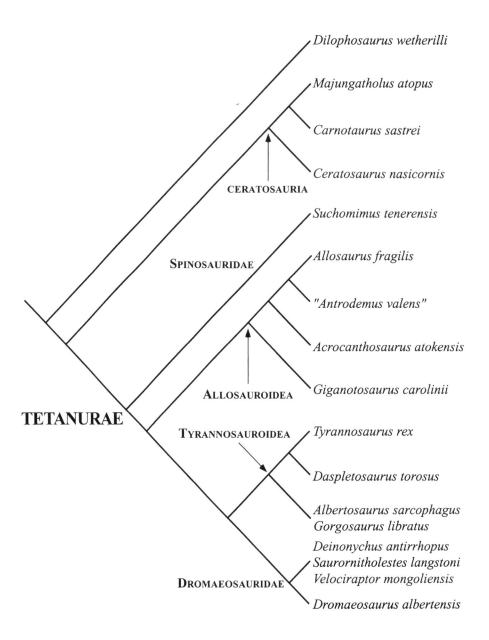

Figure 10.5. Cladogram of studied theropod taxa. Based on Sereno et al. 1998; Padian et al. 1999; Currie and Carpenter 2000; Holtz 2001; and Carrano et al. 2002.

in the symphyseal region, where "the chin is upcurved" (Welles 1984, p. 147). The mandibular symphysis is relatively long (~16% mandibular length) and nearly twice as long as it is deep (Fig. 10.3). Like that of *Dilophosaurus*, the mandible of *Ceratosaurus* is also relatively shallow, its depth decreasing gradually anteriorly. The mandibular symphysis is deeper than it is long and is oriented subvertically (Fig. 10.3). The mandible of abelisaurids is shallow, the depth of the dentary remaining relatively constant along its length, and the mandibular symphysis is slightly deeper than it is long (Fig. 10.3).

Mandibular Force Profiles. In terms of dorsoventral force profile (Fig. 10.6), *Dilophosaurus* has a strongly negative post-middentary slope and a slightly positive pre-middentary slope due to the mandible's

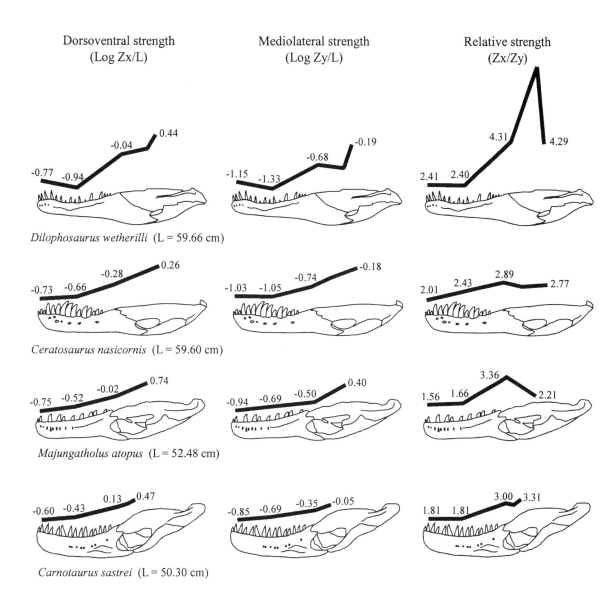

Dorsoventral strength (Log Zx/L)	Mediolateral strength (Log Zy/L)	Relative strength (Zx/Zy)

Dilophosaurus wetherilli (L = 59.66 cm)

Dorsoventral: -0.77, -0.94, -0.04, 0.44
Mediolateral: -1.15, -1.33, -0.68, -0.19
Relative: 2.41, 2.40, 4.31, 4.29

Ceratosaurus nasicornis (L = 59.60 cm)

Dorsoventral: -0.73, -0.66, -0.28, 0.26
Mediolateral: -1.03, -1.05, -0.74, -0.18
Relative: 2.01, 2.43, 2.89, 2.77

Majungatholus atopus (L = 52.48 cm)

Dorsoventral: -0.75, -0.52, -0.02, 0.74
Mediolateral: -0.94, -0.69, -0.50, 0.40
Relative: 1.56, 1.66, 3.36, 2.21

Carnotaurus sastrei (L = 50.30 cm)

Dorsoventral: -0.60, -0.43, 0.13, 0.47
Mediolateral: -0.85, -0.69, -0.35, -0.05
Relative: 1.81, 1.81, 3.00, 3.31

being relatively deeper at the second tooth alveolus than at middentary. *Ceratosaurus* has a less negative post-middentary slope than *Dilophosaurus* and a weakly negative pre-middentary slope due to a Zx/L value that is slightly lower at the second alveolus than at middentary. Abelisaurids have a post-middentary slope similar to that of *Ceratosaurus*, but their pre-middentary slope is more negative than that of the latter; in fact, the pre-middentary slope is just slightly less negative than the post-middentary, a condition reminiscent of that observed in *V. komodoensis* (see Fig. 10.4). In terms of absolute applied force, abelisaurids have stronger mandibles than both *Ceratosaurus* and *Dilophosaurus*.

The labiolingual mandibular force profile (Fig. 10.6) of *Dilophosaurus* is similar to the dorsoventral force profile, although the post-middentary slope is not as steep. In *Ceratosaurus*, the labiolingual force

Figure 10.6. Mandibular properties of ceratosaurs and Dilophosaurus. *From left to right: landmark 1 (second tooth alveolus), landmark 2 (mid-dentary), landmark 3 (upper dentary suture), and landmark 5 (maximum depth). Mandibular length L represents the distance between the articular fossa and the anterior extremity of the mandible. Because the mandible of* Dilophosaurus *was mediolaterally compressed during burial, its Zx/Zy values may be overestimated (particularly in the postdentary region).*

profile differs from the dorsoventral profile in that the pre-middentary slope is close to nil or slightly positive because the force of the mandible at the second alveolus is subequal to that at middentary. The mandible of abelisaurids, like that of varanids (Fig. 10.4), behaves as a simple lever, with labiolingual force decreasing at a nearly constant rate along the tooth row. Once again, abelisaurids have relatively stronger mandibles than the other two theropods.

All studied taxa have mandibles that are strongly buttressed dorsoventrally (Fig. 10.6). *Dilophosaurus* has the most dorsoventrally buttressed mandible, although this is most assuredly exaggerated by the mediolaterally compressed state of preservation of the specimen (especially in the postdentary region; see above). The mandible is relatively much deeper than it is wide at the superior dentary suture but rapidly increases in width toward the middentary. The anterior extremity of the mandible is approximately the same shape as at middentary. The mandible of *Ceratosaurus* has Zx/Zy values that decrease linearly from 2.89 at the upper dentary suture to 2.01 at the second alveolus, indicating that although the mandible becomes gradually rounder anteriorly, it is still at least twice as deep as it is wide. Both abelisaurids have similar Zx/Zy mandibular profiles: they exhibit a rapid decrease in Zx/Zy in the post-middentary region, attaining values inferior to 2.00 at middentary, whereas the Zx/Zy values at the middentary and the second tooth are subequal.

Interpretation. The dorsoventral and labiolingual mandibular force profiles of *Dilophosaurus* indicate a rapidly decreasing bite force along the posterior tooth row, but a relatively stronger mandible at the second alveolus than at middentary. This buttressing of the symphyseal region suggests that the extremity of the mandible may have played a significant role for prey capture, manipulation, or both in *Dilophosaurus*, possibly to resist loads exerted during these activities. Similar symphyseal properties are also characteristic of modern felids and predatory crocodilians (e.g., *Crocodylus niloticus*), which deliver a powerful bite with the anterior of their jaws to subdue their prey (Therrien, in prep.; Therrien and Ruff, in prep.). Because the mandible is mediolaterally compressed by postmortem processes, the true Zx/Zy values cannot be determined; however, the tooth-bearing portion of the mandible, made of solid bone, was considerably less affected by postmortem deformation than the postdentary portion, which consisted of a hollow cylinder. Consequently, although the Zx/Zy values for the pre-middentary region are probably artificially inflated, they were probably originally still in the vicinity of 2.00, indicating that the mandible is dorsoventrally buttressed. Interestingly, the Zx/Zy ratio is relatively constant at the middentary and the second alveolus, revealing that the mandible maintains the same shape between the two landmarks in spite of its increased strength at the symphysis. Consequently, loads exerted on the mandibles were primarily associated with biting; torsional and labiolingual loads were lower, presumably induced by prey of relatively small size. This discovery, combined with the upturned symphyseal region, the loose premaxilla-maxilla articulation that forms a subnarial notch, and the orientation of the premaxillary and anterior-most maxillary teeth (Welles 1984), suggests that *Dilophosaurus* may have hunted relatively small prey by delivering slashing bites and capturing them with the specially designed anterior extremity of its jaws, as suggested by Paul (1988). After the initial capture, prey may have been moved posteriorly, either inertially or gravitationally, into the concave portion of the mandible, in opposition with the convex margin and enlarged teeth of the maxilla (Welles 1984); and slicing bites were then rapidly delivered until the prey died. A similar behavior is commonly observed in modern crocodilians (Drongelen and Dullemeijer 1982; Busbey 1989). Possible evidence of gregarious behavior has been reported for *Dilophosaurus* in the form of three individuals found at the same locality (Rowe and Gauthier 1990). Given that our results for this theropod suggest that it probably fed on prey smaller than itself, hunting packs must have been of limited size if this theropod was indeed gregarious.

In *Ceratosaurus*, dorsoventral and labiolingual forces decrease at a lower rate along the posterior tooth than they do in *Dilophosaurus*. However, the mandible is nearly as strong at the second alveolus as it is at middentary. The dentary is dorsoventrally buttressed (Zx/Zy values fall between 2.00 and 3.00) and becomes rounder anteriorly, although it is still twice as deep as it is wide at the second alveolus ($Zx/Zy = 2.01$). These profiles suggest that loads generated on the mandibles during feeding were very well constrained within the sagittal plane, indicating that the animal was exclusively slicing flesh and that it may have used the anterior extremity of its jaws to capture and handle prey, although not to the same extent as *Dilophosaurus*. Indeed, since the Zx/Zy values at the second alveolus for *C. nasicornis* and *C. dentisulcatus* (UMNH VP5278) are in the range for *V. komodoensis*, the prey must have offered relatively little resistance when manipulated, possibly being of relatively smaller body size. Furthermore, the presence of premaxillary teeth with asymmetrical crowns (incisiform) in *C. dentisulcatus* (Therrien, pers. obs.) and ceratosaurs in general (Bakker et al. 1988), the stout nature of the premaxillae in *C. dentisulcatus* and *C. magnicornis*, and the somewhat upturned symphyseal region of the dentary in *C. dentisulcatus* (Madsen and Welles 2000) all suggest that the anterior extremity of the jaws played an important role in *Ceratosaurus*, possibly related to the capture and handling of prey smaller than itself.

In contrast with the dorsoventral and labiolingual forces in the aforementioned theropods, the forces in the two abelisaurids decrease at a lower, nearly linear rate along the entire mandible. Also, the mandible is dorsoventrally buttressed, and mandibular morphology does not vary significantly between the middentary and the second alveolus, the Zx/Zy values of these two landmarks being nearly identical. The Zx/Zy values at the second alveolus are similar to that of *V. komodoensis*, which suggests, along with the other mandibular properties akin to varanids, that abelisaurids probably delivered fast, slashing bites without grasping prey. Indeed, with its broad skull and short teeth, *Majungatholus* is reminiscent of *V. komodoensis*, and their feeding behaviors must have been very similar: attacking large prey by inflicting gashing wounds with their jaws and then waiting for their prey to die (also see "*Antrodemus*" below). There is direct evidence, in the form of tooth-marked bones, that *Majungatholus* fed on sauropods and conspecifics (Rogers et al. 2003), although it is impossible to determine whether this was the result of active predation or scavenging.

Although the skull of *Carnotaurus* differs greatly from that of the Komodo dragon (and that of *Majungatholus*) in being surprisingly narrower, Mazzetta et al. (1998) inferred similar behaviors, saying that this theropod had a "fast (rather than powerful) bite" (p. 191). They also claimed that *Ceratosaurus* must have had a relatively "stronger bite throughout the whole mandibular length" than *Carnotaurus* because of the longer moment arm of the anterior pterygoideus muscle (Mazzetta et al. 1998, p. 188). The results presented here do not support this idea, attributing a slightly more powerful bite to abelisaurids, although the mandible of *Ceratosaurus* appears to sustain *relatively* more-important dorsoventral forces than those of abelisaurids (greater Zx/Zy values). Bakker (1998) suggested that *Carnotaurus* may have been similar to allosaurids and used its upper jaw as a club; if he refers to a slashing bite, then our models also support his views. However, we will that the mandible of allosaurids does not have the same biomechanical properties as that of abelisaurids (except for "*Antrodemus*"); consequently, they probably hunted differently.

Spinosauridae. Owing to their elongate snout and slightly flattened conical teeth, somewhat reminiscent of crocodilians, spinosaurids are thought to have been primarily piscivorous, although there is evidence they may also have fed on small terrestrial prey (e.g., small iguanodontians; see Charig and Milner 1997 and also Sues et al. 1999).

The mandible of *Suchomimus tenerensis* (cast based on MNN GDF500, 501, 502, 503) was measured to determine its biomechanical properties. The mandible is relatively shallow, and its depth decreases gradually anteriorly. The anterior extremity is upcurved in a fashion reminiscent of *Dilophosaurus* and bears a rosette of teeth in opposition with similarly arranged premaxillary teeth and a premaxilla-maxilla notch (Sereno et al. 1998). The mandibular symphysis is also relatively long (~25% of mandibular length) and twice as long as it is deep (at the upcurved chin) (Fig. 10.3). Unfortunately, owing to uncertainties regarding certain mandibular dimensions (e.g., length of mandibular symphysis, contribution of splenial to mandible), we could not reconstruct the force profiles for *Baryonyx walkeri* from Charig and Milner (1990, 1997).

Mandibular Force Profiles. The various force profiles for *Suchomimus* reflect its peculiar mandibular morphology (Fig. 10.7). The post-middentary slopes of Zx/L and Zy/L are strongly to moderately negative; however, the mandible is much stronger at the second alveolus than at middentary, owing to the elongate symphysis, resulting in a strongly positive pre-middentary slope. These profiles are similar to those of *Dilophosaurus*, although they are developed to a greater degree.

The Zx/Zy profile (Fig. 10.7) decreases at a constant rate along the dentary, going from values as high as 4.11 at the upper dentary suture

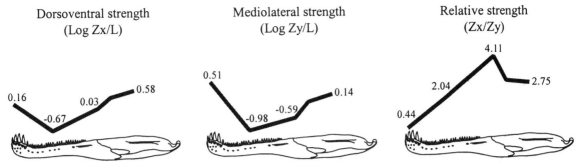

Dorsoventral strength
(Log Zx/L)

Mediolateral strength
(Log Zy/L)

Relative strength
(Zx/Zy)

Suchomimus tenerensis (L = 112.58 cm)

to as low as 0.44 at the second tooth alveolus. This profile signifies that the *effective* mandibular cross-sectional morphology of *Suchomimus* varies greatly along the tooth row, going from an ellipse four times deeper than it is wide in the back to an ellipse two times as wide as it is deep at the symphysis. It is important to remember that the *effective* cross-section at the second alveolus is oriented not mediolaterally, that is, across both hemi-mandibles or the width of the symphysis, but labiolingually through the length of the symphysis (see Fig. 10.2A).

Interpretation. The dorsoventral and labiolingual force profiles of *Suchomimus* indicate that bite force decreases gradually along the posterior half of the tooth row. However, the elongate symphysis strengthened the anterior portion of the mandible relative to the mid-dentary. As is the case for *Dilophosaurus, Suchomimus* must have used the anterior portion of its jaws to capture and manipulate prey. Unlike the mandible of *Dilophosaurus*, however, the mandible of *Suchomimus* is strongly dorsoventrally buttressed in the post-middentary region and strongly labiolingually buttressed at the second alveolus. This indicates that loads exerted on the posterior mandible were primarily in the sagittal plane and associated with biting. In contrast, torsional and labiolingual loads, presumably exerted during prey capture, must have been so much more important than those applied in the bite (sagittal) plane that the anterior portion of the mandible had to be twice as strong labiolingually as it was dorsoventrally ($Zx/Zy = 0.44$). Indeed Hylander (1984), in his studies of the patterns of stress in the mandibular symphysis, discovered that a mandibular symphysis needs to increase its anteroposterior dimension in order resist an increase in torsional moments. Consequently, these results suggest that some prey must have been large enough to offer significant resistance during capture, thus supporting the assertion of Sues et al. (1999) that spinosaurids may have fed on relatively small terrestrial prey as well as fish.

Although *Suchomimus* has a long mandibular symphysis, it is not as long as that observed in extant and extinct piscivorous crocodilians (>50% mandibular length; Therrien and Ruff, in prep.). Rather, crocodilians such as the Orinoco crocodile (*Crocodylus intermedius*), the African slender-snouted crocodile (*Crocodylus cataphractus*), and the extinct *Sebecus icaeorhinus*, for which small to medium-size terrestrial prey are known (or inferred) to be part of the diet (Molnar 1969; Pooley

Figure 10.7. Mandibular properties of the spinosaurid Suchomimus tenerensis. *From left to right: landmark 1 (second tooth alveolus), landmark 2 (mid-dentary), landmark 3 (upper dentary suture), and landmark 5 (maximum depth). Mandibular length* L *represents the distance between the articular fossa and the anterior extremity of the mandible.*

1989; Ross and Magnusson 1989), are similar to *Suchomimus* in having a moderately long mandibular symphysis (~24–30% mandibular length; Therrien and Ruff, in prep.). Furthermore, the mandibular force profiles of *Suchomimus* are extremely similar to those of *Crocodylus intermedius* and *Sebecus icaeorhinus* (Therrien and Ruff, in prep.).

Hence, it appears that *Suchomimus* may have hunted smaller prey in a manner somewhat similar to that of *Dilophosaurus* (Paul 1988), capturing prey with the rosette of teeth located at the anterior extremity of the jaws. However, unlike that theropod, spinosaurids probably held onto prey and may have shaken it in order to finish it, as suggested by the very low Zx/Zy values at the anterior extremity of the mandible and its long-rooted conical dentition (Charig and Milner 1997). The fact that the premaxilla-maxilla notch and surrounding dentition changed during spinosaur evolution (Sereno et al. 1998) supports the idea that this area played an important role during feeding, presumably for prey capture. Spinosaurids could also have hunted larger prey but would probably have captured and killed it with their robust forelimbs, which were equipped with large claws (e.g., Paul 1988; Charig and Milner 1997), rather than their bite because their skull was unable to resist important bending stresses (Henderson 2002).

Allosauroidea. Allosauroidea is known to include some of the largest theropods ever to have existed (Coria and Salgado 1995). Those allosauroids investigated are *Allosaurus fragilis* (cast of MOR 693), "*Antrodemus valens*" (USNM 4734), *Acrocanthosaurus atokensis* (cast of NCSM 14345), and *Giganotosaurus carolinii* (cast of MUCPv-CH-1). "*Antrodemus valens*" has long been considered a junior synonym of *Allosaurus fragilis* (e.g., Madsen 1976), but recently Henderson (1998) suggested that cranial and dental differences between them might warrant its recognition as a different taxon (see also Paul 1988; Bakker 1998).

The mandibles of *Allosaurus fragilis* and "*Antrodemus valens*" are moderately deep posteriorly and decrease in depth anteriorly. In *Allosaurus*, very large individuals exhibit a ventral deepening of the dentary in proximity to the mandibular symphysis (Therrien, pers. obs.). The postdentary region of *Acrocanthosaurus* is deep, and the dentary is long and shallow. The dentary deepens at the symphysis, but, unlike the dentaries of *Dilophosaurus* and *Suchomimus*, which become deeper by expanding superiorly (upturned chin), it deepens inferiorly via a ventral process (Currie and Carpenter 2000). The mandible of *Giganotosaurus* is known only from dentaries (Coria and Salgado 1995; Calvo and Coria 1998). These bones are relatively deep and become noticeably deeper at the mandibular symphysis via a ventral process similar to that in *Acrocanthosaurus* (Currie and Carpenter 2000). In all studied allosauroids, the mandibular symphysis is deeper than it is long and is subvertical (Fig. 10.3).

Mandibular Force Profiles. Although *Allosaurus* and "*Antrodemus*" have roughly the same body size, they exhibit important differences in strength profiles. Dorsoventral force decreases rapidly and nearly linearly to the extremity of the mandible in "*Antrodemus,*"

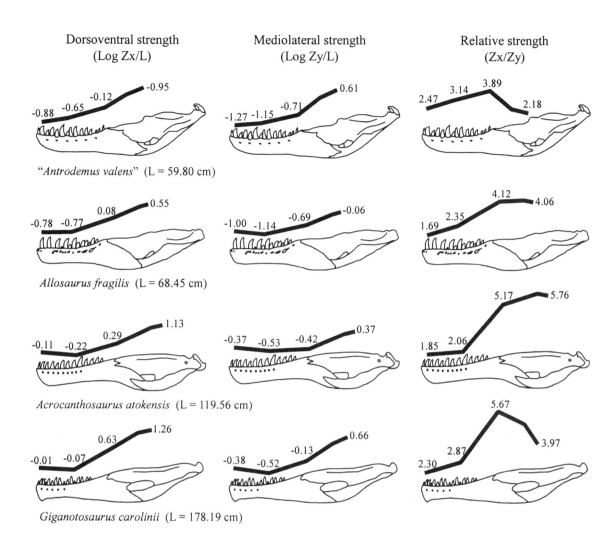

Dorsoventral strength (Log Zx/L)	Mediolateral strength (Log Zy/L)	Relative strength (Zx/Zy)

"*Antrodemus valens*" (L = 59.80 cm)

Allosaurus fragilis (L = 68.45 cm)

Acrocanthosaurus atokensis (L = 119.56 cm)

Giganotosaurus carolinii (L = 178.19 cm)

whereas the mandible is nearly as strong at the mandibular symphysis as it is at middentary in *Allosaurus*, resulting in a slightly negative pre-middentary slope (Fig. 10.8). Labiolingually, mandibular force decreases at a lower rate but still nearly linearly to the symphysis in "*Antrodemus*," whereas the mandible at the second alveolus is stronger than at middentary in *Allosaurus*, resulting in a positive pre-middentary slope (Fig. 10.8). The differences between the mandibles of these two taxa are even clearer in their *Zx/Zy* profiles: the dentary of "*Antrodemus*" is more strongly buttressed dorsoventrally than that of *Allosaurus*. The *Zx/Zy* profile of "*Antrodemus*" decreases slowly and linearly, going from a value of 3.89 at the upper dentary suture to 2.47 at the second alveolus. In contrast, the *Zx/Zy* profile decreases rapidly and nearly linearly in *Allosaurus*, going from a value of 4.12 at the upper dentary suture to 1.69 at the second alveolus (Fig. 10.8).

In spite of the large difference in body size between the two taxa, the force profiles of the larger allosauroids *Acrocanthosaurus* and *Giganotosaurus* are nearly identical. Dorsoventrally (Fig. 10.8), man-

Figure 10.8. Mandibular properties of allosauroids. From left to right: landmark 1 (second tooth alveolus), landmark 2 (middentary), landmark 3 (upper dentary suture), and landmark 5 (maximum depth). Mandibular length L represents the distance between the articular fossa and the anterior extremity of the mandible.

dibular force decreases slowly in the post-middentary region; however, the mandible is stronger at the second alveolus than at the middentary, producing a slightly positive pre-middentary slope. The labiolingual force profile is similar to the dorsoventral profile, except for a much lower post-middentary slope (Fig. 10.8). The Zx/Zy profiles of *Acrocanthosaurus* and *Giganotosaurus*, although differing in absolute values, are also very similar (Fig. 10.8): both reveal a rapid decrease, from very high values at the upper dentary suture ($Zx/Zy > 5.00$ in both taxa), in the post-middentary region and a much slower decrease in the pre-middentary region, attaining low values at the second alveolus ($Zx/Zy = 2.30$ and 1.85 in *Giganotosaurus* and *Acrocanthosaurus*, respectively).

Interpretation. The linear decrease in dorsoventral and labiolingual force along the dentary in "*Antrodemus*" is very similar to the results of the simple-lever model of varanids. Bite force thus decreases gradually along the tooth row to the anterior extremity of the mandible. The high Zx/Zy values also indicate that the mandible is strongly dorsoventrally buttressed, being much deeper than it is wide ($Zx/Zy > 3.00$), although it becomes rounder anteriorly (decreasing Zx/Zy values); however, the Zx/Zy ratio at the second alveolus is still high ($Zx/Zy = 2.47$), indicating that labiolingual loads were minimal. Thus, loads generated during feeding were well constrained in the sagittal plane, leading us to believe that "*Antrodemus*" probably delivered rapid slashing bites at prey, without holding it, in a fashion similar to *V. komodoensis*. This interpretation is compatible with the feeding behavior proposed by Bakker (1998) for his "true *Allosaurus*" (our "*Antrodemus*"), in which the upper jaw acted as a mega-serrated blade to inflict long, jagged wounds on sauropods.

Post-middentary dorsoventral and labiolingual mandibular forces decrease at similar rates in *Allosaurus* and "*Antrodemus*," but the mandible at the second alveolus is nearly as strong as or stronger than at the middentary. The mandible of *Allosaurus* is much deeper than it is wide at the upper dentary suture ($Zx/Zy = 4.12$) and becomes rapidly rounder anteriorly, reaching a Zx/Zy ratio of 1.69 at the second alveolus. This change in ratios indicates that the dentary is strongly dorsoventrally buttressed posteriorly and rapidly becomes better adapted to resist labiolingual loads anteriorly. The fact that the Zx/Zy value at the second alveolus is similar to that of *V. komodoensis* suggests that, although *Allosaurus* may have captured and manipulated prey with the extremity of its mandible, the labiolingual loads exerted on the mandible during such activities were still relatively low, possibly because the prey was of a relatively smaller body size or did not offer significant resistance. All these observations suggest that, although the mandible was adapted for slicing (e.g., bite-and-release strategy), *Allosaurus* may have used the extremity of its mandible in prey capture and manipulation to a greater extent than "*Antrodemus*." The fact that D-shaped premaxillary teeth exhibiting apical wear are observed in *Allosaurus* (Madsen 1976; see also Bakker et al. 1988) lends support to the idea that the anterior extremity of the jaws was specialized for prey capture and handling.

The force profiles of *Allosaurus*, in fact, are far more similar to those of *Ceratosaurus* (see above) than to those of *"Antrodemus."* Given that bite force seems to have been similar in the three taxa (Fig. 10.8; Table 10.1), the variation in force profiles supports the hypothesis of differentiation of feeding behaviors and niche partitioning among the Morrison theropods, as well as the hypothesis of ecological displacement of *Ceratosaurus* by *Allosaurus*, owing to a too great similarity in feeding behavior (Henderson 1998). Indeed *"Antrodemus,"* like *V. komodoensis* and *Majungatholus*, has a broad skull and short recurved teeth, whereas *Allosaurus* and *Ceratosaurus* have narrow skulls equipped with long, straight teeth (Henderson 1998).

In his work on the Komodo dragon, Auffenberg (1981) argued that a broad snout is advantageous for inflicting large gashing wounds, whereas a narrow snout is not as efficient. Furthermore, he noted that the teeth of *V. komodoensis* are too recurved to impale prey and could not have been used to prevent prey escape as previously suggested by Frazzetta (1962); rather, strongly recurved teeth are ideal to inflict "a massive wound as the teeth move through the flesh" (Auffenberg 1981, p. 214; for a differing view, see Rieppel 1979). Given these observations, it is thus not surprising that the mandibular force profiles of *"Antrodemus"* suggest a fast, slashing bite similar to that of *V. komodoensis*, whereas those of *Allosaurus* and *Ceratosaurus* suggest a slicing (and impaling) bite with more-complex techniques of prey capture and manipulation with the anterior extremity of the jaws. In light of these differences, *"Antrodemus"* may have hunted larger prey, like sauropods, whereas *Allosaurus* and *Ceratosaurus* may have preferred relatively smaller prey, leading the latter two to compete for similar resources.

In his paleoecological analysis of the Morrison Formation, Foster (2003, p. 55) established that *Allosaurus* dominated the predator population biomass, which was composed of "moderate numbers but a large biomass of *Allosaurus* and large numbers but a low amount of biomass contributed by small theropods like *Ornitholestes* and *Coelurus*." He interpreted these results as possible evidence that *Allosaurus* was the main Morrison predator, having a generalized diet consisting of a wide spectrum of medium-size prey like juvenile sauropods, adult ornithopods, and potentially stegosaurs, while the other large but rarer Morrison theropods (*Torvosaurus*, *Saurophaganax*, and *Ceratosaurus*, a group to which *"Antrodemus"* should be added) may have been "more specialized on slightly larger prey, perhaps including juvenile and subadult sauropods, than was *Allosaurus*" (Foster 2003, p. 55). As mentioned above, the mandibular models for *Ceratosaurus* are similar to those of *Allosaurus* and do not support its integration with the large-prey predators; consequently, we prefer Henderson's (1998) hypothesis of ecological displacement related to competition for similar prey to explain the scarcity of *Ceratosaurus* in the Morrison ecosystems.

Our mandibular models cannot refute either the "hatchet" slashing attack proposed by Rayfield et al. (2001, 2002) for *Allosaurus* or Bakker's (1998) mega-serrated blade model for "creosaur-type allosaurids," in which the predator opens its jaw wide and attacks by

TABLE 10.1.

Bite Force Estimates for Studied Theropods

Mandibular length corresponds to the distance between the articular fossa and the anterior extremity of the mandible. The absolute bite force value for *A. mississippiensis* is from Erickson in Meers 2002.

Taxon	Mandibular length (cm)	Zx/L mid-dentary (cm²)	Bite force relative to *V. komodoensis*	Bite force relative to *A. mississippiensis*
Varanus komodoensis	16.96	0.014	—	0.086
Alligator mississippiensis[1]	50.08	0.160	11.617	—
Velociraptor mongoliensis	17.91	0.014	1.052	0.091
Saurornitholestes langstoni	~0.019	1.376	0.118	16.96
Deinonychus antirrhopus	28.71	0.025	1.827	0.157
Gorgosaurus libratus (juvenile)	~0.039	2.832	0.244	34.66
Dromaeosaurus albertensis	18.53	0.040	2.939	0.253
Dilophosaurus wetherilli	59.66	0.114	8.254	0.710
Daspletosaurus torosus (juvenile)	~0.171	12.397	1.067	58.95
Allosaurus fragilis	68.45	0.172	12.477	1.074
Suchomimus tenerensis	112.58	0.212	15.419	1.327
Ceratosaurus nasicornis	59.60	0.218	15.825	1.362
"*Antrodemus valens*"	59.80	0.222	16.154	1.391
Majungatholus atopus	52.48	0.305	22.155	1.907
Albertosaurus sarcophagus (sub-adult)	69.62	0.340	24.683	2.125
Carnotaurus sastrei	50.30	0.369	26.781	2.305
Acrocanthosaurus atokensis	119.56	0.606	44.026	3.790
Giganotosaurus carolinii	178.19	0.858	62.304	5.363
Daspletosaurus torosus (adult)	95.59	1.155	83.879	7.221
Tyrannosaurus rex (mean)	120.39	2.546	184.876	15.915

Published estimates		Theropod bite force (N)	*A. mississippiensis* bite force (N)	Relative to *A. mississippiensis*
T. rex, maximum bite force[2]		235,123	18,912	12.432
T. rex, from crocodile regression[2]		183,389		9.697
T. rex, from mammal regression[2]		77,470		4.096
T. rex, minimum at maxillary teeth[3]		153,600		8.122
T. rex, maximum at maxillary teeth[4]		321,600		17.005
A. fragilis, muscle-driven bite[5]		2,147.88		0.114
A. fragilis, maximum force before yield[5]		18,747		0.991

[1]Therrien and Ruff, in prep..
[2]Meers 2002.
[3]Extrapolated for a bilateral bite from Erickson et al. 1996 by Meers (2002).
[4]Extrapolated for a bilateral bite from Erickson et al. 1996 following Meers's (2002) calculation.
[5]Rayfield et al. 2001.

depressing its skull, driving its upper teeth into the prey, because these interpretations are based on cranial morphology. However, our *Allosaurus* model does indicate that the mandible was involved in the attack, contrary to the aforementioned models in which the mandible serves no active role. Furthermore, our results lend support to Frazzetta and Kardong's (2002, p. 388) view that the "overengineered" aspect of the *Allosaurus* skull may be an adaptation to resist stresses "generated by struggling prey held by a very narrow-headed predator" rather than to resist loads due to a "high velocity impact of the skull into its prey" (Rayfield et al. 2001, p. 1035). A more detailed, integrative approach, in which cranial and cervical anatomy and muscle reconstructions are considered, could help resolve the uncertainty surrounding the feeding behavior of *Allosaurus*.

As mentioned above, the force profiles of the larger allosauroids *Acrocanthosaurus* and *Giganotosaurus* are nearly identical, the main difference being that *Giganotosaurus* has a stronger bite force (absolute Zx/L values). In both taxa, bite force decreases slowly in the posterior half of the tooth row. The mandible is stronger, both dorsoventrally and labiolingually, at the second alveolus than at middentary, and relatively more so than in *Allosaurus*, as indicated by the positive pre-middentary slope. The mandibles of the two large allosauroids are extremely well buttressed dorsoventrally in the posterior half of the tooth row but become much rounder in the anterior half, indicating that labiolingual loads are relatively much more important anteriorly. All these observations suggest that *Acrocanthosaurus* and *Giganotosaurus* had powerful slicing bites and captured or manipulated prey with the anterior extremity of their mandibles, in a fashion similar to *Allosaurus*. If these large allosauroids filled an ecological role similar to that of *Allosaurus* in their respective ecosystems, they may have been generalized predators feeding on a wide spectrum of prey. Given their large body size, they could have tackled prey relatively smaller than themselves as well as sub-adult sauropods. Possibly reflecting predation on larger prey than *Allosaurus,* the premaxillary teeth of *Acrocanthosaurus* are not D-shaped, and the premaxillae meet at an acute angle (Currie and Carpenter 2000); unfortunately, nothing is known about the premaxillae of *Giganotosaurus*.

Currie and Carpenter (2000) noted the presence of the ventral expansion of the mandible at the level of the symphysis in *Acrocanthosaurus* and *Giganotosaurus*, and to a lesser extent in some large tyrannosaurids, and suggested this feature may be a size-dependent character. We believe that it reflects the feeding behavior of these animals rather than their body size. In his strain gauge analysis, Hylander (1984) observed that the mandibular symphysis of *Macaca fascicularis* was loaded in bending during incision and isometric incisor biting, the superior margin undergoing compression while the inferior margin underwent tension. He then theorized that a possible way to adapt for such loads would be to maximize the second moment of area of the symphyseal section by increasing "the vertical or dorsoventral dimensions of the symphysis by adding cortical bone along the lower border of the symphysis" (Hylander 1984, p. 39). Such a feature, equivalent to the ventral process in *Acrocanthosaurus* and *Giganotosaurus,* would act as an efficient asymmetrical beam by adding bone where it was most needed, because bone is weaker under tension than compression (Evans 1973). Thus, animals exerting large bite forces at the extremity of their mandible would need to evolve adaptations to resist the ensuing stresses; and as body size increased, symphyseal morphology would have to change (Hylander 1985). Consequently, we believe that the ventral process at the symphysis in the two large allosauroids might be *indirectly* a size-dependent character, in the sense that larger animals tend to have stronger bites and would exert more important loads on the anterior extremity of their mandibles during prey capture (somewhat equivalent to "incisor biting"). Two observations seem to support this claim: (1) the largest individuals of *Allosaurus* also exhibit a deepening of the mandible in proximity to the mandibular symphysis; and (2) among all the large tyrannosaurid mandibles studied, none exhibited a ventral process comparable to that observed in *Acrocanthosaurus* and *Giganotosaurus,* probably because the mandibular symphysis of tyrannosaurids is diagonally oriented, thus indicating different patterns of stress at the mandible (e.g., Hylander 1984) and, consequently, a different feeding behavior (see Tyrannosauridae, below).

Dromaeosauridae. Dromaeosaurids are a clade of small to medium-size, gracile theropods inferred to have been highly active. The dromaeosaurids studied are the dromaeosaurine *Dromaeosaurus albertensis* (cast of AMNH 5356) and the velociraptorines *Deinonychus antirrhopus* (YPM 41147, cast), *Saurornitholestes langstoni* (TMP 88.121.39), and *Velociraptor mongoliensis* (Utah Geological Survey cast of a privately owned specimen). The mandible of *Saurornitholestes,* consisting of a partial postdentary section (articulated surangular and articular) and a nearly complete anterior half (dentary, but lacking splenial), is sufficiently complete to allow reconstruction of the mandibular force profiles for the anterior segment of the mandible. The contribution of the splenial to the mandibular depth was estimated, but its contribution to mandibular width could not be estimated; therefore, the force values at the upper dentary suture (and possibly at the middentary) are minimum values. All dromaeosaurids have a shallow mandible and a subvertical mandibular symphysis that ranges from deeper than it is long to nearly equidimensional (Fig. 10.3).

Mandibular Force Profiles. Besides differences in absolute mandibular force values and variations in slope, the mandibular force profiles are nearly identical among the studied dromaeosaurids. Both dorsoventral and labiolingual force profiles demonstrate a slow, linear decrease in force from the upper dentary suture to the second alveolus, indicating that the mandible behaves as a simple lever (Fig. 10.9). The rate of mandibular force decrease is nearly identical in all dromaeosaurids, the lower values at the upper dentary suture for *Saurornitholestes* being caused by the measurement error surrounding the missing splenial. In terms of absolute dorsoventral force values (Zx/L), *Velociraptor* and *Saurornitholestes* have similarly low values, *Deinonychus* has a slightly stronger bite, and *Dromaeosaurus* has the strongest bite. Labiolingually, *Dromaeosaurus* stands out by having rela-

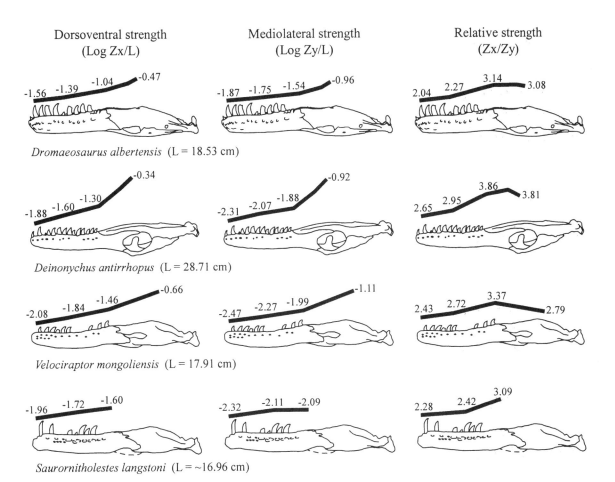

Dorsoventral strength
(Log Zx/L)

Mediolateral strength
(Log Zy/L)

Relative strength
(Zx/Zy)

Dromaeosaurus albertensis (L = 18.53 cm)

Deinonychus antirrhopus (L = 28.71 cm)

Velociraptor mongoliensis (L = 17.91 cm)

Saurornitholestes langstoni (L = ~16.96 cm)

tively higher values than the other three taxa, which cluster together, reflecting the dromaeosaurine/velociraptorine dichotomy. All four dromaeosaurids possess similar *Zx/Zy* profiles, characterized by an anterior decrease, although not linear, of approximately 1.00 between the upper dentary suture and the second alveolus (Fig. 10.9). In other words, the difference in *Zx/Zy* values between the posterior and anterior extremities of the dentary is 1.00. We also observe that *Deinonychus* and *Velociraptor* have the most strongly dorsoventrally buttressed mandibles (higher *Zx/Zy* values), whereas *Dromaeosaurus* has a rounder mandible (lower *Zx/Zy* values). *Saurornitholestes* appears to be intermediate between the two groups, although this position could be an artifact produced by the missing splenial.

Interpretation. With their simple-lever force profiles, in which bite force decreases linearly toward the mandibular symphysis, dromaeosaurids must have used their mandibles in a fashion similar to varanids, to deliver fast bites without a significant degree of prey manipulation. Also, the fact that the *Zx/Zy* values at the second alveolus in dromaeosaurids are much higher than in wild-caught *V. komodoensis* of similar mandibular length but are similar to those who lived in captivity,

Figure 10.9. Mandibular properties of dromaeosaurids. From left to right: landmark 1 (second tooth alveolus), landmark 2 (mid-dentary), and landmark 3 (upper dentary suture); if present, the next point represents landmark 5 (maximum depth). Mandibular length L represents the distance between the articular fossa and the anterior extremity of the mandible.

suggests one of the following possibilities: (1) that the prey did not offer significant resistance when it was bitten (relatively small or nearly dead prey); (2) that the predator delivered a slashing bite without holding its prey, thus avoiding the stresses exerted by struggling prey; or (3) that the mandible may not have been involved at all in prey capture but solely in food processing. These interpretations are supported by earlier claims that dromaeosaurids may have used their long, grasping hands to hold onto their prey while inflicting gashing wounds with their terrifying sickle claws (e.g., Ostrom 1969a,b 1990).

On the basis of the position of the sickle claw of *Velociraptor* in the throat region of *Protoceratops* in the famous "fighting dinosaurs" death assemblage, Carpenter (1998) suggested that dromaeosaurids might have used their sickle claws to pierce major blood vessels in the carotid sheath or the trachea rather than to slash prey. Although this appears to be valid for the "fighting dinosaurs," it is not likely to have been the habitual killing method of dromaeosaurids. Indeed, an attack to the throat requires a great deal of precision, and the only extant mammalian predators that use such a killing technique (felids) restrain their prey with retractile claws and use the sensory input provided by their vibrissae (whiskers) to carefully position their bite and crush the trachea (e.g., Kingdon 1977; Leyhausen 1979). The prey dies from strangulation, not from blood loss associated with severed blood vessels (see Schaller 1972). In fact, the only predators thought to have actively aimed at the major blood vessels of the neck as a means to kill their prey, the sabertooths, evolved complex craniodental, cervical, and postcranial adaptations to immobilize their prey and position their bite properly to cut through the carotid sheath and its contents (e.g., Martin 1980; Anyonge 1996; Turner and Antón 1996; Antón and Galobart 1999). Given the degree of complexity surrounding a throat attack, which would have been even greater if the goal was to cut the jugular vein or carotid artery, we find it unlikely that dromaeosaurids aimed specifically at the throat with their sickle claws. Rather, numerous although superficial wounds inflicted over various parts of the body could have greatly weakened the prey through blood loss before it was killed, a technique used by modern pack-hunting canids and hyenids (Kruuk and Turner 1967; Kruuk 1972; Biknevicius and Ruff 1992a).

The most intriguing conclusion drawn from our mandibular models is that *Dromaeosaurus* had a stronger bite, in absolute terms, than other dromaeosaurids. The dorsoventral and labiolingual force profiles are higher than those of *Deinonychus*, a much larger animal, indicating that the mandible of *Dromaeosaurus* exerted greater forces during bite. However, the Zx/Zy profile reveals that the loads exerted on the mandible of *Dromaeosaurus* were less well constrained in the sagittal plane than in they were in other dromaeosaurids. These observations suggest that *Dromaeosaurus*, and possibly dromaeosaurines in general, may have hunted slightly differently than velociraptorines, possibly relying more on its jaws to wound its prey. Although *Dromaeosaurus* did not use its jaws to hold and subdue prey, as indicated by the high Zx/Zy ratio at the second alveolus, it may have delivered powerful slashing bites in order to kill its prey, whereas velociraptorines may have preferentially killed by using their sickle claws.

Indeed, Paul (1987, 1988) mentioned that *Dromaeosaurus* had a more robust skull, larger teeth, and a broader front tooth arcade than other dromaeosaurids. No sickle-shaped second pedal ungal is known for *Dromaeosaurus* (see Currie 1995); however, Paul observed that the phalanges of pedal digit II were more robust than in *Velociraptor* (= *Velociraptor* + *Deinonychus* + *Saurornitholestes* sensu Paul 1988) but similar to those of a closely related taxon, *Adasaurus mongoliensis*, which had a sickle claw reduced to the size of a normal claw (Barsbold 1983). Such features, combined with the reduced cranial kinesis observed in dromaeosaurines such as *Dromaeosaurus* ("deep-skulled" dromaeosaurids of Barsbold 1983), have been interpreted as an indication of increased reliance on the jaws as opposed to claws to attack prey (Barsbold 1983; Paul 1988). Furthermore, on the basis of denticle morphology, Currie et al. (1990) suggested that dromaeosaurines, with their short and broad denticles, may have been better suited to bite through the small bones of their prey than were velociraptorines, which, with their long, hooked posterior denticles, were slicing flesh parallel to the surface of bones. If such was the case, the stronger bite demonstrated here for *Dromaeosaurus* would also have been useful in feeding.

Tyrannosauroidae. Tyrannosaurids are a radiation of large theropods characterized by enormous heads, stout teeth, and short, two-fingered hands. Though they were definitively carnivorous and appear to have ingested bones (Chin et al. 1998; Varricchio 2001), there are still arguments as to whether tyrannosaurids were active predators or scavengers (e.g., Horner and Lessem 1993; Farlow 1994; Erickson and Olson 1996). The tyrannosaurids studied here are *Albertosaurus sarcophagus* (cast of ROM 1247), *Daspletosaurus torosus* (NMC 8506), and *Tyrannosaurus rex* (CM 9380, FMNH PR 2081, and casts of MOR 555, AMNH 5027, LACM 23844, TMP 81.6.1, and BHI-3033). The larger number of individuals for *Tyrannosaurus* should provide us, for the first time, with an idea of the range of variability occurring within a species (Fig. 10.10A–C). No measurements were made on the post-dentary region of TMP 81.6.1. Also, because the first two dentary teeth in tyrannosaurids are D-shaped (Molnar 1998), relatively smaller, and closer to the anterior of the mandible than in other theropods (owing to the inclined aspect of the symphysis), mandibular dimensions at the third alveolus were substituted for those of the second in order to construct the beam models.

Two juvenile tyrannosaurids represented by nearly complete mandibles were also studied: *Gorgosaurus libratus* (TMP 94.12.155, considered *Albertosaurus libratus* by Carr [1999]) and *Daspletosaurus torosus* (TMP 94.143.1). The preserved elements of the juvenile *Gorgosaurus* mandible consist of an articulated surangular-angular-articular complex and a dentary lacking the associated splenial. The mandible of the juvenile *Daspletosaurus* consists of a dentary broken at the level of the last alveolus and an articulated surangular-angular-articular com-

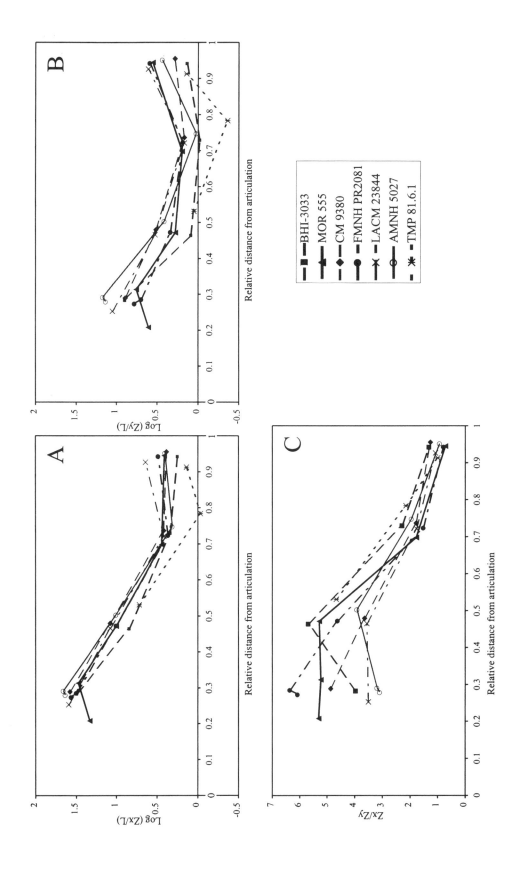

plex. Consequently, the total mandibular length had to be estimated for both juvenile individuals, but this uncertainty is minimal and should not influence the results dramatically. However, other uncertainties must be considered. First, measurements had to be taken posterior to the last alveolus rather than at the upper dentary suture in the juvenile *Daspletosaurus*; thus the third landmark on the force profiles is not the same as that of other taxa, although the values should be relatively close. Second, the missing splenial in *Gorgosaurus* prevents an accurate determination of the mandibular width at the upper dentary suture. Although less than the expected value, the width at the middentary was substituted for calculation purposes, with the consequences that the force values (Zx/L and Zy/L) at the upper dentary suture will be underestimated and the Zx/Zy values will be overestimated.

The mandible of tyrannosaurids is deeper than that of other theropods. The dentary is relatively deeper in order to sustain greater bending moments (Molnar 1998); and the larger surangular, which effectively increases the depth of the postdentary region, presumably offers a greater insertion area for the powerful mandibular adductor musculature (Molnar 1991). The dentary of *Tyrannosaurus rex* is oval and bears a lingual bar, which increases the resistance of the mandible to torsion (Molnar 1998). The orientation of the mandibular symphysis in tyrannosaurids is different from that of other theropods. Whereas the symphysis is subvertical (deeper than long) in most theropods, it is strongly inclined posteroventrally in tyrannosaurids, ranging from practically subequal to longer than it is deep (Fig. 10.3). The symphyseal surface often exhibits rugosities or bony projections at the posteroventral extremity that appear to have been interlocking with similar projections on the opposite mandible. According to Hylander (1984), extending the symphysis ventrally and lingually (posteriorly) is a solution adopted by animals whose mandibles need to sustain important torsional stresses.

Mandibular Force Profiles. Except for their absolute values, the mandibular force profiles of the four tyrannosaurid taxa studied are very similar. In *Albertosaurus*, dorsoventral force decreases at a rapid rate in the post-middentary region, and it is slightly lower at the third alveolus than at middentary (Fig. 10.11). Labiolingual force decreases rapidly again in the post-middentary region, and the mandible is slightly stronger at the second alveolus than at the middentary, creating a slightly positive pre-middentary slope (Fig. 10.11). The Zx/Zy profile demonstrates a linear decrease along the mandible, from a value of 2.63 at the upper dentary suture to a value of 1.15 at the third alveolus, revealing that the mandible of *Albertosaurus* is less well dorsoventrally buttressed than in other theropods and that it becomes nearly round anteriorly (Fig. 10.11).

In *Tyrannosaurus*, dorsoventral force decreases at a lower rate in the post-middentary region than in *Albertosaurus*; in addition, dorsoventral force at the third alveolus is nearly equal or superior to that at the middentary (Figs. 10.10A and 10.11). The variability at the third alveolus in *Tyrannosaurus* is related to differences in the posterior extent of the mandibular symphysis between individuals, as determined

Figure 10.10. (opposite page) Mandibular properties of Tyrannosaurus rex. *The abscissa represents the relative distance (in percent) from the articular fossa. (A) Dorsoventral force profiles (Zx/L) for all* T. rex *specimens studied. (B) Labiolingual force profiles (Zy/L) for all* T. rex *specimens studied. (C) Relative mandibular strength profiles (Zx/ Zy) for all* T. rex *specimens studied. From right to left: landmark 1 (second tooth alveolus), landmark 2 (middentary), and landmark 3 (upper dentary suture); if present, the next points represent landmarks 4 and 5 (lower dentary suture and maximum depth, respectively).*

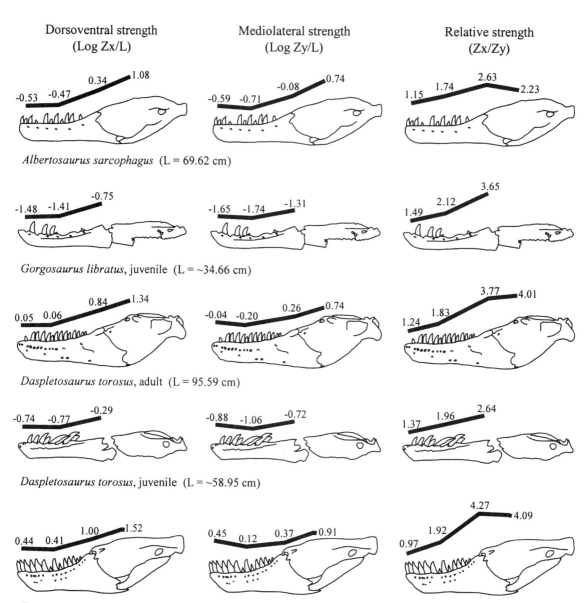

| Dorsoventral strength (Log Zx/L) | Mediolateral strength (Log Zy/L) | Relative strength (Zx/Zy) |

Albertosaurus sarcophagus (L = 69.62 cm)

Gorgosaurus libratus, juvenile (L = ~34.66 cm)

Daspletosaurus torosus, adult (L = 95.59 cm)

Daspletosaurus torosus, juvenile (L = ~58.95 cm)

Tyrannosaurus rex (mean, L = 120.39 cm)

Figure 10.11. Mandibular properties of tyrannosaurids. From left to right: landmark 1 (second tooth alveolus), landmark 2 (mid-dentary), landmark 3 (upper dentary suture), and landmark 5 (maximum depth). Mandibular length L represents the distance between the articular fossa and the anterior extremity of the mandible.

by the position of the symphyseal rugosities; however, the amount of variability does not greatly affect the interpretation of feeding behavior (see below). On average, the strength at the anterior extremity of the mandible is subequal to or greater than at the middentary. Labiolingually, force decreases slowly between the upper dentary suture and the middentary, but the mandible is stronger at the third alveolus than at the middentary, thus resulting in a positive pre-middentary slope (Figs. 10.10B and 10.11). Finally, the Zx/Zy profile (Figs. 10.10C and 10.11) reveals that the mandible is strongly dorsoventrally buttressed at the upper dentary suture (mean Zx/Zy = 4.27) and becomes rapidly rounder anteriorly, reaching values close to 1.00 at the third alveolus (mean Zx/Zy = 0.94).

In *Daspletosaurus*, dorsoventral force in the post-middentary region decreases at a rate similar to that seen in *Tyrannosaurus* (Fig. 10.11). The anterior extremity of the mandible is nearly as strong as at the middentary. The labiolingual force profile is nearly identical to that of *Tyrannosaurus* with a slow post-middentary rate of decrease and a moderate pre-middentary rate of increase (Fig. 10.11). Finally, the Zx/Zy profile (Fig. 10.11) is, again, similar to that of *Tyrannosaurus*, revealing a strongly dorsoventrally buttressed mandible at the upper suture (Zx/Zy = 3.77) that becomes rounder at the anterior extremity of the mandible (Zx/Zy = 1.24).

Interestingly, the dorsoventral and labiolingual force profiles for the juvenile tyrannosaurids differ from those of other tyrannosaurid taxa in absolute values only (Fig. 10.11). However, the Zx/Zy profile of the juvenile *Daspletosaurus* is more similar to that of *Albertosaurus* than to the adult *Daspletosaurus*, although the absolute values are slightly higher (behind last alveolus, Zx/Zy = 2.64; at the third alveolus, Zx/Zy = 1.37). Since the Zx/Zy value at the upper dentary suture is overestimated in the juvenile *Gorgosaurus* (see above), it may have been similar to that of *Albertosaurus*.

Interpretation. The greater values of dorsoventral and labiolingual force of *Tyrannosaurus* indicate clearly that its bite was stronger than that of the studied *Albertosaurus* specimen, a fact also suggested by skull bending strength (Henderson 2002). However, the absolute values of the mandibular force profiles for the adult *Daspletosaurus* (NMC 8506, mandibular length = 95.60 cm) are very close to those of the smallest adult *Tyrannosaurus* studied (TMP 81.6.1, mandibular length = 106.79 cm), suggesting a similar bite force for these two individuals (Figs. 10.10A and 10.11). When the force profiles of the tyrannosaurids are compared with those of other theropods, the bite of tyrannosaurids is seen to be more powerful than that of similar-size or larger theropods: the bite of *Albertosaurus* was only slightly weaker than that of *Acrocanthosaurus*, an animal with a skull nearly twice as long, and *Tyrannosaurus* had a bite more powerful than *Giganotosaurus!* Furthermore, the slow decrease rate of mandibular force along the tooth row in *Tyrannosaurus* is a testimony to its powerful bite, indicating that important pressure could be generated during bite along the entire tooth row (Molnar 1991).

The fact that the mandible at the third alveolus is nearly as strong as or stronger than at middentary in both dorsoventral and labiolingual planes indicates that tyrannosaurids used the extremity of their jaws for prey capture, handling, or both, even as juveniles (see *Gorgosaurus* and *Daspletosaurus* in the next section). Furthermore, the Zx/Zy profiles reveal that the mandible is strongly dorsoventrally buttressed posteriorly (mean Zx/Zy at the upper dentary suture = 4.27 for *Tyrannosaurus*; Zx/Zy = 3.77 in *Daspletosaurus*; Zx/Zy = 2.63 in *Albertosaurus*) but becomes rapidly rounder anteriorly, to the point of being equally adapted to resist dorsoventral and labiolingual loads at the third alveolus (mean of 0.94 in *Tyrannosaurus*; 1.24 and 1.15 in *Daspletosaurus* and *Albertosaurus*, respectively, but still within the range of variation of *T. rex*; 1.49 in the juvenile *Gorgosaurus;* and 1.37 in the juvenile *Daspletosaurus*). Such low values indicate that labiolingual and torsional stresses were as important (or nearly so) as those exerted in the plane of bite at the anterior extremity of the jaws. Given the powerful dorsoventral bite force generated by tyrannosaurids, and *Tyrannosaurus* in particular, these other stresses must have been extremely important. The broad snout, robust premaxillae, D-shaped premaxillary and anterior-most dentary teeth of tyrannosaurids (e.g., Paul 1987, 1988; Molnar 1998; Carr 1999), and strongly built skull (Holtz 1998; Henderson 2002) strongly support this interpretation.

Low values of Zx/Zy in the symphyseal region of the mandible have not been observed in any other theropods, with the exception of spinosaurids (see above). Although other theropods have Zx/Zy values similar to or higher than the value for the Komodo dragon, and hence probably delivered slashing bites, tyrannosaurids have values closer to the value for the molluscivorous Nile monitor, *Varanus niloticus*. As mentioned above, *V. niloticus* crushes snails, mussels, and crabs with its posterior teeth, and the beam models have revealed that its mandible is well adapted to resist the bending and torsional loads related to durophagy. Thus, tyrannosaurids may have similarly crushed bones using their posterior teeth (as suggested by Bakker et al. 1988), creating important torsional loads at the symphysis (Hylander 1981) and explaining the low Zx/Zy values observed.

The fact that the mandible is as strong as or stronger at the third alveolus than at the middentary (both Zx/L and Zy/L), however, suggests that the low Zx/Zy values observed at the symphysis are not merely an adaptation to resist the torsional and bending stresses induced while crushing bones along the posterior tooth row. Indeed, similar profiles are also characteristic of predatory crocodilians (Therrien and Ruff, in prep.). We can envision two ways in which important stresses can be induced at the anterior extremity of the mandible: (1) they may have been induced during prey capture, either by the struggles of large prey or by the shaking of the prey in order to finish it; or (2) they may be related to pulling and tugging movements of the head to detach flesh and bone from a carcass. The shape of tyrannosaurid teeth and the multidirectional microwear observed on them suggest that complex head movements were involved during feeding (Farlow et al. 1991; Abler 1992, 1997, 1999, 2001), somewhat reminiscent of the twist-feeding behavior observed in some modern crocodilians (Taylor 1987; Pooley 1989). Although the first hypothesis would be more typical of an active predator and the second of a scavenger, it is impossible to determine which type of feeding behavior is responsible for the force profiles observed in tyrannosaurids. In fact, the two hypotheses are not mutually exclusive, and both behaviors may very well have been practiced by tyrannosaurids. *Tyrannosaurus* might have captured large prey by holding it with its powerful jaws, shaken it to bring it to the ground or break its neck, bitten deeply through flesh and bone, and pulled and tugged at the carcass to detach a chunk of flesh.

The beam models may not answer the question of whether tyrannosaurids were active predators or scavengers, as they may have been both. However, the models do reveal that they were unique among theropods: rather than being adapted to deliver slashing, superficial bites, the mandible of adult tyrannosaurids was adapted to sustain important torsional and bending stresses related to deep bites, allowing them to both subdue large prey and bite through flesh and bone alike, as supported by cranial morphology (Molnar 1973, 1991, 1998; Carr 1999; Henderson 2002). It must also be noted that, owing to their reduced forelimbs, the jaws were the only means to capture prey for tyrannosaurids; other theropods had longer forelimbs armed with sharp claws that were most assuredly used in prey capture. The lower Zx/Zy values at the anterior extremity of the mandible in tyrannosaurids clearly demonstrate this difference in hunting strategy.

Given the importance of torsional stresses induced while feeding, the intramandibular joint would have been a plane of weakness in the tyrannosaurid mandible that may have led to important stress dissipation (rendering the bite less effective) or even catastrophic failure. To avoid these problems, tyrannosaurids appear to have secondarily lost the ability to move their dentaries relative to the postdentary elements by interlocking the bones forming the intramandibular joint (Hurum and Currie 2000; Bluhm 2002).

Bite Force Estimates of Studied Theropods

Because theropods are often pictured as "big, mean, killing machines," the amount of force they could exert in a bite has intrigued and fascinated generations of children and paleontologists alike. Knowledge about the bite force of an animal can provide insights into its behavioral ecology and the prey it may have hunted (see Meers 2002). The amount of force an animal can exert during its bite depends on the interplay between cranial morphology and jaw adductor muscle architecture, more precisely the position of insertion and origin relative to the mandibular articulation (i.e., lever arm) and the cross-sectional area (i.e., number of fibers) of each muscle. As a result, determination of bite force absolute intensity requires information that is rarely preserved in fossils and that can only be partly gained from closely related extant species or modern analogues.

Over the past few years, numerous authors have attempted to determine the absolute bite force of various theropod taxa. Because direct methods, such as the use of strain gauges or the measurement of jaw adductor muscle cross-sectional areas, could not be employed to estimate bite force in theropods, indirect but scientifically valid approaches were taken. On the basis tooth mark impressions left on *Triceratops* bones, Erickson et al. (1996) estimated that *Tyrannosaurus rex* could have exerted between 6,410 and 13,400 N of force with one of its long caniniform teeth (fourth through seventh maxillary teeth). Meers (2002), using the lowest force estimate obtained by Erickson et al. (1996) as a mean value for each maxillary tooth, calculated that *T.*

rex could have exerted as much as 153,600 N of force in a bilateral bite (twelve maxillary teeth per side). However, this value is much lower than the bite force estimate he derived from a method based on correlations observed between the body mass of extant predators and their maximum bite force. Indeed, Meers (2002) estimated an astounding 235,123 N bite force for *T. rex* by extrapolating the regression line of the combined reptilian and mammalian predator sample to the estimated body mass of the large tyrannosaurid. Bite force estimates obtained from regressions calculated for crocodilians and mammals independently were significantly lower than the one obtained from the combined sample, but the estimate from the crocodilian regression comes close to the value derived from the results of Erickson et al. (1996) (Table 10.1). Finally, using an advanced engineering modeling technique and computer software, Rayfield et al. (2001) conducted a finite element analysis of a CT-scanned *Allosaurus fragilis* skull (MOR 693) to study stress resistance. Their results suggested that this allosaurid, using solely its jaw adductors, could have exerted a bilateral bite force (at the third through fifth maxillary teeth) of 2,148 N and that the maximum bilateral force of the skull before yield would have been 18,748 N.

Can the results of the mandibular beam models presented above give insight into the bite force of the various theropod taxa? The answer is "only in relative terms." As mentioned earlier, the assumptions regarding the solid nature of the mandible and the constancy of safety factors of vertebrate mandibles allow us to study only relative changes in biomechanical properties and not their absolute values. It is therefore impossible to derive a bite force estimate in absolute values (i.e., Newtons) from our results. Bite force comparisons among theropod taxa have been discussed in terms of the force profiles above, but they do not convey a concept of actual bite intensity. However, comparison of Zx/L values at similar landmarks with extant taxa can provide an idea of *relative* bite force, assuming that all mandibles are equally approximated by a solid beam model, that is, that the amount of cortical bone contributing to the real mandibular cross-section is proportional in the compared taxa. Unfortunately, no extensive study of variation in mandibular cortical bone thickness over a wide range of individuals and taxa has ever been conducted. An inspection of CT scans of the mandibles of *V. komodoensis* (Bluhm 2002) and *A. mississippiensis* (Rowe et al. 1999) reveals that the dentaries of these animals are not truly solid structures: they possess a hollow core for the passage of blood vessels and nerves (Fig. 10.12A,B). Therefore, the solid beam model will somewhat overestimate the Zx/L values of these taxa. In contrast, the dentaries of *T. rex* are solid, interrupted only by tooth alveoli (Fig. 10.12C–D; Molnar 1998; Brochu and Ketcham 2002), so their cross-sectional properties should be well approximated by the solid beam model. Consequently, if the dentaries of all theropods can be considered solid, as claimed by Van Valkenburgh and Molnar (2002), the bite force estimates relative to our modern (non-solid) analogues (ratio Zx/L_{theropod} over $Zx/L_{\text{extant taxon}}$; see Table 10.1) should be considered as minimal values.

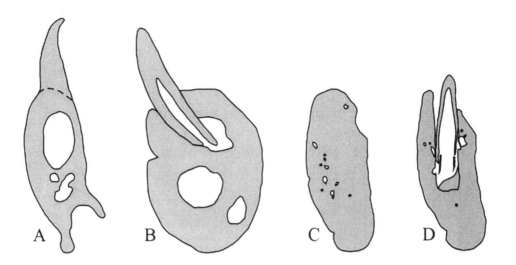

Middentary Zx/L values of theropods are compared to those obtained for *Varanus komodoensis* and *Alligator mississippiensis* (Table 10.1). Owing to the paucity of data surrounding the bite force of the giant varanid and because previous theropod bite force studies have involved data for *A. mississippiensis*, comparison with this crocodilian is warranted. The $Zx/L_{middentary}$ value for *A. mississippiensis* was derived from seven adult individuals with a mean mandibular length of 50.08 cm (Therrien and Ruff, in prep.). Finally, our relative bite force estimates for theropods are compared to those obtained from published absolute bite force values (Table 10.1).

The dichotomy between velociraptorines and dromaeosaurines discussed above is even clearer when bite forces are estimated as a function of the bite forces of extant predators: although velociraptorines had a maximum bite force close to that of *V. komodoensis*, that of *Dromaeosaurus* was three times as strong (Table 10.1). Surprisingly, the juvenile *Gorgosaurus* studied had a bite force nearly identical to that of *Dromaeosaurus* even though it was a much larger animal (mandible twice as long), suggesting that competition between dromaeosaurines and juvenile tyrannosaurids may have been intense. In spite of having similar mandibular lengths, *Dilophosaurus* appears to have had a slightly weaker bite than *A. mississippiensis* (Table 10.1). In light of the results obtained from the mandibular force profiles, we infer that this theropod may have specialized in prey relatively smaller than itself.

Suchomimus, Ceratosaurus, Allosaurus, "*Antrodemus*," and the juvenile *Daspletosaurus* all seem to have been capable of exerting bite forces close to that of *A. mississippiensis*. Of particular interest is our value obtained for the mandible of *Allosaurus* (1.074 relative to *A. mississippiensis*), which is comparable to the maximum bilateral bite force before yield obtained by Rayfield et al. (2001) for the skull of the same specimen (0.991 relative to *A. mississippiensis*) (Table 10.1). The close correspondence of these results validates our methods, indicating

Figure 10.12. Coronal section through mandibles near the middentary landmark (from CT scans): (A) Varanus komodoensis (Bluhm 2002 [p. 14], exact location unspecified); (B) Alligator mississippiensis (Rowe et al. 1999, CT slice 627); (C) Tyrannosaurus rex, through interalveolar space (Brochu and Ketcham, CT slice 277); (D) Tyrannosaurus rex, through tooth alveolus (Brochu and Ketcham, CT slice 288). Mandibles have been standardized to a common depth. Assuming that teeth and alveoli do not have a significant effect on beam models in a comparative context (Daegling et al. 1992; Daegling and Hylander 1998), the mandible of T. rex can be considered a truly solid ellipse. In contrast, the mandibles of V. komodoensis and A. mississippiensis have a hollow core. Consequently, their mandibular cross-sectional properties will be somewhat overestimated by a solid beam model and the ensuing theropod bite force estimates (ratio of theropod Zx/L to extant taxon Zx/L) will be minimum values.

that simple beam models can provide results as valid as those of more complex biomechanical techniques.

The results presented in Table 10.1 also reveal that the specimen of *Albertosaurus,* presumably a sub-adult (Carr 1999), and the abelisaurids *Majungatholus* and *Carnotaurus* had bites twice as powerful as the bite of *A. mississippiensis.* These results come as a surprise given the generally slender appearance of the abelisaurid mandible; however, they confirm the role of abelisaurids as ferocious predators and invite further study of their paleoecology.

Finally, in the largest theropods, the large allosauroids *Acrocanthosaurus* and *Giganotosaurus* had extremely powerful bites, nearly four and five times as strong as the bite of *A. mississippiensis,* respectively; they are surpassed by tyrannosaurids (Table 10.1). The estimates indicate that *Daspletosaurus* had a bite force seven times as high as that of *A. mississippiensis,* whereas that of the average *Tyrannosaurus rex* was nearly sixteen times as high! These values are a testimony of the great power of these theropods and are consistent with published values indicative of bone-cracking abilities (e.g., Erickson et al. 1996; Meers 2002). Surprisingly, although the estimate for *Daspletosaurus* approaches the minimum bite force estimated for *T. rex* by Erickson et al. (1996), the bite force estimate for *T. rex* surpasses the highest estimate obtained by Meers (2002). Indeed, our estimate (15.914 relative to *A. mississippiensis*) is quite close to the value for a bilateral bite extrapolated from the *maximum* force of Erickson et al. (1996) following Meers's (2002) calculation methods (17.005 relative to *A. mississippiensis*) (Table 10.1).

Ontogenetic Changes in Feeding Behavior in Theropods

The mandibular models presented above, and the interpretations drawn from them, apply to adult theropods (unless specified otherwise). However, as is the case in crocodilians (e.g., Cott 1961; Fogarty and Albury 1968; Pooley 1989) and varanids (e.g., Rieppel and Labhardt 1979; Auffenberg 1981, 1988, 1994), feeding behavior and diet probably changed during theropod ontogeny. Indeed, this is supported by juveniles with body proportions different from those of adults (Russell 1970) and by studies of ontogenetic changes in skull and postcranial morphology (Currie 1998, 2003; Smith 1998; Carr 1999).

Growth series for *Allosaurus fragilis, Albertosaurus sarcophagus* (sensu Horseshoe Canyon Formation species; Holtz 2001; P. Currie, pers. comm. 2002), *Gorgosaurus libratus* (sensu Dinosaur Park Formation species; Holtz 2001; P. Currie, pers. comm. 2002), and *Tyrannosaurus rex* (including "*Nanotyrannus lancensis*"; see Carr, 1999) were studied in order to determine whether mandibular biomechanical properties suggested differences in feeding behavior at various ages. Unfortunately, complete mandibles were not available for all individuals, the vast majority being represented solely by dentaries. Thus, it was impossible to compare the complete mandibular force profiles at different ontogenetic stages. However, we were able to determine the section modulus (Z, see equation 4 in Methods), an indicator of bending rigidity, from the external dimensions of the dentary. We have calculated the bending rigidity in the plane of bite (Zx) and the relative mandibular shape (Zx/Zy) at the middentary (tenth tooth in *Allosaurus*, ninth in tyrannosaurids) and at the second alveolus (third in tyrannosaurids) (Table 10.2). The distance of each landmark from the anterosuperior extremity of the dentary was measured as a mandibular length indicator, which should reflect differences in body size within a particular species. In other words, the distance between the middentary and the anterior extremity of the mandible increases as the individual becomes larger; thus, this distance can be used as an ontogenetic index to differentiate younger (smaller) and older (larger) individuals.

Although numerous *Allosaurus* specimens were studied, the sample size for the tyrannosaurid taxa was much smaller. Consequently, it is difficult to determine how significant the patterns observed truly are. Nevertheless, we present the results for these taxa in the hope that future discoveries will complement this database and that researchers will benefit from it and use it to either confirm or falsify our interpretations. Statistical analyses were conducted with the statistical software SPSS 10.0.7.

Ontogenetic Changes in Allosaurus fragilis. A total of thirty-four dentaries from juvenile to adult individuals (including MOR 693) were measured, the majority of which were from the Cleveland-Lloyd Quarry (UMNH, UUVP, BYU, and YPM-PU specimens). Previous studies (Smith 1998) have revealed that the thickness (i.e., width) of the dentary may vary isometrically with dentary depth at the twelfth alveolus and that smaller individuals are more robust than larger ones. Furthermore, no size-independent characters could be identified that distinguished juvenile from adult animals (Smith 1998, p. 140).

The Zx values presented in Table 10.2 clearly demonstrate that bending rigidity increases by more than an order of magnitude during ontogeny in *Allosaurus* (compare UMNH VP6475 and BYU 2028). Also apparent is a slight increase in bending rigidity at the second alveolus relative to that at the middentary, a trend confirmed by the ratio of $Zx_{middentary}/Zx_{2nd\ alveolus}$ (Fig. 10.13A). Although the predictive value of the least-squares linear regression of $Zx_{middentary}/Zx_{2nd\ alveolus}$ is poor (slope = -0.02, R^2 = 0.12), there is a significant correlation ($p < 0.05$) between the ratio of Zx values and body size, as expressed by the distance between the middentary and the anterior extremity of the mandible. Thus, the decreasing Zx ratio associated with the increasing body size indicates that the mandible gradually becomes stronger in bending at the second alveolus relative to the middentary during ontogeny.

From the results presented in Table 10.2, we observe that, with the exception of UMNH VP9347 and UMNH VP9362, all individuals regardless of size (i.e., distance to middentary) exhibit an anterior decrease in Zx/Zy, going from a mean of 1.89 (s = 0.2) at the middentary to a mean of 1.50 (s = 0.2) at the second alveolus. In other words, the mandible becomes slightly rounder anteriorly because labiolingual stresses are relatively more important. Furthermore, the ratio of the Zx/Zy values at middentary and at the second alveolus reveals that the mandible preserves the same relative shape throughout ontogeny (see

TABLE 10.2.

Mandibular Properties of Growth Series in *Allosaurus fragilis*, *Albertosaurus sarcophagus*, *Gorgosaurus libratus*, and *Tyrannosaurus rex*

The distance between the mid-dentary and the anterior extremity of the mandible acts as an ontogenetic index (see text for explanation).

Taxon	Distance, mid-dentary to anterior extremity of mandible (cm)	Zx mid-dentary	Zx 2nd alveolus	Zx/Zy mid-dentary	Zx/Zy 2nd alveolus
Allosaurus fragilis					
UMNH VP6475	9.50	1.67	1.45	2.12	1.76
UMNH CLDQ 01-130	10.24	3.72	3.40	1.87	1.46
UMNH VP9336	10.34	2.79	2.46	1.68	1.27
UMNH VP9362	10.69	3.24	3.30	1.81	2.05
UMNH VP9357	11.05	3.48	3.37	1.75	1.50
UMNH VP9333	11.14	4.08	3.61	1.85	1.32
UMNH VP9342	11.42	3.04	2.69	1.86	1.25
UMNH VP9370	11.62	4.54	3.56	1.60	1.23
UMNH VP9334	11.63	3.24	2.13	1.96	1.43
UUVP 2903	11.95	5.73	5.83	1.78	1.48
UMNH VP9349	12.09	5.52	5.44	2.09	1.59
UMNH VP6474	12.18	4.14	2.82	2.03	1.57
UMNH VP9365	12.36	6.52	5.33	1.84	1.58
UMNH VP9343	12.49	5.01	4.18	1.79	1.48
UMNH VP9369	12.65	4.64	3.81	1.79	1.27
UMNH VP9338	12.90	5.75	5.04	1.89	1.35
YPM-PU 72	13.93	3.95	3.57	2.05	1.68
UMNH VP0038 (left)	13.94	8.85	8.26	1.76	1.32
UMNH VP6473	14.32	9.02	10.67	1.91	1.46
YPM-PU D12	14.74	13.65	15.09	2.28	1.77
UMNH VP0038 (right)	14.76	8.74	11.54	1.73	1.36
UMNH VP6480	15.38	14.69	12.53	1.71	1.39
Cast of MOR 693	15.45	9.11	11.00	2.35	1.69
UMNH VP9348	16.67	15.72	12.21	1.68	1.48
YPM-PU 14554	16.70	10.78	12.01	2.01	1.51
UMNH VP9351	16.70	12.75	12.04	2.01	1.18
UMNH VP9347	16.74	10.60	9.44	1.64	1.67
UMNH VP9344	17.47	14.52	11.17	1.77	1.60
UMNH VP9345	17.82	16.82	12.57	1.88	1.49
UMNH VP9337	18.04	12.64	13.30	1.76	1.46
UMNH VP6476	18.95	21.36	23.14	1.74	1.46
BYU 759/2028	19.14	27.20	33.67	1.99	1.53
BYU 725/17125	22.00	21.49	23.77	2.45	2.02

TABLE 10.2. *(cont.)*

Mandibular Properties of Growth Series in *Allosaurus fragilis*, *Albertosaurus sarcophagus*, *Gorgosaurus libratus*, and *Tyrannosaurus rex*

Taxon	Distance, mid-dentary to anterior extremity of mandible (cm)	Zx mid-dentary	Zx 2nd alveolus	Zx/Zy mid-dentary	Zx/Zy 2nd alveolus
Albertosaurus sarcophagus					
TMP 99.50.40	15.6	9.37	14.78	1.89	1.33
Cast of ROM 1247	18.57	17.36	19.43	1.74	1.15
USNM 12814	20.05	40.53	49.46	1.71	1.33
TMP 94.25.6	22.8	49.39	45.91	2.23	1.20
TMP 96.25.6	24.6	61.53	85.82	2.12	1.30
TMP 67.9.164	26.41	56.09	56.67	2.63	1.33
Gorgosaurus libratus					
TMP 94.12.155	9.75	0.97	1.07	2.12	1.49
TMP 86.144.1	13.51	4.27	5.76	2.09	1.34
TMP 99.55.170	17.03	14.65	19.29	2.14	1.01
TMP 95.5.1	17.91	24.50	36.01	2.05	1.20
TMP 82.28.1	20.09	44.61	61.98	1.95	1.31
TMP 83.36.134	21.89	53.39	56.10	1.91	1.22
TMP 94.12.602	22.12	49.58	60.41	2.02	0.96
Tyrannosaurus rex					
TMP 96.5.7 9	17.73	16.04	20.27	1.76	0.74
Cast of CMNH 7541	?	?	26.57	?	0.86
Cast of TMP 81.6.1	23.14	77.15	134.02	2.15	0.99
Cast of AMNH 5027	28.52	173.71	270.72	1.95	0.92
CM 9380	30.32	221.89	266.44	1.78	1.26
Cast of BHI-3033	33.12	194.72	204.91	2.28	1.30
Cast of LACM 23844	33.54	225.50	489.17	1.76	1.08
Cast of MOR 555	35.50	226.92	284.82	1.77	0.72
FMNH PR 2081	36.68	302.73	381.84	2.06	0.78
Mean, adult *T. rex*	32.95	222.66	312.95	1.92	0.97

regression in Fig. 10.13B), thus indicating that the ontogenetic change in Zx at the second alveolus described above is accompanied by an equivalent increase in Zy. These results seem to support Smith's (1998) allegation that juveniles cannot be distinguished from adults on a morphological basis. However, the nearly constant Zx/Zy ratios, regardless of body size, do not indicate that larger individuals are more gracile (i.e., that they have a smaller mandibular width relative to depth) than smaller ones as Smith (1998, p. 131) suggested.

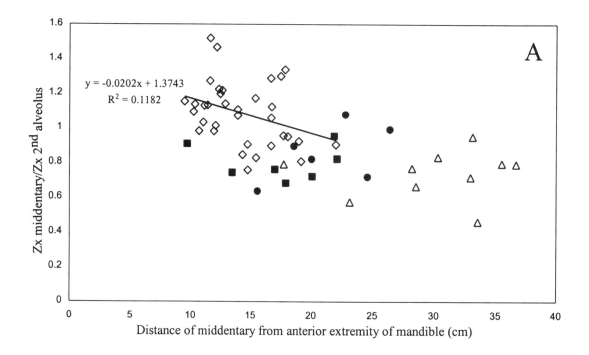

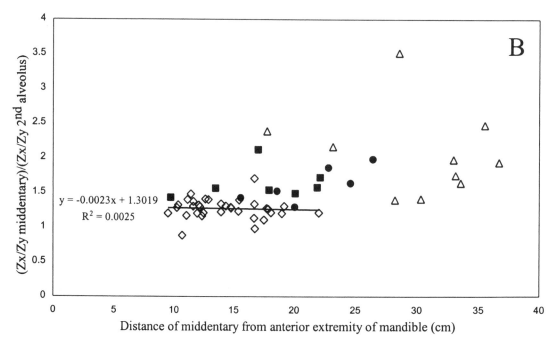

But how can we interpret these data in terms of feeding behavior? The simultaneous increase in dorsoventral and labiolingual bending rigidity at the second alveolus relative to the middentary during ontogeny suggests that the force applied at the extremity of the jaw for prey capture and handling increased in importance with age. Younger (and smaller) individuals apparently did not require a strengthened mandibular symphysis to capture prey. Several possible scenarios could explain such an observation. First, if small individuals hunted small prey, a strong symphyseal region may not have been necessary to subdue it; prey would just have been captured with the jaws and swallowed whole. Alternatively, small individuals may have hunted prey that was relatively large with respect to their body size (prey that was thus too large for them to have held and subdued as adults did) by delivering slashing bites and waiting for the prey to die (an ambush technique similar to that of Komodo dragons), by hunting in small packs, or both.

The nearly constant Zx/Zy values throughout ontogeny, in spite of changing Zx (and Zy) values, strongly suggests that, although the *absolute* stress magnitude exerted on the mandible may have varied, the *relative* stress magnitude did not. In other words, Zx/Zy values could have remained constant either because prey selection changed with age or because younger and smaller individuals had different hunting techniques than adults. By attacking small prey, which exerted *absolutely* smaller but *relatively* equivalent loads on the mandibles as larger prey in adults, or by attacking in packs, juveniles maintained similar Zx/Zy values as adults. As the individuals grew larger, the *absolutely* larger prey of their early years then became *relatively* smaller. These prey could now be grasped and subdued as described above (see Allosauroidea), and because dorsoventral and labiolingual bending rigidity (Zx and Zy) at the second alveolus increased proportionally, the Zx/Zy ratio was preserved.

Recently, Bakker (1997) described fossil localities where large, disarticulated herbivore remains exhibiting abundant tooth marks were found associated with monospecific tooth assemblages representing *Allosaurus* ontogenetic series. He interpreted these fossil assemblages as evidence of parental care in this theropod, where adults carried carcasses back to the lair to feed their young. Do our results give insight into the possibility of parental care? It is tempting to interpret the increase in bending rigidity at the anterior extremity of the mandible in the *Allosaurus* ontogenetic series as evidence for the gradual learning and perfecting of capture techniques leading to a greater involvement of juveniles in prey capture under parental supervision. However, this pattern could also have developed in the absence of parental care, as explained above. Furthermore, juveniles of extant carnivoran species having a long period of dependence (e.g., lions) do not differ significantly from the adults in the shape of their mandibular force profiles (Zx/L and Zy/L), but do differ in their Zx/Zy values, especially at the canine (Therrien, in prep.). This difference in values reflects the fact that the juveniles do not participate in prey capture, instead feeding on prey killed by the parents or the pack. In contrast, juveniles of extant carnivoran species who have to participate in the hunt and compete for access to a carcass at an early age (e.g., spotted hyenas) have mandibular force profiles and Zx/Zy_{canine} values that do not differ significantly from those of adults (Therrien, in prep.). Given the various ways in which the observed ontogenetic changes in mandibular properties could

Figure 10.13. (opposite page) Relative changes in mandibular properties during theropod ontogeny. The distance separating the mid-dentary and the anterior extremity of the mandible acts as an ontogenetic index within a given species (hollow diamonds, Allosaurus fragilis; solid circles, Albertosaurus sarcophagus; solid squares, Gorgosaurus libratus; hollow triangles, Tyrannosaurus rex). (A) Changes in mandibular bending rigidity ($Zx_{mid-dentary}/Zx_{2nd\ alveolus}$). There appears to be a statistically significant decrease in this ratio in Allosaurus with body size, indicating that the anterior extremity of the mandible becomes stronger relative to the mid-dentary with age. No significant change in this ratio is observed in tyrannosaurids. (B) Changes in relative mandibular strength ($Zx_{mid-dentary}/Zx_{2nd\ alveolus}$). No significant changes are observed in any theropods, although the range of variation in Tyrannosaurus is important.

have arisen in *Allosaurus*, it is unfortunately impossible to resolve the question of parental care in this theropod from these data.

Ontogenetic Changes in Albertosaurus sarcophagus. Six dentaries, from juvenile to large adult individuals, referable to *Albertosaurus sarcophagus* (including ROM 1247), have been studied. Previous studies (Carr 1999) have shown that the mandible of small juveniles was equidimensional (depth = width) and that early during ontogeny it became deeper than it was wide. Also, all tyrannosaurids were shown to have dentary height (i.e., depth) that scales positively with body size, whereas mandibular length scales isometrically, and to have dentary length and dentary tooth row that scale negatively with skull length and maxillary tooth row, respectively (Currie 2003).

The Zx values reported in Table 10.2 indicate that a ten-fold increase in bending rigidity occurs as the animal grows, reflecting its more powerful bite force (compare TMP 99.50.40 and TMP 96.25.6). However, there appears to be a change in bending rigidity profile between juveniles and adults (individuals larger than ROM 1247, sensu Carr 1999): bending rigidity is higher at the third alveolus than at the middentary in juveniles, whereas it is subequal in adults. A one-sample *t*-test of $Zx_{middentary}/Zx_{3rd\ alveolus}$ (Fig. 10.13A) suggests that adult *Albertosaurus* may have had a significantly different profile from the juvenile TMP 99.50.40, but a least-squares regression failed to find a significant correlation between body size and this ratio. Needless to say, caution is necessary given the small sample size.

Although the Zx/Zy value (Table 10.2) at the third alveolus is relatively high in the juvenile ($Zx/Zy = 1.33$), it is within the range of adults (mean = 1.26, s = 0.08). On the other hand, the Zx/Zy values suggest that the shape of the mandibular corpus at the middentary changes during ontogeny. Individuals larger than USNM 12814 have higher Zx/Zy values at the middentary than smaller individuals (Table 10.2). In other words, the mandible appears to become deeper than it is wide at the middentary during ontogeny, as previously described by Carr (1999). The ratios of the Zx/Zy values at the middentary over those at the third alveolus (Fig. 10.13B) suggest the existence of two clusters (USNM 12814 and smaller individuals versus TMP 94.25.6 and larger). Unfortunately, the sample size is too small to determine whether this difference is really significant (significantly different if equal variances assumed, not significantly different if equal variances not assumed).

In terms of feeding behavior, these results suggest that the anterior extremity of the jaw played an important role in prey capture in *Albertosaurus* throughout ontogeny. Such an adaptation would be necessary if juveniles were active predators that needed to capture prey themselves, as opposed to feeding on carrion or prey captured by a parent, and could explain why their mandibles are rounder at the middentary: the stresses induced by a struggling prey would be relatively more important on the mandible of small individuals than on that of adults. As the individual grew larger, the prey would become relatively smaller, thus leading to a reduction of torsional and labiolingual stresses and allowing the mandible to become dorsoventrally buttressed to resist bending due to increased bite force. However, if the discovery and study of further juveniles were to reveal that the Zx/Zy value at the third alveolus is significantly higher in juveniles than in adults, it may suggest slightly different feeding habits, where the juveniles may have delivered slashing bites, as suggested by their laterally compressed teeth.

Ontogenetic Changes in Gorgosaurus libratus. Seven individuals, ranging from juvenile to large adult individuals, identified as *Gorgosaurus libratus* were studied. Two juveniles (including TMP 94.12.155) are included and should give insight into changes associated with increase in body size.

The Zx values reported in Table 10.2 clearly indicate that dorsoventral bending rigidity increases immensely, by nearly two orders of magnitude, during ontogeny. Surprisingly, the profile does not change with increase in body size; bending rigidity is consistently higher at the third alveolus than at the middentary, as indicated by the relative constancy of the proportion of $Zx_{middentary}$ to $Zx_{3rd\ alveolus}$ throughout the growth series (mean = 0.80, s = 0.10; Fig. 10.13A).

The Zx/Zy values (Table 10.2) represent the changes in mandibular shape during ontogeny. The Zx/Zy values at the middentary remain constant throughout the growth series (mean = 2.04, s = 0.09), but those at the third alveolus are more variable. As in *Albertosaurus,* the Zx/Zy values at the third alveolus of juveniles are slightly higher (mean = 1.42, s = 0.10) than those of adults (mean = 1.14, s = 0.15), but a *t*-test does not detect a significant difference ($p < 0.067$)—although caution is required given the small sample size. It thus appears that the relative shape of the mandible did not change with an increase in body size, as indicated by the relative constancy of the proportion between Zx/Zy at the middentary and Zx/Zy at the third alveolus (Fig. 10.13B).

In terms of feeding behavior, the anterior extremity of the mandible played an important role in prey capture and handling throughout ontogeny, as in *Albertosaurus* (see above). Unlike the middentary region in *Albertosaurus,* however, that region in *Gorgosaurus* appears to maintain the same shape throughout ontogeny. Unless the Zx/Zy values at the third alveolus in juveniles can be demonstrated to be significantly different from those of adults through the discovery and study of additional specimens, *Gorgosaurus* appears to have been a shape-conservative theropod in its mandibular morphology, juveniles being scaled-down versions of adults. Behaviorally, this suggests that feeding habits probably did not change greatly during ontogeny. The way these animals captured and handled their prey exerted important stresses— torsional and labiolingual—on the mandible, as seen in the low Zx/Zy values. This suggests that juveniles must, therefore, have learned to hunt and subdue prey by themselves at an early age, although the prey was correspondingly smaller. In contrast, if the difference in Zx/Zy value is statistically significant, then hunting behavior may have been different in juveniles. Stresses induced on the mandible by prey were relatively lower than in adults, so juveniles may have killed by delivering slashing bites, as suggested by their laterally compressed teeth (Currie et al. 1990; Carr 1999).

Ontogenetic Changes in Tyrannosaurus rex. Nine dentaries of *Tyrannosaurus* were compared: the seven specimens used above for the mandibular force profiles and two juveniles (a cast of "*Nanotyrannus lancensis*" CMNH 7541 and TMP 96.5.7, possibly attributable to the same taxon; P. Currie, pers. comm. 2002). Because the mandible of CMNH 7541 is preserved in occlusion with the skull, it was impossible to determine the mandibular depth at middentary. However, it was possible to estimate the mandibular depth and width at the third alveolus, and these results are reported here. Previous studies revealed that *Tyrannosaurus* underwent an increase in robustness during ontogeny (Carr 1999) and that two morphs, a robust and a gracile, can be recognized among adults (Carpenter 1990).

The Zx values reported in Table 10.2 demonstrate that bending rigidity increases by more than an order of magnitude during ontogeny (compare TMP 96.5.7 and FMNH PR 2081). In addition, bending rigidity at the third alveolus appears to be always subequal to or greater than that at the middentary. The proportion between $Zx_{middentary}$ and $Zx_{3rd\,alveolus}$ remains relatively constant throughout ontogeny (Fig. 10.13A), seemingly supporting this idea (mean = 0.74, s = 0.14).

Relative mandibular shape (Zx/Zy, Table 10.2) appears to be almost constant in the *Tyrannosaurus* growth series. The mandible is nearly round at the third alveolus (mean Zx/Zy = 0.95, s = 0.22) but is nearly twice as deep as it is wide at the middentary (mean Zx/Zy = 1.98, s = 0.39). Although highly variable, the ratio between the Zx/Zy values at the middentary and at the third alveolus (Fig. 10.13B) does not differ between juveniles and adults (mean = 2.08, s = 0.67).

Given the degree of cranial modifications that *Tyrannosaurus* is thought to have undergone during ontogeny (Carr 1999; also see Currie 2003), it is surprising to discover that the mandibular force profiles of juveniles are so similar to those of adults. This similarity suggests that, even though the *absolute* stress magnitude exerted on the mandible differed between juveniles and adults, the *relative* stress intensity along the mandible did not vary. Since the mandible is still relatively stronger at the symphysis than at the middentary in juveniles, one must conclude that the anterior extremity of the jaw played an important role in prey capture and handling even in juveniles. The low Zx/Zy values at the third alveolus indicate that the struggling prey generated important torsional and labiolingual stresses or that complex head movements were made to extract a chunk of flesh. The laterally compressed dentition of CMNH 7541 indicates that juvenile *Tyrannosaurus* did not process bone, but the incisiform premaxillary teeth support the conclusion that juveniles may have subdued live prey (Bakker et al. 1988).

Thus, it appears that tyrannosaurids captured and subdued prey at a relatively young age, as indicated by the high Zx values at the third alveolus of juvenile *Albertosaurus, Gorgosaurus,* and *Tyrannosaurus* and their low Zx/Zy values at the third alveolus, values that were not significantly different from those of adults (Table 10.2). This similarity in values suggests that juveniles were active predators, captured their own prey, and did not feed exclusively on prey killed by their parents or on carrion. Extended periods of parental care were probably not typical in tyrannosaurids. Indeed, as mentioned above, juveniles of extant carnivoran species relying on their parents to kill prey differ significantly from adults in their Zx/Zy values at the anterior extremity of the mandible, whereas those who have to hunt and capture prey at an early age do not.

Another interesting issue is that of competition with dromaeosaurids. Juvenile tyrannosaurids would have competed for prey with the contemporaneous dromaeosaurids. The difference in mandibular force profiles between tyrannosaurids and dromaeosaurids indicates that these theropods hunted differently. Whereas tyrannosaurids subdued a prey animal by biting and possibly shaking it in order to finish it, dromaeosaurids may have hunted in packs and killed prey by inflicting gashing wounds with their claws and delivering slashing bites. Associated with these different killing methods were probably different kinds of prey: tyrannosaurids would have attacked prey they could contain (no larger than themselves), whereas dromaeosaurids could have attacked prey much larger than themselves. This ecological partitioning may have minimized the competition between the two taxa during the early ontogeny of tyrannosaurids. By the time prey size overlapped, tyrannosaurids would have been much larger than their rivals, and a solitary hunter (or a small group, see Currie 1998) would have fared better against a pack of dromaeosaurids.

Summary

Beam theory has shown itself to be useful for interpreting variation in the cross-sectional properties of the mandibles of varanids and theropods. Comparison of biomechanical models for the mandibles of these animals has demonstrated similarities and differences in mandibular force profiles that can be interpreted in terms of feeding behavior. Because the theropod feeding behaviors were interpreted through comparison with mandibular models for extant taxa of well-documented feeding behavior, we feel that our interpretations of *possible* feeding behaviors are scientifically valid. The strength of our approach lies in its quantitative nature, which makes it possible to compare a wide variety of taxa and to test the results and interpretations against those obtained from different methods.

The varanid mandible behaves as a simple lever, where bite force decreases at a constant rate along the tooth row, reflecting a mandible designed to deliver slashing bites without specialization to hold or handle prey. However, specialization toward a specific feeding behavior can be observed in the overall shape of the mandibular corpus (Zx/Zy). The mandible of *Varanus komodoensis* is deeper than it is wide ($Zx/Zy > 2.00$) along the tooth row but becomes slightly rounder ($Zx/Zy = 1.63$) near the symphysis, indicating that the principal stresses along the tooth row are related to dorsoventral bending induced while slicing flesh but that labiolingual stresses become more important anteriorly owing to the impact with prey during the bite. In contrast, the mandible of the molluscivorous *Varanus niloticus* is nearly round along the tooth row and becomes perfectly round at the anterior extremity (*Zx/Zy* ~1.00), an adaptation to resist the strong torsional moments generated while crushing shells.

Theropods exhibit a great diversity of feeding behaviors. The use of the largest, non-mammalian, terrestrial predator alive today, the Komodo dragon, as a modern analogue to infer theropod behavior is not applicable to all taxa. In fact, only the allosauroid "*Antrodemus valens*" and the abelisaurid *Majungatholus atopus* appear to have been similar to *V. komodoensis* in their mandibular strength profiles and in their craniodental morphology (broad skull and short, recurved teeth); hence, these theropods were well suited to attack larger prey and probably practiced an ambush attack strategy. *Carnotaurus sastrei* also appears to have delivered slashing bites, as suggested by the simple-lever mandibular force profiles, but its deep, narrow skull suggests that it could neither inflict large gashing wounds nor detach large chunks of flesh from its prey, something that the broad skull of *V. komodoensis* allows it to do.

Dromaeosaurids all possess mandibles behaving as simple levers, but dromaeosaurines (*Dromaeosaurus albertensis*) can be distinguished from velociraptorines (*Deinonychus antirrhopus, Velociraptor mongoliensis,* and *Saurornitholestes langstoni*) on the basis of their mandibular strength profiles. Dromaeosaurines had a more powerful bite and a rounder mandible than velociraptorines, indicating that stresses induced by prey were more important in the former. The prey stresses suggest that dromaeosaurines may have relied more on their bite to kill their prey, whereas velociraptorines probably killed using their sickle claws. Thus, this difference may represent ecological partitioning among contemporaneous dromaeosaurid taxa: dromaeosaurines would have preyed on relatively smaller prey than velociraptorines, for which a strong bite would have been sufficient, whereas the latter could have taken down prey much larger than themselves by attacking in packs and using their sickle claws to kill it.

Spinosaurids and *Dilophosaurus wetherilli* probably hunted prey relatively smaller than themselves. The anterior extremity of their mandible was specialized to capture small live prey, as indicated by the upturned chin, the rosette of teeth, and the greatly strengthened nature of the symphyseal region to resist the important stresses induced by struggling prey (high *Zx/L* and *Zy/L,* and low *Zx/Zy*). The nearly conical teeth of spinosaurids would have been well designed to impale and hold prey, a common feeding habit indicated by the low *Zx/Zy* values at the second alveolus, their shape allowing them to withstand bending loads applied from all directions. In contrast, *Dilophosaurus* had laterally compressed lateral teeth, indicating that it still had a slicing bite and did not hold its prey to the same extent as spinosaurids. The animals may have delivered slashing bites to wound their prey, capturing it with the extremity of their jaws only when it was weakened to the point of not offering great resistance, an interpretation supported by their relatively higher *Zx/Zy* values at the second tooth alveolus.

The extremity of the mandible also appears to have played an important role for prey capture and handling in *Ceratosaurus nasicornis*, *Allosaurus fragilis*, *Acrocanthosaurus atokensis*, and *Giganotosaurus carolinii*, but not to the extent seen in spinosaurids and *Dilophosaurus*. The former group still practiced a bite-and-release strategy (high Zx/Zy values), but the symphyseal region was strengthened to sustain high bending stresses (high Zx/L and Zy/L values), possibly for pulling the prey to the ground or delivering the final blow when it was down. The largest allosauroids, *Acrocanthosaurus* and *Giganotosaurus*, and some large *Allosaurus* individuals exhibit a ventral process at the symphysis, suggesting growth of bone where tensile stresses were high, which is consistent with a powerful bite force delivered against prey with the extremity of the jaws.

Tyrannosaurids were unique among theropods in having an effectively round mandibular cross-section at the anterior extremity of the mandible, where labiolingual loads were as important as dorsoventral ones. This indicates that the mandible of tyrannosaurids was suited (1) to employ a bite-and-hold strategy, in which the struggling prey would exert important stresses on the mandible from many directions; (2) to sustain the stresses related to a deep bite, potentially encountering bone, and complex head movements to detach pieces of flesh; and (3) to resist the important twisting moments generated when crushing hard objects (bones), as in *Varanus niloticus*.

Comparison of the theropod beam models to those of extant taxa, namely, *Varanus komodoensis* and *Alligator mississippiensis*, gives insight into the relative maximum bite force of the extinct predators, although the results must be considered with caution. Velociraptorines appear to have had maximum bite forces similar to those of *V. komodoensis*, whereas the bite force of *Dromaeosaurus* was three times as great. *Suchomimus*, *Allosaurus*, "*Antrodemus*," and *Ceratosaurus* were capable of exerting maximum bite forces as great as *A. mississippiensis*, whereas those of abelisaurids and *Albertosaurus* were twice as powerful. Among the largest theropods, *Acrocanthosaurus* and *Giganotosaurus* were surpassed by *Daspletosaurus* and *Tyrannosaurus*. The high estimates obtained for tyrannosaurids are consistent with previously published values suggesting bone-cracking abilities.

The study of growth series for *Allosaurus fragilis* and three tyrannosaurid taxa has given insight into possible changes in feeding behavior occurring during ontogeny. Mandibular bending strength at the second alveolus has been shown to vary relative to that at the middentary with an increase in body size in *Allosaurus*: from lesser than at the middentary in juveniles to greater than at the middentary in adults. However, Zx/Zy values remained constant throughout ontogeny. If the interpretation that the stronger mandibular extremity of adults is related to grasping and finishing prey, then juveniles probably did not kill this way and may, instead, have delivered simple slashing bites. In contrast, no relative changes in bending strength appear to occur during tyrannosaurid ontogeny: mandibular bending strength at the third alveolus is greater than at the middentary in the same proportion in juveniles as in adults. Also, as in *Allosaurus, Zx/Zy* values appear to remain constant throughout ontogeny, although some potential differences would require a greater sample size in order to determine their significance. Thus, the capability to grasp and subdue prey with the anterior extremity of the mandible was pronounced at all ages during tyrannosaurid ontogeny. Just like adults, juveniles were capable of capturing and holding live prey rather than relying on carrion or parental care for subsistence. However, juvenile tyrannosaurids probably delivered more superficial bites and did not crush through bones while feeding, as suggested by their laterally compressed dentition.

Acknowledgments. We thank all the personnel of the institutions visited for their willingness to help and their patience during the (often numerous) visits to their collections, particularly Michael Brett-Surman, Philip Currie, Denny Diveley, Pat Holroyd, Jim Kirkland, Vien Lam, Lyndon Murray, and Robert Purdy. We also thank Amy Chew, Mason Meers, Matt O'Neill, David Weishampel, and Shawn Zack for their help at various stages of this research. We are grateful for the comments of Ronan Allain and Ken Carpenter on a previous version of the manuscript. Finally, we wish to thank The Jurassic Foundation and the Sam and Doris Welles Fund (UCMP, University of California, Berkeley) for financial support of this research. The translation of Barsbold 1983 was done by C. Siskron and S. P. Welles and obtained courtesy of the Polyglot Paleontologist Web site (http://ravenel.si.edu/paleo/paleoglot/index.cfm).

References Cited

Abler, W. L. 1992. The serrated teeth of tyrannosaurid dinosaurs, and biting structures in other animals. *Paleobiology* 18: 161–183.

Abler, W. L. 1997. Tooth serrations in carnivorous dinosaurs. In P. J. Currie and K. Padian (eds.), *Encyclopedia of Dinosaurs*, pp. 740–743. San Diego: Academic Press.

Abler, W. L. 1999. The teeth of the tyrannosaurs. *Scientific American* 281: 50–51.

Abler, W. L. 2001. A kerf-and-drill model of tyrannosaur tooth serrations. In D. H. Tanke and K. Carpenter (eds.), *Mesozoic Vertebrate Life*, pp. 84–89. Bloomington: Indiana University Press.

Antón, M., and À. Galobart. 1999. Neck function and predatory behavior in the scimitar toothed cat *Homotherium latidens* (Owen). *Journal of Vertebrate Paleontology* 19: 771–784.

Anyonge, W. 1996. Locomotor behaviour in Plio-Pleistocene sabre-tooth cats: A biomechanical analysis. *Journal of Zoology, London* 238: 395–413.

Auffenberg, W. 1981. *The Behavioral Ecology of the Komodo Monitor.* Gainesville: University Presses of Florida.

Auffenberg, W. 1988. *Gray's Monitor Lizard.* Gainesville: University of Florida Press.

Auffenberg, W. 1994. *The Bengal Monitor.* Gainesville: University of Florida Press.

Bakker, R. T. 1997. Raptor family values: Allosaur parents brought giant carcasses into their lair to feed their young. In D. L. Wolberg, E. Stump, and G. D. Rosenberg (eds.), *Dinofest International,* pp. 51–63. Philadelphia: Academy of Natural Sciences.

Bakker, R. T. 1998. Brontosaur killer: Late Jurassic allosaurids as saber-tooth cat analogues. In B. P. Pérez-Moreno, T. Holtz Jr., J. L. Sanz, and J. Moratalla (eds.), *Gaia: Aspects of Theropod Paleobiology,* vol. 15, pp. 145–158. Lisbon: Museu Nacional de História Natural.

Bakker, R. T., M. Williams, and P. J. Currie. 1988. *Nanotyrannus,* a new genus of pygmy tyrannosaur, from the latest Cretaceous of Montana. *Hunteria* 1 (5): 1–30.

Barsbold, R. 1983. [Carnivorous dinosaurs from the Cretaceous of Mongolia]. [*Transactions, Joint Soviet-Mongolian Paleontological Expedition*] 19: 5–119. [In Russian.]

Biewener, A. A. 1992. Overview of structural mechanics. In A. A. Biewener (ed.), *Biomechanics, Structures and Systems: A Practical Approach,* pp. 1–20. Oxford: Oxford University Press.

Biknevicius, A. R., and C. B. Ruff. 1992a. The structure of the mandibular corpus and its relationship to feeding behaviours in extant carnivorans. *Journal of Zoology, London* 228: 479–507.

Biknevicius, A. R., and C. B. Ruff. 1992b. Use of biplanar radiographs for estimating cross-sectional geometric properties of mandibles. *Anatomical Record* 232: 157–163.

Biknevicius, A. R., and B. Van Valkenburgh. 1996. Design for killing: Craniodental adaptations of predators. In J. L. Gittleman (ed.), *Carnivore Behavior, Ecology, and Evolution,* vol. 2, pp. 393–428.

Bluhm, M. E. 2002. The intramandibular joint in varanid lizards: Insights into non-avian theropod morphology and behavior. Master's thesis, Johns Hopkins University, Baltimore.

Brochu, C. A., and R. A. Ketcham. 2002. Computed tomographic analysis of the skull of *Tyrannosaurus rex. Journal of Vertebrate Paleontology, Memoir 7,* 22 (4 suppl.) Supplemental X-ray CT data on CD-ROM.

Busbey, A. B. 1989. Form and function of the feeding apparatus of *Alligator mississippiensis. Journal of Morphology* 202: 99–127.

Busbey, A. B. 1995. The structural consequences of skull flattening in crocodilians. In J. Thomason (ed.), *Functional Morphology in Vertebrate Paleontology,* pp. 173–192. Melbourne: Cambridge University Press.

Calvo, J. O., and R. Coria. 1998. New specimen of *Giganotosaurus carolinii* (Coria & Salgado, 1995), supports it as the largest theropod ever found. In B. P. Pérez-Moreno, T. Holtz Jr., J. L. Sanz, and J. Moratalla (eds.), *Gaia: Aspects of Theropod Paleobiology,* vol. 15, pp. 117–122. Lisbon: Museu Nacional de História Natural.

Carpenter, K. 1990. Variation in *Tyrannosaurus rex.* In K. Carpenter and P. J. Currie (eds.), *Dinosaur Systematics: Approaches and Perspectives,* pp. 141–145. Cambridge: Cambridge University Press.

Carpenter, K. 1998. Evidence of predatory behavior by carnivorous dinosaurs. In B. P. Pérez-Moreno, T. Holtz Jr., J. L. Sanz, and J. Moratalla (eds.), *Gaia: Aspects of Theropod Paleobiology,* vol. 15, pp. 135–144. Lisbon: Museu Nacional de História Natural.

Carr, T. D. 1999. Craniofacial ontogeny in Tyrannosauridae (Dinosauria: Coelurosauria). *Journal of Vertebrate Paleontology* 19: 497–520.

Carrano, M. T., S. D. Sampson, and C. A. Forster. 2002. The osteology of *Masiakasaurus knopfleri,* a small abelisauroid (Dinosauria: Theropoda) from the Late Cretaceous of Madagascar. *Journal of Vertebrate Paleontology* 22: 510–534.

Charig, A. J., and A. C. Milner. 1990. The systematic position of *Baryonyx walkeri,* in the light of Gauthier's reclassification of the Theropoda. In K. Carpenter and P. J. Currie (eds.), *Dinosaur Systematics: Approaches and Perspectives,* pp. 127–140. Cambridge: Cambridge University Press.

Charig, A. J., and A. C. Milner. 1997. *Baryonyx walkeri,* a fish-eating dinosaur from the Wealden of Surrey. *Bulletin of the Natural History Museum, Geology Series,* 53: 11–70.

Chen P., Dong Z., and Zhen S. 1998. An exceptionally well-preserved theropod dinosaur from the Yixian Formation of China. *Nature* 391: 147–152.

Chin, K., T. T. Tokaryk, G. M. Erickson, and L. C. Calk. 1998. A king-sized theropod coprolite. *Nature* 393: 680–682.

Chure, D. J., A. R. Fiorillo, and A. Jacobsen. 1998. Prey bone utilization in the Late Jurassic of North America, with comments on prey bone use by dinosaurs throughout the Mesozoic. In B. P. Pérez-Moreno, T. Holtz Jr., J. L. Sanz, and J. Moratalla (eds.), *Gaia: Aspects of Theropod Paleobiology,* vol. 15, pp. 227–232. Lisbon: Museu Nacional de História Natural.

Cleuren, J., P. Aerts, and F. De Vree. 1995. Bite and joint force analysis in *Caiman crocodilus. Belgian Journal of Zoology* 125: 79–94.

Colbert, E. H. 1989. *The Triassic Dinosaur* Coelophysis. Museum of Northern Arizona Bulletin, no. 57. Flagstaff: Museum of Northern Arizona Press.

Coria, R. A., and L. Salgado. 1995. A new giant carnivorous dinosaur from the Cretaceous of Patagonia. *Nature* 377: 224–226.

Cott, H. B. 1961. Scientific results of an inquiry into the ecology and economic status of the Nile crocodile (*Crocodilus niloticus*) in Uganda and Northern Rhodesia. *Transactions of the Zoological Society of London* 29: 211–357.

Cowin, S. C. 1989. The mechanical properties of cortical bone tissue. In S. C. Cowin (ed.), *Bone Mechanics,* pp. 97–127. Boca Raton, Fla.: CRC Press.

Crompton, A. W. 1995. Masticatory function in nonmammalian cynodonts and early mammals. In J. Thomason (ed.), *Functional Morphology in Vertebrate Paleontology,* pp. 55–75. Melbourne: Cambridge University Press.

Currie, P. J. 1995. New information on the anatomy and relationships of *Dromaeosaurus albertensis* (Dinosauria: Theropoda). *Journal of Vertebrate Paleontology* 15: 576–591.

Currie, P. J. 1998. Possible evidence of gregarious behavior in tyrannosaurids. In B. P. Pérez-Moreno, T. Holtz Jr., J. L. Sanz, and J. Moratalla (eds.), *Gaia: Aspects of Theropod Paleobiology,* vol. 15, pp. 271–277. Lisbon: Museu Nacional de História Natural.

Currie, P. J. 2003. Allometric growth in tyrannosaurids (Dinosauria: Theropoda) from the Upper Cretaceous of North America and Asia. *Canadian Journal of Earth Sciences* 40: 651–665.

Currie, P. J., and K. Carpenter. 2000. A new specimen of *Acrocanthosaurus atokensis* (Theropoda, Dinosauria) from the Lower Cretaceous Antlers Formation (Lower Cretaceous, Aptian) of Oklahoma, USA. *Geodiversitas* 22: 207–246.

Currie, P. J., J. K. Rigby Jr., and R. E. Sloan. 1990. Theropod teeth from the Judith River Formation of southern Alberta, Canada. In K. Carpenter and P. J. Currie (eds.), *Dinosaur Systematics: Approaches and Perspectives,* pp. 107–125. Cambridge: Cambridge University Press.

Daegling, D. J., and W. L. Hylander. 1998. Biomechanics of torsion in the human mandible. *American Journal of Physical Anthropology* 105: 73–87.

Daegling, D. J., M. J. Ravosa, K. R. Johnson, and W. L. Hylander. 1992. Influence of teeth, alveoli, and periodontal ligaments on torsional rigidity in human mandibles. *American Journal of Physical Anthropology* 89: 59–72.

Dessem, D. 1985. The transmission of muscle force across the unfused symphysis in mammalian carnivores. In H.-R. Duncker and G. Fleischer (eds.), *Vertebrate Morphology*, pp. 289–291. New York: Gustav Fischer Verlag.

Drongelen, W. v., and P. Dullemeijer. 1982. The feeding apparatus of *Caiman crocodilus:* A functional-morphological study. *Anatomischer Anzeiger* 151: 337–366.

Erickson, G. M., and K. H. Olson. 1996. Bite marks attributable to *Tyrannosaurus rex:* Preliminary description and implications. *Journal of Vertebrate Paleontology* 16: 175–178.

Erickson, G. M., S. D. Van Kirk, J. Su, M. E. Levenston, W. E. Caler, and D. R. Carter. 1996. Bite-force estimation for *Tyrannosaurus rex* from tooth-marked bones. *Nature* 382: 706–708.

Evans, F. G. 1973. *Mechanical Properties of Bone*. Springfield, Ill.: Charles C. Thomas.

Farlow, J. O. 1976. Speculations about the diet and foraging behavior of large carnivorous dinosaurs. *American Midland Naturalist* 95: 186–191.

Farlow, J. O. 1994. Speculations about the carrion-locating ability of tyrannosaurs. *Historical Biology* 7: 159–165.

Farlow, J. O., D. L. Brinkman, W. L. Abler, and P. J. Currie. 1991. Size, shape, and serration density of theropod dinosaur lateral teeth. *Modern Geology* 16: 161–198.

Fiorillo, A. R. 1991. Prey bone utilization by predatory dinosaurs. *Palaeogeography, Palaeoclimatology, Palaeoecology* 88: 157–166.

Fogarty, M. J., and J. D. Albury. 1968. Late summer foods of young alligators in Florida. *Proceedings of the 21st Southeastern Association of Game and Fish Community* 1967: 220–222.

Foster, J. R. 2003. *Paleoecological Analysis of the Vertebrate Fauna of the Morrison Formation (Upper Jurassic), Rocky Mountain Region, U.S.A.* New Mexico Museum of Natural History and Science Bulletin, no. 23. Albuquerque: New Mexico Museum of Natural History and Science.

Frazzetta, T. H. 1962. A functional consideration of cranial kinesis in lizards. *Journal of Morphology* 111: 287–319.

Frazzetta, T. H., and K. V. Kardong. 2002. Prey attack by a large theropod dinosaur. *Nature* 416: 387–388.

Greaves, W. S. 1995. Functional predictions from theoretical models of the skull and jaws in reptiles and mammals. In J. Thomason (ed.), *Functional Morphology in Vertebrate Paleontology*, pp. 99–115. Melbourne: Cambridge University Press.

Haas, G. 1973. Muscles of the jaws and associated structures in the Rhynchocephalia and Squamata. In C. Gans and T. S. Parsons (eds.), *Biology of the Reptilia*, vol. 4, *Morphology D*, pp. 285–490. New York: Academic Press.

Henderson, D. M. 1998. Skull and tooth morphology as indicators of niche partitioning in sympatric Morrison Formation theropods. In B. P. Pérez-Moreno, T. Holtz Jr., J. L. Sanz, and J. Moratalla (eds.), *Gaia: Aspects of Theropod Paleobiology*, vol. 15, pp. 219–226. Lisbon: Museu Nacional de História Natural.

Henderson, D. M. 2002. The eyes have it: The sizes, shapes, and orientations of theropod orbits as indicators of skull strength and bite force. *Journal of Vertebrate Paleontology* 22: 766–778.

Holtz, T. R., Jr. 1998. Large theropod comparative cranial function: A new "twist" for tyrannosaurs. *Journal of Vertebrate Paleontology* 18 (3 suppl.): 51A.

Holtz, T. R., Jr. 2001. The phylogeny and taxonomy of the Tyrannosauridae. In D. H. Tanke and K. Carpenter (eds.), *Mesozoic Vertebrate Life,* pp. 64–83. Bloomington: Indiana University Press.

Horner, J. R., and D. Lessem 1993. *The Complete* T. rex. New York: Simon and Schuster.

Hurum, J. H., and P. J. Currie. 2000. The crushing bite of tyrannosaurids. *Journal of Vertebrate Paleontology* 20: 619–621.

Hylander, W. L. 1979. Mandibular function in *Galago crassicaudatus* and *Macaca fascicularis:* An *in vivo* approach to stress analysis of the mandible. *Journal of Morphology* 159: 253–296.

Hylander, W. L. 1981. Patterns of stress and strain in the macaque mandible. In D. S. Carlson (eds.), *Craniofacial Biology,* pp. 1–35. Monograph no. 10. Ann Arbor: University of Michigan Press.

Hylander, W. L. 1984. Stress and strain in the mandibular symphysis of primates: A test of competing hypotheses. *American Journal of Physical Anthropology* 64: 1–46.

Hylander, W. L. 1985. Mandibular function and biomechanical stress and scaling. *American Zoologist* 25: 315–330.

Iordansky, N. N. 1964. The jaw muscles of the crocodiles and some relating structures of the crocodilian skull. *Anatomischer Anzeiger* 115: 256–280.

Iordansky, N. N. 1970. Structure and biomechanical analysis of functions of the jaw muscles in the lizards. *Anatomischer Anzeiger* 127: 383–413.

Jacobsen, A. R. 1998. Feeding behaviour of carnivorous dinosaurs as determined by tooth marks on dinosaur bones. *Historical Biology* 13: 17–26.

Jacobsen, A. R. 2001. Tooth-marked small theropod bone: An extremely rare trace. In D. H. Tanke and K. Carpenter (eds.), *Mesozoic Vertebrate Life,* pp. 58–63. Bloomington: Indiana University Press.

Kingdon, J. 1989. *East African Mammals: An Atlas of Evolution in Africa.* Vol. 3, pt. A, *Carnivores.* Chicago: University of Chicago Press.

Kruuk, H. 1972. *The Spotted Hyaena.* Chicago: University of Chicago Press.

Kruuk, H., and M. Turner. 1967. Comparative notes on predation by lion, leopard, cheetah and wild dog in the Serengeti area, East Africa. *Mammalia* 31: 1–27.

Leyhausen, P. 1979. *Cat Behavior.* New York: Garland STPM Press.

Losos, J. B., and H. W. Greene. 1988. Ecological and evolutionary implications of diet in monitor lizards. *Biological Journal of the Linnean Society* 35: 379–407.

Madsen, J. H., Jr. 1976. Allosaurus fragilis: *A Revised Osteology.* Utah Geological and Mineral Survey Bulletin, no. 109. Salt Lake City: Utah Geological and Mineral Survey, Utah Department of Natural Resources.

Madsen, J. H., Jr., and S. P. Welles. 2000. *Ceratosaurus* (Dinosauria, Theropoda), a revised osteology. *Utah Geological Survey Miscellaneous Publication* 00-2: 1–80.

Martin, L. D. 1980. Functional morphology and the evolution of cats. *Transactions of the Nebraska Academy of Sciences* 8: 141–154.

Mazzetta, G. V., R. A. Fariña, and S. F. Viscaíno. 1998. On the palaeo-biology of the South American horned theropod *Carnotaurus sastrei* Bonaparte. In B. P. Pérez-Moreno, T. Holtz Jr., J. L. Sanz, and J. Moratalla (eds.), *Gaia: Aspects of Theropod Paleobiology,* vol. 15, pp. 185–192. Lisbon: Museu Nacional de História Natural.

McAllister, G. B., and D. D. Moyle. 1983. Some mechanical properties of goose femoral cortical bone. *Journal of Biomechanics* 16: 577–589.

Meers, M. B. 2002. Maximum bite force and prey size of *Tyrannosaurus rex* and their relationships to the inference of feeding behavior. *Historical Biology* 16: 1–12.

Meyer, E. R. 1984. Crocodiles as living fossils. In S. M. Stanley and N. Eldridge (eds.), *Living Fossils,* pp. 105–131. New York: Springer Verlag.

Molnar, R. E. 1969. Jaw musculature and jaw mechanics of the Eocene crocodilian *Sebecus icaeorhinus.* Master's thesis, University of Texas at Austin.

Molnar, R. E. 1973. The cranial morphology and mechanics of *Tyrannosaurus rex* (Reptilia:Saurischia). Ph.D. dissertation, University of California at Los Angeles.

Molnar, R. E. 1991. The cranial morphology of *Tyrannosaurus rex. Palaeontographica Abt. A* 217: 137–176.

Molnar, R. E. 1998. Mechanical factors in the design of the skull of *Tyrannosaurus rex* (Osborn, 1905). In B. P. Pérez-Moreno, T. Holtz Jr., J. L. Sanz, and J. Moratalla (eds.), *Gaia: Aspects of Theropod Paleobiology,* vol. 15, pp. 193–218. Lisbon: Museu Nacional de História Natural.

Molnar, R. E., and J. O. Farlow. 1990. Carnosaur paleobiology. In D. B. Weishampel, P. Dodson, and H. Osmólska (eds.), *The Dinosauria,* pp. 210–224. Berkeley: University of California Press.

Norman, D. B. 1984. On the cranial morphology and evolution of ornithopod dinosaurs. *Zoological Symposium* 52: 521–547.

Ostrom, J. H. 1961. Cranial morphology of the hadrosaurian dinosaurs of North America. *Bulletin of the American Museum of Natural History* 122: 1–186.

Ostrom, J. H. 1964. A functional analysis of jaw mechanics of the dinosaur *Triceratops. Postilla* 88: 1–35.

Ostrom, J. H. 1969a. A new theropod dinosaur from the Lower Cretaceous of Montana. *Postilla* 128: 1–17.

Ostrom, J. H. 1969b. Osteology of *Deinonychus antirrhopus,* an unusual theropod from the Lower Cretaceous of Montana. *Bulletin of the Peabody Museum of Natural History* 30: 1–165.

Ostrom, J. H. 1978. The osteology of *Compsognathus longipes* Wagner. *Zitteliana* 4: 73–118.

Ostrom, J. H. 1990. Dromaeosauridae. In D. B. Weishampel, P. Dodson, and H. Osmólska (eds.), *The Dinosauria,* pp. 269–279. Berkeley: University of California Press.

Padian, K., J. R. Hutchinson, and T. R. Holtz Jr. 1999. Phylogenetic definitions and nomenclature of the major taxonomic categories of the carnivorous Dinosauria (Theropoda). *Journal of Vertebrate Paleontology* 19: 69–80.

Paul, G. S. 1987. Predation in the meat-eating dinosaurs. In P. J. Currie and E. H. Koster (eds.), *Fourth Symposium on Mesozoic Terrestrial Ecosystems, Short Papers,* pp. 171–176. Occasional Papers of the Royal Tyrrell Museum of Palaeontology 3. Drumheller, Alberta: Royal Tyrrell Museum of Palaeontology.

Paul, G. S. 1988. *Predatory Dinosaurs of the World: A Complete Illustrated Guide.* New York: Simon and Schuster.

Pooley, A. C. 1989. Food and feeding habits. In C. A. Ross and S. Garnet (eds.), *Crocodiles and Alligators,* pp. 76–91. New York: Facts On File.

Rayfield, E. J., D. B. Norman, C. C. Horner, J. R. Horner, P. M. Smith, J. J. Thomason, and P. Upchurch. 2001. Cranial design and function in a large theropod dinosaur. *Nature* 409: 1033–1037.

Rayfield, E. J., D. B. Norman, and P. Upchurch. 2002. Prey attack by a large theropod dinosaur—Reply. *Nature* 416: 388.

Rieppel, O. 1979. A functional interpretation of the varanid dentition (Reptilia, Lacertilia, Varanidae). *Gegenbaurs Morphologisches Jahrbuch, Leipzig* 125: 797–817.

Rieppel, O., and L. Labhardt. 1979. Mandibular mechanics in *Varanus niloticus* (Reptilia: Lacertilia). *Herpetologica* 35: 158–163.

Rieppel, O., and H. Zaher. 2000. The intramandibular joint in squamates, and the phylogenetic relationships of the fossil snake *Pachyrhachis problematicus* Haas. *Fieldiana—Geology* 43: 1–69.

Rogers, R. R., D. W. Krause, and K. C. Rogers. 2003. Cannibalism in the Madagascan dinosaur *Majungatholus atopus. Nature* 422: 515–518.

Ross, C. A., and W. E. Magnusson. 1989. Living crocodilians. In C. A. Ross and S. Garnet (eds.), *Crocodiles and Alligators,* pp. 58–73. New York: Facts On File.

Rowe, T., and J. A. Gauthier. 1990. Ceratosauria. In D. B. Weishampel, P. Dodson, and H. Osmólska (eds.), *The Dinosauria,* pp. 151–168. Berkeley: University of California Press.

Rowe, T., C. A. Brochu, M. Colbert, J. W. Merck Jr., K. Kishi, E. Saglamer, and S. Warren. 1999. Alligator: Digital atlas of the skull. *Journal of Vertebrate Paleontology, Memoir 6,* 19 (2 suppl.). Supplemental X-ray CT data on CD-ROM.

Russell, D. A. 1970. *Tyrannosaurs from the Late Cretaceous of Western Canada.* National Museum of Natural Sciences, Publications in Palaeontology, no. 1. Ottawa: [Queen's Printer].

Scapino, R. 1981. Morphological investigation into functions of the jaw symphysis in carnivorans. *Journal of Morphology* 167: 339–375.

Schaller, G. B. 1972. *The Serengeti Lion.* Chicago: University of Chicago Press.

Schumacher, G.-H. 1973. The head muscles and hyolaryngeal skeleton or turtles and crocodilians. In C. Gans and T. S. Parsons (eds.), *Biology of the Reptilia,* vol. 4, *Morphology D,* pp. 101–199. New York: Academic Press.

Sereno, P. C., C. A. Forster, R. R. Rogers, and A. M. Monetta. 1993. Primitive dinosaur skeleton from Argentina and the early evolution of Dinosauria. *Nature* 361: 64–66.

Sereno, P. C., J. A. Wilson, H. C. E. Larsson, D. B. Dutheil, and H.-D. Sues. 1994. Early Cretaceous dinosaurs from the Sahara. *Science* 265: 267–271.

Sereno, P. C., A. L. Beck, D. B. Dutheil, B. Gado, H. C. E. Larsson, G. H. Lyon, J. D. Marcot, O. W. M. Rauhut, R. W. Sadleir, C. A. Sidor, D. D. Varricchio, G. P. Wilson, and J. A. Wilson. 1998. A long-snouted predatory dinosaur from Africa and the evolution of spinosaurids. *Science* 282: 1298–1302.

Sinclair, A. G., and R. M. Alexander. 1987. Estimates of forces exerted by the jaw muscles of some reptiles. *Journal of Zoology, London* 213: 107–115.

Smith, D. K. 1998. A morphometric analysis of *Allosaurus*. *Journal of Vertebrate Paleontology* 18: 126–142.

Smith, R. J. 1984. Allometric scaling in comparative biology: problems of concept and method. *American Journal of Physiology* 246: R152–R160.

Smith, R. J. 1993. Logarithmic transformation bias in allometry. *American Journal of Physical Anthropology* 90: 215–228.

Sues, H.-D., E. Frey, and D. M. Martill. 1999. The skull of *Irritator challengeri* (Dinosauria:Theropoda:Spinosauridae). *Journal of Vertebrate Paleontology* 19 (3 suppl.): 79A.

Taylor, M. A. 1987. How tetrapods feed in water: a functional analysis by paradigm. *Zoological Journal of the Linnean Society* 91: 171–195.

Timoshenko, S. P., and J. M. Gere. 1972. *Mechanics of Materials*. New York: Van Nostrand Reinhold.

Turner, A., and M. Antón. 1996. *The Big Cats and Their Fossil Relatives*. New York: Columbia University Press.

Van Valkenburgh, B., and R. E. Molnar. 2002. Dinosaurian and mammalian predators compared. *Paleobiology* 28: 527–543.

Van Valkenburgh, B., and C. B. Ruff. 1987. Canine tooth strength and killing behavior in large carnivores. *Journal of Zoology, London* 212: 379–397.

Varricchio, D. J. 2001. Gut contents from a Cretaceous tyrannosaurid: Implications for theropod dinosaur digestive tracts. *Journal of Paleontology* 75: 401–406.

Weishampel, D. B. 1983. Hadrosaurid jaw mechanics. *Paleontologica* 28: 271–280.

Weishampel, D. B. 1984. *Evolution of Jaw Mechanisms in Ornithopod Dinosaurs*. New York: Springer-Verlag.

Welles, S. P. 1984. *Dilophosaurus wetherilli* (Dinosauria, Theropoda): Osteology and comparisons. *Palaeontographica Abt. A*. 185: 85–180.

Witmer, L. M., and K. D. Rose. 1991. Biomechanics of the jaw apparatus of the gigantic Eocene bird *Diatryma*: Implications for diet and mode of life. *Paleobiology* 17: 95–120.

11. Body and Tail Posture in Theropod Dinosaurs

GREGORY S. PAUL

Abstract

Theropods have been restored with their bodies held in positions ranging from subvertical to horizontal. The former posture was most common until the recent "new look" favored the latter, more avian pose, but there has been a recent attempt to revisit a human-like erect carriage, which may have permitted better turning ability owing to a reduction in distal mass. However, the anatomical arrangement of theropod pelves and legs prevented the adoption of an erect body, owing to muscle stretch factors and the limitations of the articular surfaces of the hip socket. The dinosaurs may have reduced their distal inertial mass when turning by retracting the neck and arm and lifting the tail. The few dinosaurs that did carry their bodies erect had specialized, retroverted pelves that allowed the legs to function when the body was tilted up.

Introduction

The body posture of theropod and other bipedal dinosaurs has been variably restored since the first attempts to mount their skeletons and portray their life appearance. Until the late twentieth century, bipedal dinosaurs, especially large examples, were commonly restored with erect bodies, rather like humans or standing kangaroos, and with tails dragging on the ground (Osborn 1917; Knight 1946; Scheele 1954; Augusta and Burian 1958; Watson and Zallinger 1960; Jackson and Matternes 1972). There were exceptions, however, including Charles Knight's classic painting of *Tyrannosaurus* confronting *Triceratops*. Little supporting data, anatomical or otherwise, was presented in favor of the erect body posture, which was largely intuitive. The

Figure 11.1. (opposite page) (A) Allosaurus *running with erect body and tail, after Carrier et al. (2001). (B) Same subject with horizontal body, neck pulled as posteriorly as possible, arms shown in normal and posteriorly directed turning position, and tail in normal horizontal and erect turning position. (C) Composite-derived therizinosaur, vertebral column and pelvis after* Nanshiungosaurus *(Dong 1979), with erect trunk and horizontal pelvis and tail. (D)* Brachiosaurus *with semi-erect trunk and horizontal pelvis and tail.*

238

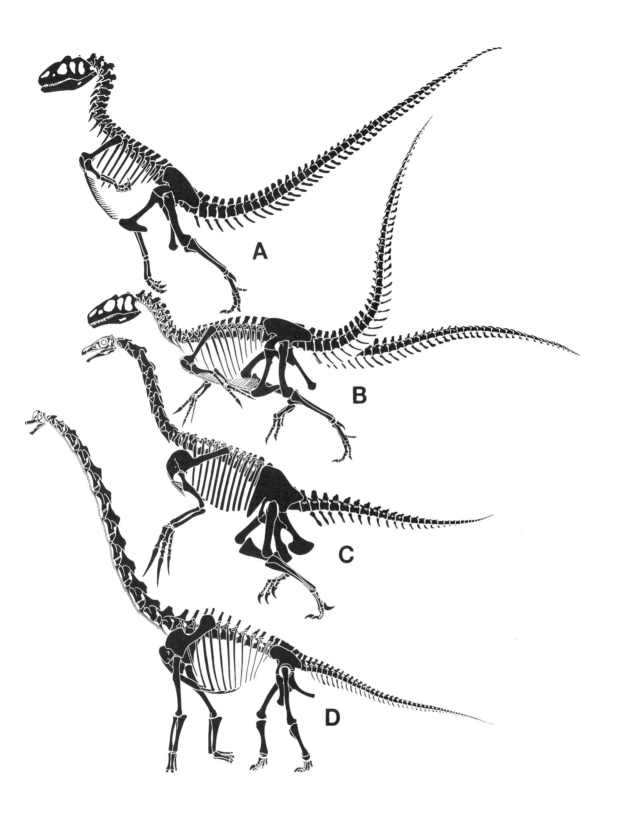

A

B

C

D

revolution in dinosaur paleontology in the last third of the twentieth century was the basis for the "new look" of dinosaur restoration, in which bipedal examples of all sizes were posed with more bird-like horizontal bodies and tails. The tail was held clear of the ground and so did not leave drag marks, which are absent from all but a few theropod trackways (Ostrom 1969; Galton 1970; Newman 1970; Padian 1986; Paul 1987, 1988, 2000; Carpenter et al. 1994). Recently, however, Carrier et al. (2001) claimed that the horizontal posture is based on limited evidence. They proposed a novel alternative in which both the body and the tail were held semi-erect, in the manner of bipedal running lizards, as a means of reducing distal inertial mass and improving maneuverability. But Carrier et al. (2001) failed to consider previously published information that supports a horizontal posture, and additional analysis below favors the latter in most bipedal dinosaurs. An exception are the peculiar therizinosaurs that appear to have combined a semi-erect trunk with a horizontal pelvis and tail.

Muscle Stretch Factors and Pelvic Orientation

Muscle stretch factors reflect the difference between the length of a muscle and its fibers at maximum stretch, at rest, and at maximum contraction. Voluntary muscles are fully functional only when their stretch factors do not exceed about one-third of resting length (Schmidt-Nielsen 1984). If this stretch limit is exceeded, the muscle becomes slack and sags after maximal contraction or becomes overstretched and subject to damage at maximum extension. Limb retractors can fulfill their proper function only if they are anchored posterior to the maximum retraction point of the muscle insertion throughout the limb stroke, unless the path of the muscle is strongly modified so that it takes a suitable looping course. In tetrapods, including crocodilians and birds, and as consistently restored in dinosaurs, important hindlimb retractors are anchored on the ischium (Romer 1923; Perle 1985; Paul 1988; Hutchinson 2001). These ischium-based retractors do not follow a looping path, so the ischium must be significantly posterior to the femur and proximal tibia throughout the propulsive stroke. If the ventral pelvis is proverted until the ischium is rotated anteriorly between the femora, the stretch factors of the ischial-based retractors become excessively large, and the muscles are no longer able to retract the limb for the latter portion of the retractive swing of the limb. A posterior placement of the ischium relative to the hindlimb is especially important to erect-limbed tetrapods in which the excursion arc of the femur is inherently close to the ischium.

In the restoration of a running theropod by Carrier et al. (2001), the ventral part of the pelvis is so strongly proverted that the ischium is nearly vertical and is anterior to the retracted left femur (Fig. 11.1A). The result of such a posture would be highly dysfunctional stretch factors for the ischial-based retractors, and thus these muscles would be unable to retract the hindlimb during the last phase of the propulsive stroke (Fig. 11.2B). This arrangement is so unworkable that it could not have occurred. In all dinosaurs, including birds, the ischium is placed sufficiently posterior to the proximal leg elements throughout the limb

Figure 11.2. (opposite page) Posture of the pelvis and hip joint function in Allosaurus.
(A) Normal femoral rotation and muscle action with pelvis horizontal and the femur not retracting past vertical.
(B) Abnormal function of the ischial-based retractors with the pelvis tilted up, compare to Fig. 11.1A. (C) Full articulation of the acetabular and femoral articular surfaces with the pelvis horizontal and the femur not retracting past vertical.
(D) Incomplete articulation of the femoral head with the acetabulum with the pelvis tilted up.

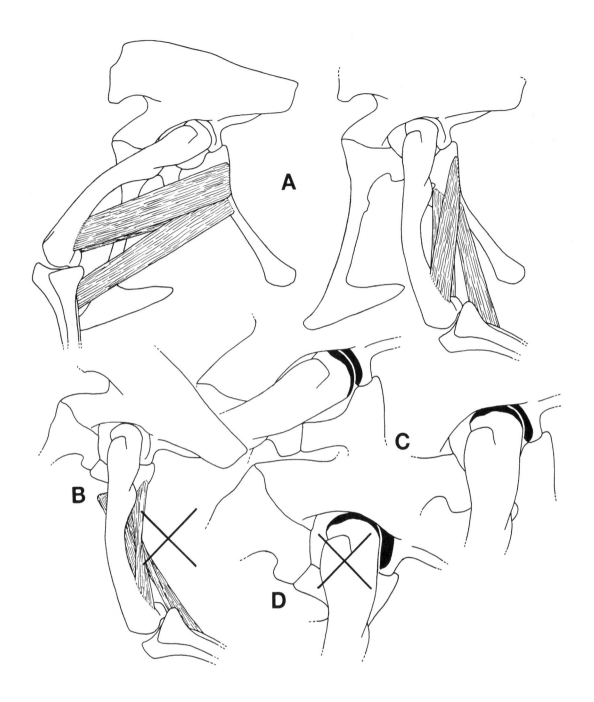

stroke only if the pelvis is close to horizontal, with the ischium project-
ing ventroposteriorly from the acetabulum (Figs. 11.1B, 11.2A; Paul
1988). This follows the normal tetrapod pattern and is preferred to the
Carrier et al. (2001) alternative.

In bipedal ornithischians and most theropods, the dorsal, sacral,
and caudal series are in approximately the same line with one another
and with the long axis of the ilium, although in some ornithopods there
is a gentle downsweep at the base of the tail (Paul 1987, 2000).
Wellnhofer (1993) argued that prosauropod tail bases were so strongly
arced ventrally that they dragged when the animals were quadrupedal,

but the lack of beveling of the centra to allow the resulting dorsal curve at mid-tail, as well as the scarcity of tail drag marks in prosauropod trackways, suggests that Wellnhofer's suggestion is not correct. The tail base of theropods may have been able to flex dorsally in the manner restored by Carrier et al. (2001), but the articulation of the caudal centra when they are parallel to one another, and when the zygapophyses are in a neutral overlapping position, suggests that the tail base was normally straight. A horizontal pelvis, therefore, results in a horizontal body and tail in most bipedal dinosaurs (Fig. 11.1B). Padian (1986) argued that theropods normally carried their bodies tilted dorsally at a modest 20° above horizontal, but the anatomy of these extinct creatures cannot be restored to the level of accuracy needed to estimate body orientation to this fine level (Paul 1988). Body posture may have varied within this near-horizontal range within the group, and even among individuals.

An exception to the standard dinosaur horizontal body posture is seen in therizinosaurs, especially the derived taxa (Fig. 11.1C). The entire pelvis is strongly retroverted relative to the dorsosacral series, so the latter is tilted dorsally when the pelvis is horizontal (Russell and Russell 1993). The tail is flexed dorsally relative to the dorsosacral series, so it remains horizontal with the pelvis. A similar arrangement was necessary in the high-shouldered brachiosaur sauropods (Fig. 11.1D). The retroverted pelvis of brachiosaurs and therizinosaurs allowed them to carry their trunks semi-erect while maintaining the pelvis and tail in the horizontal posture needed for normal locomotion. The absence of pelvic retroversion in other theropods is additional evidence against a habitual semi-erect body posture.

Hip Socket Stressing and Articulation

Carrier et al. (2001) considered the most robust part of the theropod acetabulum to be near and on the pubic peduncle, although they presented no supporting data or figures. In their view, this pattern of acetabular strengthening is logical only if the pelvis was so strongly proverted that the strongest part of the acetabulum was dorsal to the femoral head, as it is in humans. This interpretation contrasts with that of Hotton (1980), who concluded that theropod acetabula were reinforced posteriorly.

Strengthening of the acetabulum via thickening of the surrounding regions of the ilium is inconsistent in tetrapods (Fig. 11.3). In most mammals, the strongest aspect of the acetabulum is anterodorsal because this is the direction of the anterior ilial process that articulates with the sacrum. In birds the strongest region is posterodorsal to the acetabulum, owing to the presence of a massive antitrochanter and the absence of a robust, subvertical pubis anchored on the pubic peduncle. The robustness of the theropod pubic peduncle probably reflects its function as the anchor for the pubis, which bore much of the animal's weight when it was resting on its belly, even in taxa whose pubes were strongly retroverted (Paul 1988). In general, theropod acetabula were well strengthened in anterior, dorsal, and posterior aspects, although dorsal bracing was stronger in theropods with a supra-acetabular shelf than in those without (Fig. 11.3C,D). A thorough assessment of the

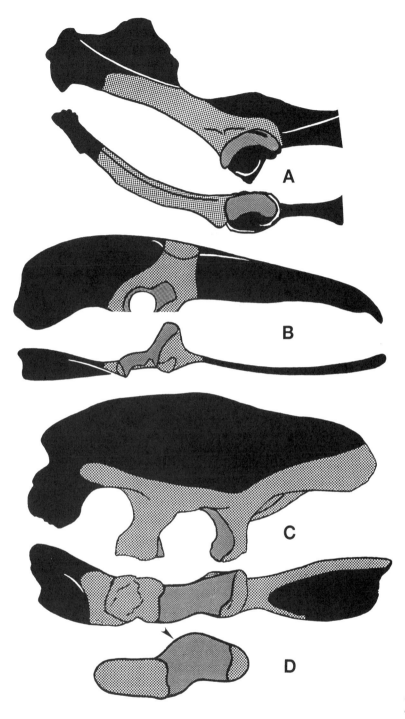

Figure 11.3. Left ilia and acetabula in lateral and ventral or ventrolateral views comparing patterns of bone strengthening (stippled) relative to thinner bone (black). (A) mammal, (B) ostrich, (C) Tyrannosaurus, and (D) Allosaurus, acetabulum only in ventral view, arrow indicates supra-acetabular crest.

reinforcement of theropod and other tetrapod hip sockets requires a detailed analysis and comparison of the external morphology and internal microstructure. At this time, the differential strengthening of tetrapod acetabula appears to be the result of variable factors not always directly related to locomotion, and does not provide reliable evidence of pelvic posture (Paul 1987, 1988).

As explained in Paul 1988, the articular surface of the femoral head remains in full articulation with the corresponding surfaces of the acetabulum when the femur does not retract strongly posterior to the long axis of the ilium (Fig. 11.2C). If the femur retracts more posteriorly, then the posterior articular surface of the acetabulum articulates with the non-articular surface of the posterior face of the proximal femur (Fig. 11.2D). Such an arrangement would be so dysfunctional when the limb was bearing high mass and force loads during locomotion that it must be ruled out. The hip socket is restored in such an improper position by Carrier et al. (2001). If the ilium is tilted strongly dorsally along with the rest of the body, the femur can work only in a semihorizontal arc, which does not produce an effective propulsive stroke. Only if the ilium is horizontal or close to it can the femur swing though a normal propulsive stroke, with the femur being at or just posterior to vertical at maximum retraction (Paul 1987, 1988, 1998).

Discussion and Conclusions

The running theropods restored by Carrier et al. (2001) show the left knees too straight to avoid dislocation (Paul 1987, 1988, 1998, 2000), an error shared by many restorations. Carrier et al. restore the right ankle as nearly straight, but the ankle strongly folds up during the recovery stroke in long-footed tetrapods, such as theropods (Paul 1987, 1988). In the erect-bodied theropod offered by Carrier et al. (Fig. 11.1A), the left elbow is over extended, and the wrist is disarticulated so that the palm of the hand faces incorrectly posteriorly.

The overall visual impression of the erect-bodied and erect-tailed running theropod is implausibly awkward. Carrier et al. (2001) noted that the high-held body and tail would reduce stability. Nontherizinosaur theropods probably ran with the body and tail horizontal and the neck strongly S curved. Reducing distal mass distribution when turning by raising the body would result in a major and dysfunctional alteration of the locomotory complex. Instead, the head and neck may have been pulled back as posterior as possible, and the tail may have been lifted dorsally to reduce distal inertial mass during turns. Although the resulting anterior reduction of distal mass would have been less than in the Carrier et al. restoration, the posterior reduction would have been greater because the tail could be lifted more anterodorsally when the body was horizontal. The arms may have been directed backward to reduce distal mass when turning, but since the arms and hands were usually predatory grasping organs, this may not have been the habitual posture. In contrast, the arms of bipedally running lizards can be directed posteriorly because they have no purpose to serve when not used for locomotion. Because the arms of ornithomimosaurs were long and nonraptorial, they may have been directed posteriorly more often than those of their predatory relatives. Carrier et al. (2001) suggested that the folding arms of derived, bird-like theropods evolved as a means to reduce distal inertial mass, but distal mass reduction is even greater if the distal arm is directed posteriorly rather than tucked up near the shoulder. The folding mechanism may have instead evolved in the context of flight and have been retained in secondarily flightless dinosaurs (Paul 2002).

Although the big-headed, long-tailed theropods probably were not as maneuverable as smaller-headed and less-elongated ground birds, the difference may have not been as great as it may appear, because the horizontal bodies of birds are substantially longer fore and aft than are the vertically oriented humans used by Carrier et al. (2001) as the model for a short-bodied biped. The pelvis is more posteriorly elongated in birds than in other theropods, and this condition may have made up for some of the loss of caudal inertial mass. The rigid beam used by Carrier et al. to estimate the inertial mass of horizontal bodies and tails may not be sufficiently realistic either. Theropods should have been able to improve maneuverability by leaning into turns the way narrow trackway bicycles and humans do. Wide-track lizards cannot turn sharply by banking, so they must reduce distal inertial mass by erection of the body and tail when running bipedally.

Snively and Henderson (2002) observed that the normal allometric scaling of body proportions, in which bodies and tails become shorter relative to total mass as robustness increases, improved the turning potential of giant theropods by reducing their rotational inertia compared to isometric scaling—an advantage to active predators. In contrast, the big-bellied, broad-hipped, short-footed therizinosaurs were slow herbivores. Their erect bodies could have enhanced maneuverability when they tried to escape from or fight faster predatory theropods, but probably evolved primarily as a means to increase vertical browsing reach, as well as to display to and intimidate other individuals.

Carrier et al. (2001) claim that elephants are limited to level ground, but in fact they are remarkably adept at traversing steep terrain (Carrington 1959). Giant dinosaurs should have been able to do the same. Nevertheless, Carrier et al. are correct that large adult theropods were largely creatures of the flats. The juveniles and small adults were probably good climbers (Paul 2002).

Further resolution of the issue requires sophisticated digital simulation of running and turning saltorial mammals, humans, birds, and theropods, simulation that incorporates the dynamic flexibility of pertinent body parts.

References Cited

Augusta, J., and Z. Burian. 1958. *Prehistoric Animals*. London: Spring Books.

Carpenter, K., J. Madsen, and A. Lewis. 1994. Mounting of fossil vertebrate skeletons. In P. Leiggi and P. May (eds.), *Vertebrate Paleontological Techniques*, pp. 285–322. New York: Cambridge University Press.

Carrier, D. R., R. M. Walter, and D. V. Lee. 2001. Influence of rotational inertia on turning performance of theropod dinosaurs: Clues from humans with increased rotational inertia. *Journal of Experimental Biology* 204: 3917–3926.

Carrington, R. 1959. *Elephants*. New York: Basic Books.

Galton, P. M. 1970. The posture of hadrosaurian dinosaurs. *Journal of Paleontology* 44: 464–473.

Hotton, N. 1980. An alternative to dinosaur endothermy: The happy wanderers. In R. D. K. Thomas and E. C. Olson (eds.), *A Cold Look at the Warm-Blooded Dinosaurs,* pp. 311–350. Washington, D.C.: AAAS.

Hutchinson, J. R. 2001. The evolution of pelvic osteology and soft tissues on the line to extant birds. *Zoological Journal of the Linnean Society* 131: 123–168.

Jackson, K., and J. H. Matternes. 1972. *Dinosaurs.* Washington, D.C.: National Geographic Society.

Knight, C. 1946. *Life through the Ages.* New York: Alfred A. Knopf.

Newman, B. H. 1970. Stance and gait in the flesh-eating dinosaur *Tyrannosaurus. Biological Journal of the Linnean Society* 2: 119–123.

Osborn, H. F. 1916. Skeletal adaptations of *Ornitholestes, Struthiomimus* and *Tyrannosaurus. Bulletin of the American Museum of Natural History* 35: 733–771.

Ostrom, J. H. 1969. Osteology of *Deinonychus antirrhopus,* an unusual theropod from the Lower Cretaceous of Montana. *Bulletin of the Peabody Museum of Natural History* 30: 1–165.

Padian, K. 1986. On the type material of *Coelophysis* and a new specimen from the Petrified Forest of Arizona. In K. Padian (ed.), *The Beginning of the Age of Dinosaurs,* pp. 46–60. Cambridge: Cambridge University Press.

Paul, G. S. 1987. The science and art of restoring the life appearance of dinosaurs and their relatives. In S. J. Czerkas and E. C. Olson (eds.), *Dinosaurs Past and Present,* vol. 2, pp. 4–49. Los Angeles: Natural History Museum of Los Angeles County.

Paul, G. S. 1988. *Predatory Dinosaurs of the World: A Complete Illustrated Guide.* New York: Simon and Schuster.

Paul, G. S. 1998. Limb design, function and running performance in ostrich-mimics and tyrannosaurs. B. P. Pérez-Moreno, T. Holtz Jr., J. L. Sanz, and J. Moratalla (eds.), *Gaia: Aspects of Theropod Paleobiology,* vol. 15, pp. 257–270. Lisbon: Museu Nacional de História Natural.

Paul, G. S. 2000. Restoring the life appearance of dinosaurs. In G. S. Paul (ed.), *The Scientific American Book of the Dinosaur,* pp. 78–106. New York: St. Martin's Press.

Paul, G. S. 2002. *Dinosaurs of the Air.* Baltimore: Johns Hopkins University Press.

Perle, A. 1985. Comparative myology of the pelvic-femoral region in the bipedal dinosaurs. *Paleontological Journal* 1: 105–109.

Romer, A. S. 1923. The pelvic musculature of the saurischian dinosaurs. *Bulletin of the American Museum of Natural History* 48: 533–552.

Russell, D. A., and D. E. Russell. 1993. Evolutionary convergence between a mammalian and a dinosaurian clawed herbivore. *National Geographic Research and Exploration* 9 (1): 70–79.

Scheele, W. E. 1954. *Prehistoric Animals.* Cleveland: World Publishing Co.

Schmidt-Nielsen, K. 1984. *Scaling: Why Is Animal Size So Important?* Cambridge: Cambridge University Press.

Snively, E., and D. M. Henderson. 2002. *Albertosaurus* en pointe: Allometric scaling of theropod rotational inertia and turning performance. *Journal of Vertebrate Paleontology* 22: 109A.

Watson, J. W., and R. F. Zallinger. 1960. *Dinosaurs.* New York: Golden Press.

Wellnhofer, P. 1993. Prosauropod dinosaurs from the Feuerletten of Ellingen near Weissenburg in Bavaria. *Revue de Paleobiologie* 7: 263–271.

12. Furcula of *Tyrannosaurus rex*

PETER LARSON AND J. KEITH RIGBY JR.

Abstract

A furcula for the Upper Cretaceous theropod *Tyrannosaurus rex* has been discovered for the first time associated with a new skeleton from the Hell Creek Formation in Perkins County, South Dakota. The furcula is formed by the clavicles fusing into a boomerang, or broad U-shape. The specimen is nearly complete, missing only the left epicleidium, and is estimated to have been 29 cm wide and 14 cm deep. Overall, this specimen is comparable to other known theropod furculae. This discovery has led to the recognition of two other *T. rex* furculae in extant collections.

Introduction

Furculae are now recognized in a variety of avian and non-avian theropods, including *Allosaurus* (Chure and Madsen 1996), *Oviraptor* (Barsbold 1983; Barsbold et al. 1990), *Ingenia* (Barsbold 1983; Barsbold et al. 1990), *Bambiraptor* (Burnham et al. 1997, 2000), *Velociraptor* (Norell and Clark 1997; Clark et al. 1999), *Protoarchaeopteryx* (Ji et al. 1998), *Caudipteryx* (Ji et al. 1998), *Scipionyx* (Dal Sasso and Signore 1998), *Beipeiiosaurus* (Xu et al. 1999a), and *Sinornithosaurus* (Xu et al. 1999b). They are also known in the tyrannosaurids: *Albertosaurus, Gorgosaurus,* and *Daspletosaurus* (Makovicky and Currie 1998). One has been purported in *Tyrannosaurus rex* (Brochu 2003), but this specimen does not match well with the furculae of other tyrannosaurids, as is discussed below.

A definitive furcula has recently been discovered associated with a *Tyrannosaurus* skeleton collected in the Hell Creek Formation of

Perkins County, South Dakota. The specimen, nicknamed "Bucky," was discovered by Bucky Derflinger in November 1998 on land owned by his parents, Wade and Lorena Derflinger. Excavation did not begin until April 18, 2001, by the Black Hills Institute on behalf of The Children's Museum, Indianapolis, Indiana, where the skeleton is curated. Because no skull elements are known, the identification of the specimen as *Tyrannosaurus rex* is based solely upon the massive postcranial skeleton, particularly the morphology of the ulna, ischium, and cervical vertebrae. On May 23, 2001, an unusual U-shaped bone was discovered at the site. This element is comparable to furculae reported for other tyrannosaurids (e.g., Makovicky and Currie 1998) and is described below.

Institutional Abbreviations: AMNH, American Museum of Natural History, New York; BHI, Black Hills Institute of Geological Research, Hill City, S.D.; BMNH, British Museum of Natural History, London; CMI, The Children's Museum, Indianapolis, Indiana; CMN, Canadian Museum of Nature, Ottawa, Ontario; DINO, Dinosaur National Monument, Dinosaur, Utah; FMNH, Field Museum of Natural History, Chicago; IGM, Mongolian Institute of Geology, Ulaan Baatar, Mongolia; MOR, Museum of the Rockies, Bozeman, Mont. (curatorial duties shared with Fort Peck Paleontology, Inc., Fort Peck, Mont.); TMP, Royal Tyrrell Museum of Palaeontology, Drumheller, Alberta.

Description

The furcula of CMI 2001.90.1 is nearly complete, missing only the distal one-third of the left ramus (Fig. 12.1); measurements and their locations are presented in Figure 12.2. Restored, it would form a slightly asymmetrical boomerang or broad U shape. This asymmetry may be correlated with the pathology at the base of the left ramus (see below). Although the left ramus is incomplete (possibly broken in life), it is clear that both rami taper in anterior and lateral views (Fig. 12.1A,C); this tapering is more pronounced than in other theropods (Fig. 12.3), including other tyrannosaurids (see also Makovicky and Currie 1998). In lateral view, the furcula is slightly bowed posteriorly, and the distal end is wedge-shaped (Fig. 12.1C). On the posterior surface of the epicleidium is a roughened sutural scar (epicleideal facet) about 40 mm long for a ligamentous attachment with the acromion of the scapula as in oviraptorids (Fig. 12.4; see also Barsbold 1981, 1983). In addition, there is a shallow, rough groove along the posterior side (Fig. 12.1B) that may be where the furcula overlaps the coracoid as in oviraptorids (Fig. 12.4). This surface of the furcula is also rough, suggesting a ligamentous attachment, although this contradicts the conclusion of Makovicky and Currie (1998, p. 146). The apex is rounded and lacks the prominent hypocleidium of *Daspletosaurus* and *Albertosaurus,* although another tyrannosaur, *Gorgosaurus,* also lacks a hypocleidium (see Makovicky and Currie 1998). The anterior surface of the furcula, where it is wedge-shaped, is rough, possibly for a broad tendinous membrane of the M. supracoracoideus rather than the M. deltoideus minor as suggested by Carpenter and Smith (2001).

Figure 12.1. (opposite page top) Furcula of Tyrannosaurus rex CMI 2001.90.1 in (A) anterior, (B) posterior, (C) lateral, and (D) ventral views. Arrow indicates abrupt change in slope of ramus. Abbreviations: ant—anterior side; ap—apex; ep—epicleidium; epfa—epicleideal facet; gr—groove; pa—pathology; pas—pathological scar; r—ramus; sp—bone spicule or spur. Scale in cm.

Figure 12.2. (opposite page bottom) Measurements of Tyrannosaurus rex furcula, CMI 2001.90.1.

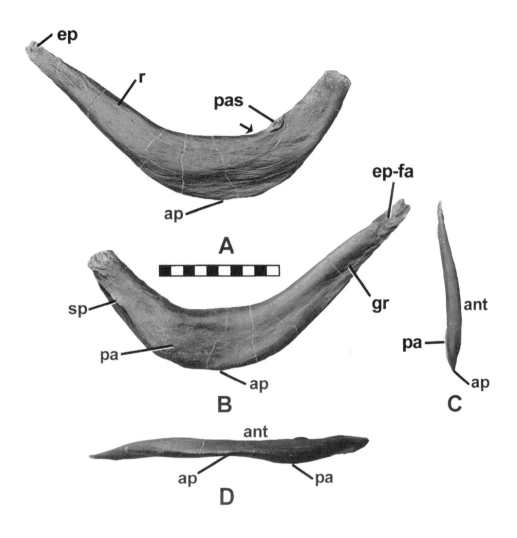

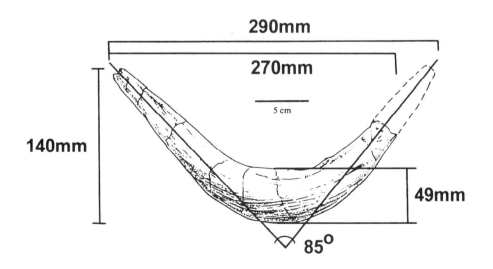

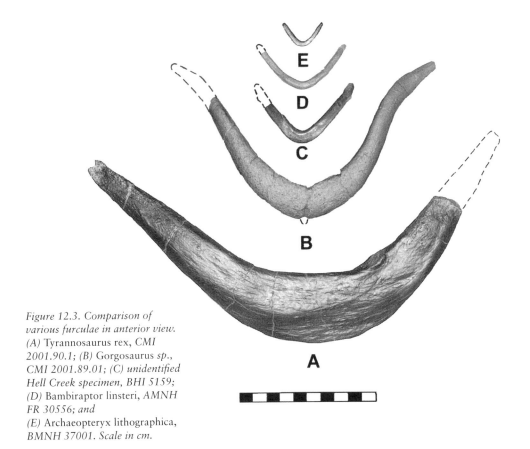

Figure 12.3. Comparison of various furculae in anterior view. (A) Tyrannosaurus rex, CMI 2001.90.1; (B) Gorgosaurus sp., CMI 2001.89.01; (C) unidentified Hell Creek specimen, BHI 5159; (D) Bambiraptor linsteri, AMNH FR 30556; and (E) Archaeopteryx lithographica, BMNH 37001. Scale in cm.

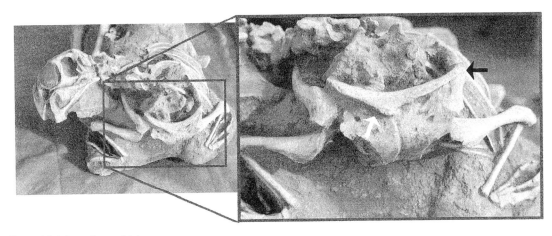

Figure 12.4. Ingenia yanshini (cast) showing furcula in place. Black arrow shows contact epicleidium and acromion process of the scapula. White arrow shows furcula in contact with coracoid.

TABLE 12.1.

Measurements for Various Fossil Furculae

Specimen	Interclavicle angle	Maximum height (mm)	Maximum width (mm)	Shape	Hypocleidium-present?
*Archaeopteryx** BMNH 37001	70°	17	30	U-shaped	No
Confuciusornis BHI 6215	75°	16	30	U-shaped	No
*Bambiraptor** AMNH FR 30556	80°	31	64	U-shaped	No
Unidentified Hell Creek BHI 5159	80°	45	74	U-shaped	No
Gorgosaurus CMI 2001.89.1	80°	10.7	19.5	Sinusoidal	Damaged
Tyrannosaurus **CMI 2001.90.1** FMNH PR2081 MOR 980	**85°** 85° 80°	**140** 135 130	2.90	**U-shaped** U-shaped U-Shaped	**No** No No
Oviraptorid** Clark et al. 1999, IGM 100/979	100°	40	70	U-shaped	Yes
*Velociraptor** Norell and Makovicky 1999, IGM100/976	105°	32	85	V-shaped	Yes
*Albertosaurus** Makovicky and Currie 1998, TMP 86.64.1	110°	17	19	Sinusoidal	No
*Daspletosaurus** Makovicky and Currie 1998, CMN 8056	120°	80	250	U-shaped	No
*Allosaurus** Chure and Madsen 1996, DINO 11541	130°	60	220	V-shaped	No

*Measurements made from cast.
**Measurements made from illustration.

On the posterior side, near the base of the left ramus, is an asymmetrical swelling of a pathology (Fig. 12.1B). This swelling is associated with a bony spur or spicule on the posterior side, as well as an oval, rugose scar along the dorsal surface with an overhanging lip on the anterior side (Fig. 12.1A,B). The left ramus changes slope abruptly near this upper pathology. The swelling is probably the result of a calcified or resolving bone hematoma (Larson 2001), possibly due to a greenstick or stress fracture.

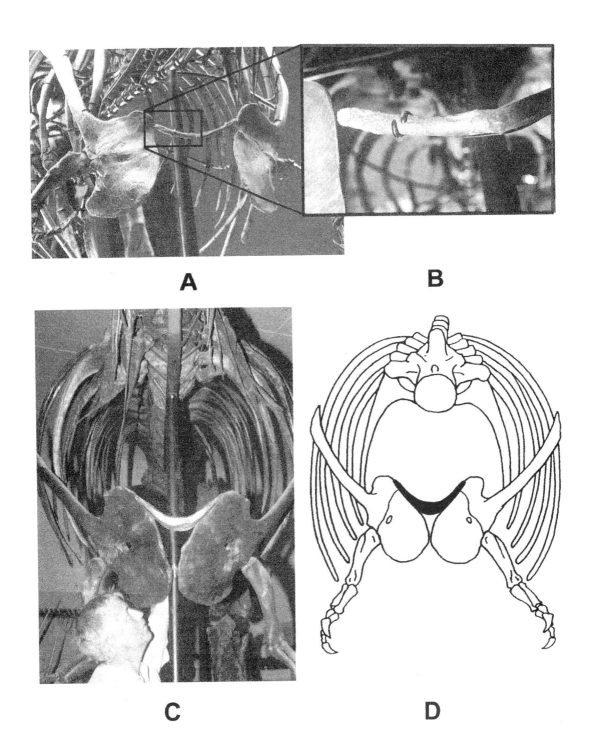

A

B

C

D

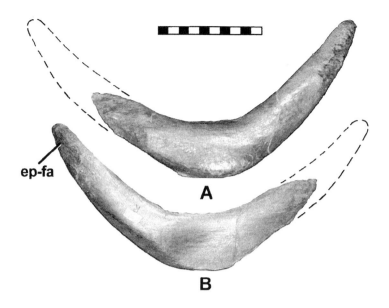

ep-fa

A

B

Discussion

The discovery of the furcula for *Tyrannosaurus rex* verifies the prediction of one by Carpenter and Smith (2001) and increases the number of taxa with this structure (Table 12.1). There is growing evidence suggesting that perhaps all theropods possessed clavicles, although these can be difficult to recognize, especially if the furcula is pathological and particularly if the bone is incomplete. Such a mis-identification occurred with the *T. rex* skeleton known as "Sue" (BHI 2033 = FMNH PR2081). We believe that what was reported to be the first *Tyrannosaurus rex* furcula (Brochu 2003) is actually a pathologic gastralia (Fig. 12.5A,B). The fragment is approximately 15 cm long, 2.5 cm wide throughout, and 1.5 cm deep, tapering to less than 1 cm deep. In shape, it is virtually identical to the distal end of a mid-line gastralia. There is, however, an element associated with FMNH PR2081 that is nearly identical to the furcula found with CMI 2001.90.1 (Neal Larson, pers. comm., 2001). This element is also pathologic, which obscured its identification. Originally identified tentatively as the twelfth dorsal rib, the specimen resembles the furcula of CMI 2000.90.1 in its boomerang shape and in having an angle between the two rami of about 85° (Fig. 12.6). The specimen shows an interesting pathology that creates an asymmetry. The right ramus is only two-thirds the length of the left. The left ramus shows an epicleideal facet, but the right does not. Presumably the articulation with the right scapula was lost (fractured and not reattached) in life or never present. Sue's right scapula and humerus are also pathologic (Brochu 2003).

A third specimen surfaced during our examination of a *T. rex* skeleton excavated near Fort Peck, Montana. This specimen, MOR 980, is also boomerang-shaped with an angle between the two rami of approximately 80°. It is nearly the same size and shape as the two

Figure 12.5. (opposite page) (A) "Furcula" of FMNH PR2081; note wide spacing of coracoids. (B) Detail of same; more than half of the specimen was restored. (C) Furcula of CMI 2001.90.01 held in position in Tyrannosaurus rex *skeleton; note coracoids almost in contact. (D) Chest cavity of* Tyrannosaurus rex, *showing the correct articulation and orientation of the furcula and pectoral girdle.*

Figure 12.6. (above) Pathologic furcula of FMNH PR2081 in (A) anterior and (B) posterior views. Scale in cm. Abbreviation: ep-fa—epicleideal facet.

previous specimens but differs from those two specimens in that it is more or less symmetrical and slightly smaller.

Makovicky and Currie (1998) demonstrated that the furcula in the tyrannosaurids *Albertosaurus, Gorgosaurus,* and *Daspletosaurus* could articulate with the acromion process of the scapula only if the coracoids almost touched along the mid-line. Such an interpretation is supported by articulated theropod skeletons (Barsbold 1983; Chure and Madsen 1996; Currie, pers. comm.; pers. obs.) and may be seen in the small oviraptorid in Figure 12.4. A similar contact of coracoids was true of *Tyrannosaurus* as well, as may be seen in Figure 12.5C,D.

Conclusion

The discovery of a furcula with a *Tyrannosaurus rex* skeleton increases the number of theropod genera known to possess fused clavicles, or furculae. The missed identification of the furcula in FMNH PR2081 suggests that other theropod furculae may be unrecognized in museum collections.

Acknowledgments. Thanks to Bucky Derflinger of Faith, South Dakota, for the original discovery of the *Tyrannosaurus rex* site; to Dallas Evans and Jeff Patchen at The Children's Museum for their support of this research; to Neal Larson of the Black Hills Institute for recognizing the furcula found with Sue; to Bill Simpson and Peter Makovicky at the Field Museum of Natural History for their assistance and photographs; to the board of directors of Fort Peck Paleontology, Inc., and to the Museum of the Rockies for access to their collections; and to Phil Currie at the Royal Tyrrell Museum of Palaeontology for his input; to Dorothy Sigler Norton of Bend, Oregon, for preparation of the illustrations; and to Bob Baker and Kenneth Carpenter for their ideas, encouragement, and editing of the manuscript.

References Cited

Barsbold, R. 1981. [Toothless carnivorous dinosaurs of Mongolia]. [*Transactions, Joint Soviet-Mongolian Paleontological Expedition*] 15: 28–39. [In Russian.]

Barsbold, R. 1983. [Carnivorous dinosaurs from the Cretaceous of Mongolia]. [*Transactions, Joint Soviet-Mongolian Paleontological Expedition*] 19: 5–119. [In Russian.]

Barsbold, R., T. MaryaDska, and H. Osmólska. 1990. Oviraptorosauria. In D. B. Weishampel, P. Dodson, and H. Osmólska (eds.), *The Dinosauria*, pp. 249–258. Berkley: University of California Press.

Brochu, C. A. 2003. Osteology of *Tyrannosaurus rex:* Insights from a nearly complete skeleton and high-resolution computed tomographic analysis of the skull. *Journal of Vertebrate Paleontology, Memoir* 7: 1–138.

Burnham, D. A., K. L. Derstler, and C. J. Linster. 1997. A new specimen of *Velociraptor* (Dinosauria: Theropoda) from the Two Medicine Formation of Montana. In D. L. Wolberg, E. Stump, and G. Rosenberg (eds.), *Dinofest International: Proceedings of a Symposium Held at Arizona State University*, pp. 73–75. Philadelphia: Academy of Natural Sciences.

Burnham, D. A., K. L. Derstler, P. J. Currie, R. T. Bakker, Zhou Z., and J. H. Ostrom. 2000. Remarkable new birdlike dinosaur (Theropoda: Maniraptora) from the Upper Cretaceous of Montana. *University of Kansas Paleontological Contributions,* no. 13: 1–14.

Carpenter, K., and M. Smith. 2001. Forelimb osteology and biomechanics of *Tyrannosaurus rex*. In D. H. Tanke and K. Carpenter (eds.), *Mesozoic Vertebrate Life,* pp. 90–116. Bloomington: Indiana University Press.

Chure, D. J., and J. H. Madsen. 1996. On the presence of furculae in some non-maniraptoran theropods. *Journal of Vertebrate Paleontology* 16: 573–577.

Clark, J. M., M. A. Norell, and L. M. Chiappe. 1999. An oviraptorid skeleton from the Late Cretaceous of Ukhaa Tolgod, Mongolia, preserved in an avian-like brooding position over an oviraptorid nest. *American Museum Novitates,* no. 3265: 1–36.

Dal Sasso, C., and M. Signore. 1998. Exceptional soft-tissue preservation in a theropod from Italy. *Nature* 392: 383–389.

Ji Q., P. J. Currie, M. A. Norell, and Ji S.-A. 1998. Two feathered dinosaurs from Northeastern China. *Nature* 393: 753–761.

Larson, P. L. 2001. Paleopathologies in *Tyrannosaurus rex:* Snapshots of a killer's life. *Journal of Vertebrate Paleontology* 21: 71A–72A.

Makovicky, P. J., and P. J. Currie. 1998. The presence of furcula in Tyrannosaurid theropods, and its phylogenetic and functional implications. *Journal of Vertebrate Paleontology* 18: 143–149.

Norell, M. A., and J. M. Clark. 1997. A *Velociraptor* wishbone. *Nature* 389: 447.

Norell, M. A., and P. J. Makovicky. 1999. Important features of the Dromaeosaurid skeleton II: Information from the newly collected specimens of *Velociraptor mongoliensis*. *American Museum Novitates* 3282: 1–45.

Xu X., Tang Z.-L., and Wang X.-L. 1999a. A therizinosaurid dinosaur with integumentary structures from China. *Nature* 390: 350–354.

Xu X., Wang X.-L., and Wu X.-C. 1999b. A dromaeosaurid dinosaur with a filamentous integument from the Yixian Formation of China. *Nature* 401: 262–266.

13. The Pectoral Girdle and the Forelimb of *Heyuannia* (Dinosauria: Oviraptorosauria)

JUNCHANG LÜ, DONG HUANG,
AND LICHENG QIU

Abstract

Material of *Heyuannia* is described. The pectoral glenoid faces laterally, the coracoid is elongate, the scapula parallels the dorsal vertebral column, the semilunate bone is fused with metacarpals I and II, and metacarpals I and II are proximally fused. The preserved angle between the forearm and manus is less than 90°, which is similar to the angle in birds. The scapulocoracoid and sternal characters indicate a ratite-like morphology. This material provides evidence to support the hypothesis that Oviraptorosauria were secondarily flightless birds.

The postcranial evolution of birds more derived than *Archaeopteryx* may have followed two main paths. A path of increasing body size and loss of flight is indicated by an obtuse angle between the scapula and coracoid, a coracoid with a well-developed biceps tubercle, a glenoid that faces laterally, a sternum without a keel, metacarpals that are incompletely fused, and a relatively long tail. A second path involved reduced body size and active flight, characterized by an acute angle between the scapula and the coracoid, a coracoid with a well-developed acrocoracoid process, a glenoid that faces dorsolaterally, a keeled sternum, completely fused metacarpals, and a tail reduced to a pygostyle.

Introduction

Oviraptorosaurs are generally regarded as non-avian theropod dinosaurs (Osborn 1924; Barsbold 1976, 1997; Gauthier 1986; Clark et al. 1999, 2001; Sereno 1999; Barsbold et al. 2000a,b; Holtz 2000, 2001; Norell et al. 2001; Xu et al. 2002). They are medium-size theropods characterized by short, high skulls with toothless jaws in derived forms and toothed jaws in primitive forms, such as *Caudipteryx* and *Avimimus;* large external mandibular and antorbital fenestrae; deep hypapophyses in the cervicodorsal vertebrae; pneumatized caudal vertebrae in derived forms; anteriorly concave pubic shafts in Oviraptoridae; and a posteriorly curved ischium (Barsbold and Osmólska 1990; Barsbold et al. 2000b; Makovicky and Sues 1998; Maryańska et al. 2002; Lü et al. 2002). Oviraptorosauria includes three families: Oviraptoridae, Caenagnathidae, and Caudipterygidae (Barsbold 1976; Sternberg 1940; Currie 2000; Clark et al. 2001; Ji et al. 1998; Zhou and Wang 2000; Zhou et al. 2000). *Caenagnathus* was once regarded as a large bird (Sternberg 1940; Cracraft 1971) but was attributed to oviraptorosaurs by Osmólska (1976).

The hypothesis that oviraptorosaurs are secondarily flightless birds, more advanced than *Archaeopteryx,* has been proposed (Paul 1988; Olshevsky 1991; Elżanowski 1999; Lü 2000) and has been supported by phylogenetic analysis (Maryańska et al. 2002; Lü et al. 2002). Although the analyses used largely different taxa and characters, the results were similar in that oviraptorosaurs fall within Aves. Here, we report new evidence on the structures of the pectoral girdle and wrist from *Heyuannia* (Lü 2003) that supports this hypothesis.

Archaeopteryx is widely accepted as a powered flier (Feduccia and Tordoff 1979; Olson and Feduccia 1979; Martin 1991, 1995; Ostrom 1997). The glenoid in *Archaeopteryx* is generally considered to face laterally (Ostrom 1976a,b; Martin 1983; Jenkins 1993); it was once regarded as facing downward (Bakker and Galton 1974) or postero-ventrally (Tarsitano and Hecht 1980), but these interpretations were due to poor preservation. The pectoral girdle and wrist structures of *Heyuannia* are similar to those of *Archaeopteryx,* although the coracoid is more derived than that of *Archaeopteryx.*

Description

The pectoral girdle and right front limb of *Heyuannia* (HYMV1-2, Heyuan Museum, Guangdong Province, China) are partially preserved (Fig. 13.1), including the scapula; a partial coracoid; a complete humerus, ulna, and radius; a semilunate; a radiale; metacarpals I–III; parts of phalanges I-1 and II-1; and phalanx II-2. The scapula is elongate, missing only a small portion of its distal end. Its blade is slender, slightly curved, and costolaterally flattened. The dorsal margin of the blade is thick and round, whereas its ventral margin is thin and sharp. There is a well-developed acromial process for articulation with the furcula.

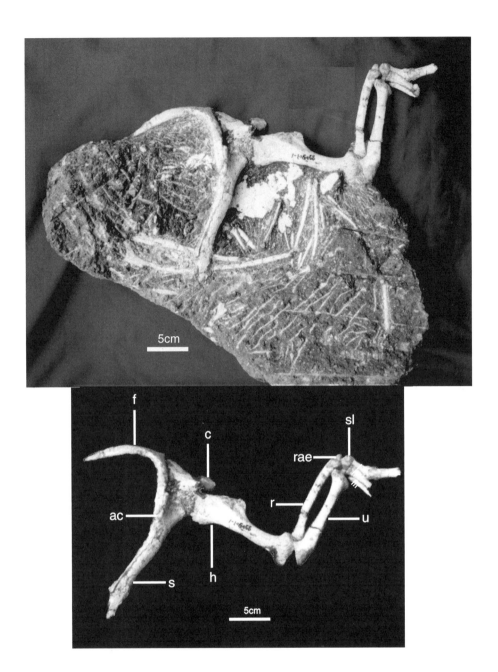

Figure 13.1. Pectoral and the right front limb of Heyuannia *(HYMV1-2). Abbreviations: ac—acromial process; c—coracoid; f—furcula; h—humerus; r—radius; rae—radiale; u—ulna; s—scapula; sl—semilunate; I–III—metacarpals I–III. Scale bar = 5 cm.*

The anterior end of the acromial process is level with the posterior margin of the glenoid.

In another specimen (HYMV1-5) from the same quarry, the scapula, dorsal vertebrae, and dorsal ribs are preserved in nearly natural articulation. The scapular blade is identical to that of HYMV1-2. The scapula is nearly parallel to the vertebral column, and its distal end is slightly expanded. The coracoid is elongate. Its medial portion is relatively thick with a convex anterior surface and a concave inner surface. Ventral to the glenoid rim, there is a conspicuous coracoid tubercle

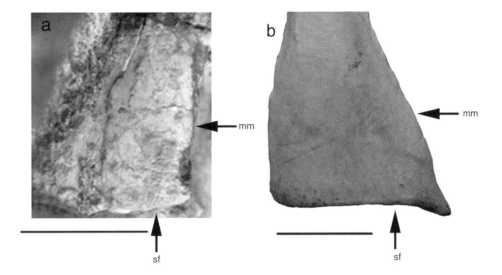

Figure 13.2. The shape of the sternal end of the right coracoid in Heyuannia *(a) and ostrich (b) in anterior view. Abbreviations: mm—medial margin; sf—sternal facet. Scale bar = 2.5 cm.*

(biceps = acrocoracoid; Ostrom 1976a). The supracoracoid foramen is situated dorsomedial to the coracoid tubercle. The distal shaft of the right coracoid is thin and plate-like, its anterior surface is convex, and the posterior surface is slightly concave. The lateral and medial margins of the sternal part are straight and nearly parallel. The margin (sternal facet), which articulates with the sternum, is also nearly straight (Fig. 13.2a). The scapula and coracoid are fused, and they join at an angle of approximately 145°. The ratio of coracoid length to scapula length is approximately 0.3.

The clavicles are fused, with no indication of a mid-line suture, into a forward facing, broadly open, U-shaped furcula (Fig. 13.1). The surface of the anteroventral part is smooth. Both branches of the furcula are slightly curved posterodorsally and become flat antero-posteriorly. The cross-section is oval near the middle the shaft. A well-developed hypocleidium is present on the caudoventral edge of the furcula. The sternum is thin, and only a small lateral portion is pre-served, which shows three distinct articular facets for sternal ribs.

The humerus is slightly twisted, with a well-developed deltopec-toral crest that lies slightly above the middle part of the shaft. The margin between the top of the deltopectoral crest and the proximal end is nearly straight. The rough area of the bone surface near the humeral head may have been caused by weak ossification. There is no sign of either the external or the internal tuberosity, nor is there a bicipital crest. The inner margin of the proximal end is round, and distally be-comes gradually depressed to form a large concave surface. In posterior view, distal to the humeral head, there is a depression on the lateral surface, which extends to the level of the deltopectoral crest. There is a slight depression for the articulation with the ulna near the distal end of the humerus. The outer surface is flat and meets the posterior sur-face at 90°.

The ulna is nearly as long as the humerus and has a distinct olecranon. The ulna is stouter than the radius. In proximal view, the ulna is robust and triangular, and in lateral view the shaft is bowed posteriorly. Its ventral surface is flat, and its dorsal surface is rounded, with a weak medial ridge; thus its cross-section is an isosceles triangle with an obtuse vertex angle. This medial ridge extends to the distal end but not to the proximal end, and it becomes gradually thinner from the middle part of the shaft toward the proximal end. The distal end of the ulna is slightly expanded. It is flat dorsoventrally and wide medio-laterally. A triangular depression, the radial notch, on the medial surface of the distal end forms the articulation with the radius. A shallow depression on the lateral surface of the distal end may contact the ulnare. The proximal and distal articular surfaces are not in the same plane.

Both ends of the radius are slightly expanded. The shaft is bowed away from the ulna, thus leaving a space between the radius and the ulna. The semilunate is the largest carpal. Its mediolateral width is slightly larger than that of metacarpal I. The semilunate is fused with metacarpals I and II (Fig. 13.3). There is a deep, straight groove on its convex articular surface, which forms a trochlear articulation proximally. Metacarpal I wraps around metacarpal II. In ventral view, the mediolateral width of the semilunate carpal is equal to that of metacarpal I. The semilunate carpal articulates with metacarpal I, not with metacarpal II. The radiale is a small carpal with an irregular quadrangular flexor surface. The radiale has a clearly movable articular surface with the radius. The ulnare was not found; however, an obvious articular surface on the distolateral surface of the ulna suggests that the ulnare was present.

The first metacarpal is complete, short, and well developed. It is the shortest and stoutest of the metacarpals. The articular surface of the distal end is ginglymoid as in other theropods. The shafts of metacarpals II and III are relatively slender. The proximal shaft of metacarpal II is flat and wide. The proximal end of metacarpal III is slightly wider than its shaft. The cross-sections of the shafts of metacarpals I and II are oval. Metacarpal III is tightly attached to the lateral side of metacarpal II, but its proximal end is distal to (lower than) the proximal end of the metacarpal II (the same is true for *Velociraptor*), thus leaving a gap between metacarpal III and the distal carpals. This space may have been occupied by cartilage.

Discussion

Figure 13.3. (opposite page) The wrist of Heyuannia *in ventral view (A), dorsal view (B), and close-up of the wrist part, showing the relationship between metacarpal I and metacarpal II (C). Abbreviations: I–III— metacarpals I–III; rae—radiale; sl—semilunate carpal.*

The distal end of the scapula tapers in most birds. The slightly expanded distal end of the scapula of *Heyuannia* is similar to that of *Archaeopteryx lithographica* (Ostrom 1976a; Martin et al. 1998) and clearly different from that of most non-avian theropod dinosaurs, which usually have strongly expanded distal ends. The narrow, elongate scapula in HYMV1-4 lies parallel to the vertebral column (Fig. 13.4), similar to that in *Archaeopteryx* but differing from that in most non-avian theropod dinosaurs, which have a long narrow scapula that crosses the rib cage diagonally (Martin 1995).

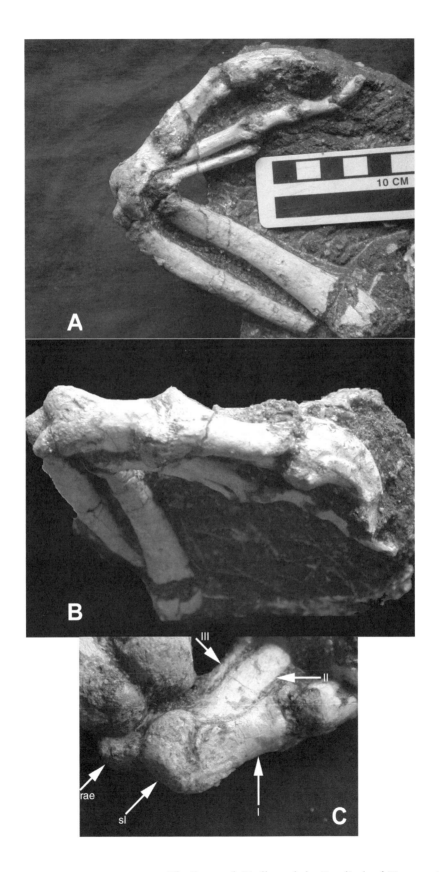

Figure 13.4. HYMV1-4, showing the shaft of the scapula parallel to the vertebral column. Abbreviations: dv—dorsal vertebra; lj—lower jaw; ls—left scapula; rs—right scapula; s— spine; vc—vertebral column.

The coracoid of *Heyuannia* is relatively long, rather than subcircular or more subrectangular as in most non-avian theropod dinosaurs (Ostrom 1974; Madsen 1976; Sereno 1994; Padian 1997), although it is not so strut-like as in *Ornithothoraces* (Chiappe 1996). The shape of the sternal contact of the coracoid is similar to that of the extinct flightless bird *Phororhacos inflatus* from Tertiary deposits of Patagonia (Andrews 1901) and to that of the ostrich (Fig. 13.2b). The shape of the contact is different from that of *Archaeopteryx* (Wellnhofer 1993) and *Struthiomimus* (Barsbold and Osmólska 1990) and differs strongly from that of *Gallimimus* (Barsbold and Osmólska 1990) and *Mononykus* (Novas 1996), in both of which the relatively short coracoid has a pointed sternal part. The coracoid in *Gallimimus* is not significantly different from the coracoids in other ornithomimids; it has an extended "sternal" margin (but its sternum, if present, has never been found; Osmólska, pers. comm.) and a pointed posteroventral process. A similar sternum-coracoid relationship occurs in *Velociraptor* (Norell and Makovicky 1999). The distal part of the coracoid is slightly expanded, and its margin is straight, similar to that of the ostrich but different from that of *Archaeopteryx*, in which the sternal end of the coracoid is rounded. Compared with the coracoids of most theropod dinosaurs, the coracoid in *Heyuannia* is longer, similar to the condition in *Archaeopteryx* (Ostrom 1976a). The coracoid tubercle is situated just ventral to the glenoid rim and immediately ventral to the coracoid foramen, in almost exactly the same position as in *Deinonychus* and *Archaeopteryx* (Ostrom 1974).

The well-preserved sternals of *Oviraptor* and *Ingenia* (Barsbold 1986) are plate-like and sometimes fused along their mid-line, with a groove along the anterior margin for reception of the coracoid, as in *Velociraptor* (Norell and Makovicky 1999), birds, and pterosaurs (Barsbold 1983a,b; Padian 1997). An obtuse angle (145°) formed between the fused scapula and coracoid in oviraptorosaurs is different from the angle in most non-avian theropod dinosaurs, in which the scapula and the coracoid are in the same plane (Sereno 1994; Chiappe 1996). An exception is *Gallimimus* (and presumably other ornithomimids; Osmólska, pers. comm. 2002), in which the scapula-coracoid angle is about 140° (Nicholls and Russell 1981; see Osmólska et al. 1972). In some dromaeosaurids and troodontids, the angle is significantly smaller (Norell and Makovicky 1999; Xu et al. 2002). The angle is similar to that in flightless birds, whereas in most modern flying birds the angle between the scapula and the coracoid is acute (Pycraft 1901).

The obtuse angle between the coracoid and the scapula has long been recognized as a characteristic of flightless birds (Huxley 1867; Parker 1882). The decreased angle in a flying bird correlates with the increased size of the sternum and pectoral muscles, resulting in the mass being brought under the bird's center of gravity to aid balance and probably to confer mechanical advantage in flight (Olson 1973). In *Confuciusornis*, the angle between the scapula and the coracoid is about 90° (Zhou and Farlow 2001). In *Avimimus*, the angle between the scapula and the coracoid is close to 160°, which is similar to the angle in ratite birds (Kurzanov 1983). *Avimimus* is regarded by some as a flightless bird from the Late Cretaceous of Mongolia (Maryańska et al. 2002). The obtuse angle between the scapula and the coracoid in the scapulocoracoid of *Oviraptor* (Barsbold and Osmólska 1990) is similar to that of a modern ostrich (Fig. 13.5).

The sterna of oviraptorosaurs are elongate, sometimes fused (as in birds), but different from those of most non-avian theropod dinosaurs. Barsbold (1983a) was the first to mention that the rectangular ossified sternum in *Oviraptor* is quite similar to that of a ratite. The sternum in *Concornis* is also similar to that in oviraptorosaurs mentioned by Sanz and Buscalioni (1992).

The U-shaped furcula is basically similar to those of *Archaeopteryx*, *Confuciusornis* (Martin et al. 1998), flightless Palaeognathae (Pycraft 1901), and other oviraptorosaurs (Barsbold and Osmólska 1990). In non-avian theropod dinosaurs, the furcula has two forms: U-shaped and V-shaped. It is V-shaped in *Velociraptor mongoliensis*, and both clavicles are relatively straight and rod-like (Norell et al. 1997). The furcula is also V-shaped in some allosaurids (Chure and Madsen 1996). It forms a caudoventrally, convex arc in tyrannosaurids (Makovicky and Currie 1998; Larson and Rigby, this volume). A U-shaped furcula is found in *Sinornithosaurus* (Xu et al. 1999). U- and V-shaped furculae are also present in birds. Burnham and Zhou (1999)

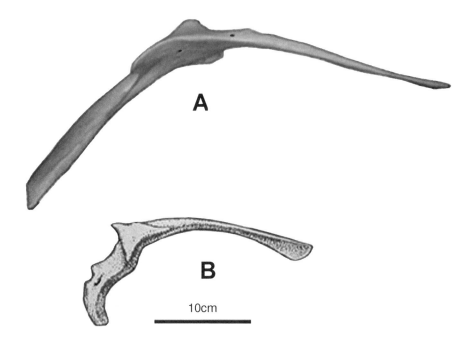

A

B

10cm

Figure 13.5. The structure of the scapula and the coracoid in the scapulocoracoid of Oviraptor philoceratops (A) and ostrich (B). A is modified from Barsbold et al. 1990.

suggested that birds with U-shaped furculae have strong flight ability. The U-shaped furcula found in oviraptorosaurs is similar to that of *Archaeopteryx* in that the upper parts of both branches are flat antero-posteriorly, and they articulate with the acromial process of the scapula. The similar shapes of the furculae of *Archaeopteryx* and oviraptoro-saurs indicate that the shape of the furculae among birds may not correlate with flight capability, and the furcula is not associated with the origin of flight as Norell et al. (1997) thought. It may be convergent or be a retained primitive character. Therefore, the hypothesis that the ostrich is derived from ancestors that had lost the clavicle (Lowe 1928, 1935; De Beer 1956), a hypothesis based on the absence of the rudiment of the clavicle in the embryo of the ostrich, may not be correct.

The length of the humerus is usually less than or equal to the length of the ulna in Ornithurae, but in dromaeosaurids and *Archaeopteryx* the humerus is longer than the ulna (Chiappe 1996). The humerus and the ulna are nearly equal in *Heyuannia,* a condition similar to that of Ornithurae.

The pronounced separation of the radius and ulna shafts is charac-teristic of modern birds (Ostrom 1972), although this condition ap-pears in some small dromaeosaurid dinosaurs (Ji et al. 2001). In primi-tive non-avian theropod dinosaurs, such as *Herrerasaurus*, the radius and the ulna are usually parallel to each other, but both bones are often twisted (Sereno 1994).

The proximal ends of metacarpals I and II are fused in *Heyuannia,* a condition that differs from that of *Deinonychus* and *Archaeopteryx,* in which they are not fused. The proximal end of metacarpal I wraps over metacarpal II in *Heyuannia.* The proximal end of metacarpal III does not contact the distal carpal, as it does in *Archaeopteryx* (Ostrom 1991) and *Khaan* (Clark et al. 2001). The proximal end of metacarpal III is tightly attached to the lateral side of metacarpal II, and this character is present in *Archaeopteryx* and modern birds but absent in *Deinonychus* (Zhou and Martin 1999).

The major alteration of the pectoral girdle in *Archaeopteryx*, in comparison to the girdles of deinonychosaurs and other non-avian theropod dinosaurs, is the reorientation of the glenoid to face dorso-laterally (Jenkins 1993); this is regarded as a preadaptation for avian flight (Novas and Puerta 1997; Burnham et al. 2000). A laterally facing glenoid in *Archaeopteryx* suggests that it was capable of extensive wing elevation (upstroke) as well as fixing the humerus against the body when the wing was held folded (Novas and Puerta 1997). This condition (a somewhat intermediate position between those of most non-avian theropod relatives and later birds) of *Archaeopteryx* is generally accepted as associated with flight (Ostrom 1976a; Jenkins 1993; Padian and Chiappe 1998; Martin 1995). In order for the glenoid to face laterally, the coracoid must lie on the front of the chest with the elongated scapula along (parallel to) the vertebral column (Martin 1997). This condition can be observed in the partially preserved individual (HYMV1-4) (Fig. 13.4). The laterally facing glenoid similar to that in *Archaeopteryx* is retained during the evolution of flightlessness in oviraptorosaurs. As the size and weight of their bodies increased, they used their legs more and their wings less (De Beer 1956), as in the evolution of the modern ratites. *Unenlagia* (Novas and Puerta 1997) is reported to have a laterally facing glenoid and is thought to represent a lineage that is closest to birds (Novas and Puerta 1997), although this orientation was questioned by Carpenter (2002). Novas and Puerta (1997) interpreted the laterally facing glenoid as filling a gap in the sequence of the evolution of the motion of the shoulder joint toward the condition necessary for flight, seen in *Archaeopteryx* and all other birds (Gauthier and Padian 1985). The laterally facing glenoid present in oviraptorosaurs may indicate that they retained this character, which first appeared in *Archaeopteryx*. This glenoid type also may indicate that the ancestors of oviraptorosaurs were capable of flight, if *Archaeopteryx* is regarded as an active flier. Oviraptorosaurs are secondarily flightless birds, reinforcing the results obtained from phylogenetic analysis (Maryańska et al. 2002; Lü et al. 2002).

The semilunate carpal is fused with metacarpals I and II, and forms a pulley-like trochlea on the carpometacarpus as in modern birds—increasing the degree of flexion possible at the wrist. This condition is similar to that in *Archaeopteryx*. The distinctive trochlea carpalis of the carpometacarpus and its unique action are characteristic of all flying birds (Ostrom 1976a; Ostrom et al. 1999). In the new oviraptorosaur, the proximal part of the first metacarpal wrapped over the second metacarpal, indicating that the wrist of *Heyuannia* is stronger than that of *Archaeopteryx*. The wrist is more derived than that of *Archaeopteryx* but more primitive than that of modern birds, in which all the metacarpals and carpals are fused into a carpometacarpus.

The preserved angle between the manus and the forearm is proximately 45°, which is more bird-like than in non-avian theropods. This angle is approximate 90° in *Archaeopteryx* (Wellnhofer 1992), *Confuciusornis* (Chiappe et al. 1999, figs. 8, 48), and *Concornis lacustris* (Sanz and Buscalioni 1992); and it is less than 70° in *Shenzhouraptor* (Ji et al. 2002), *Jeholornis* (Zhou and Zhang 2002), *Eoalulavis hoyasi* (Sanz et al. 1996), *Noguerornis gonzalezi* (Chiappe and Lacasa-Ruiz 2002), *Neuquenornis volans* (Chiappe and Calvo 1994), *Khaan mckennai* (Clark et al. 2001), and *Microraptor* (Hwang et al. 2002). But it is greater than 90° in *Caudipteryx* (Ji et al. 1998; Zhou and Wang 2000) and other non-avian theropod dinosaurs. The preserved configuration of the lower forearm in *Heyuannia* is clearly similar to that of basal birds, but it is different from that of most non-avian theropod dinosaurs in which the wrist parts are preserved. We infer that the morphological function of the manus and the forearm in oviraptorids is closer to that of birds than to that of non-avian theropod dinosaurs.

As Sanz and Buscalioni (1992) suggested, in the evolutionary succession from non-avian theropods to *Archaeopteryx*—early birds and then to later birds, size was constrained by flight. There is a size decrease from non-avian theropods to *Archaeopteryx*. This decrease is greatest from *Archaeopteryx* to the full-powered flight birds, which have the novelties of flight apparatus, such as the strut-like coracoid, keeled sternum, furcula, and pygostyle (Sanz and Buscalioni 1992). Considering the relatively large size of flightless birds, such as *Caudipteryx*, and oviraptorosaurs, these flightless birds used their legs more; and the major components associated with flight became less developed. These components include metacarpals that are not completely fused, a sternum that lacks a keel, the loss of the pygostyle, and the increase in the number of caudal vertebrae to more than in most flight birds but fewer than in non-avian theropod dinosaurs.

The restored pectoral structure of *Oviraptor* (Barsbold 1983b) (Fig. 13.6a) was questioned by Martin (1991), who stated that there are no articular surfaces between the coracoid and the sternum indicated by Barsbold. In fact, there are articular surfaces (Padian 1997). Barsbold's restoration of the pectoral structure is unacceptable because he did not show the obtuse angle between the scapula and the coracoid. The new material indicates that the furcula articulated with the scapula, not with the coracoid (Figs. 13.1, 13.6). Although the sternum is not complete in *Heyuannia,* the distal end of the coracoid is similar in shape and size to that of other oviraptorosaurs. On the basis of the well-preserved sternals of *Oviraptor* and *Ingenia* (Barsbold 1986) from Mongolia, the sternal end of the coracoid in *Heyuannia* probably also articulated with anterior margin of the sternum. When the distal margin of the coracoid articulates with the sternum, because the scapula and the coracoid are fused, the glenoid is reoriented to face laterally (Fig. 13.6b).

Although furculae have been found in many non-avian theropod dinosaurs (Thulborn 1984; Bryant and Russell 1993; Chure and Madsen 1996; Norell et al. 1997), their articulations with the acromial

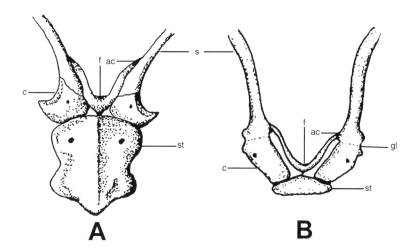

process of the scapula are different from those of *Archaeopteryx* and oviraptorosaurs. In *Archaeopteryx* and oviraptorosaurs, there is a large articulation area on the scapula (acromial process of the scapula) articulating with the epicleidium of the furcula. Most non-avian theropod dinosaurs, such as *Velociraptor,* show the primitive reptilian articulation between the acromial process of the scapula and the furculae (Barsbold, pers. comm.; Lü et al. 2002).

Owing to the uncertain preservation of the pectoral girdle of the *Archaeopteryx*, two hypotheses about the articular relationships of furculae in the pectoral girdle region were put forward. Martin (1995) thought that the furcula should articulate to the coracoid tubercle, on the basis of the idea that the biceps tubercle is the homologue of the brachial tuberosity where the furcula usually attaches in modern birds (Ostrom 1976a,b). The other is that the furcula articulated with the acromial process of the scapula (Petronievics and Woodward 1917; Wellnhofer 1993). In the emu, the scapula has a strongly marked articular facet for the vestigial furcula (Pycraft 1901). In modern flightless struthious birds, the furcula still articulates with the scapula. Thus, the furcula articulated with the scapula found in *Oviraptor* (Barsbold and Osmólska 1990) and *Heyuannia* is consistent with the furcula articulating with scapula in *Archaeopteryx*. The furcula lies in front of the chest and extends posteroventrally. Therefore, if the furcula is placed so that it articulates with the acromial process of the scapula, the glenoid will face laterally (Fig. 13.6b). This position corresponds to adjusting the articulation between the coracoid and the sternum. The laterally facing glenoid causes the arm to be extended out to the side. This placement is a prerequisite for a wing (Martin 1995) that is found in some more specialized small non-avian theropod dinosaurs, such as *Bambiraptor* (Burnham et al. 2000), *Velociraptor* (Norell and Makovicky 1999), *Sinovenator* (Xu et al. 2002), and *Microraptor* (Hwang et al. 2002), if they are not treated as birds (i.e., avian theropods).

Figure 13.6. The restoration of the pectoral structure of Oviraptor *(Barsbold 1983b) (A) and present study (B) based on* Heyuannia *in anterior view. Abbreviations: ac—acromial process; c—coracoid; f—furcula; s—scapula; st—sternum; gl—glenoid.*

Conclusion and Summary

The relatively large size (less than 3 m long) and the moderately long tails (fewer than thirty caudal vertebrae; compare with the short-tailed birds) of oviraptorosaurs indicate that they were cursorial animals. The laterally facing glenoid, and the wide articulation of coracoid with the sternum, indicate that they are derived from flying ancestors (that is, secondarily flightless birds). The laterally facing glenoid also suggests that they may have evolved independently from *Archaeopteryx* during bird evolution and that their ancestors may have been living in trees. *Heyuannia* also shows ratite-like morphologies.

Acknowledgments. Special thanks to Drs. Dale A. Winkler and Louis L. Jacobs (SMU), H. Osmólska (Poland), and R. Barsbold (Mongolia), who made critical comments and suggestions on the manuscript, and to Drs. P. Currie and K. Carpenter for comments that greatly improved the manuscript. Thanks to Mr. Shusaku Suzuki and Dr. Yoichi Azuma for their kind help in this project. Thanks also to Mrs. Weiqiang Yuan, Pinbin Zhong, Yi Liu, Zhiqing Huang, Xuejia Huang, and Qingsheng Wen (Heyuan Museum), who joined the fieldwork and repaired the specimens. This work was supported by the Jurassic Foundation, the Chang Ying-Chien Science Grant for USA-China Collaborative Field Research, the Institute for the Study of Earth and Man at Southern Methodist University, and the Graduate Student Development Program of Dedman College at Southern Methodist University to J. Lü.

References Cited

Bakker, R. T., and P. M. Galton. 1974. Dinosaur monophyly and a new class of vertebrates. *Nature* 248: 168–172.

Barsbold, R. 1976. [On a new Late Cretaceous family of small theropods (Oviraptoridae fam. n.) of Mongolia]. *Doklady Akademiia Nauk SSSR* 226 (3): 685–688. [In Russian.]

Barsbold, R. 1983a. [Carnivorous dinosaurs from the Cretaceous of Mongolia]. [*Transactions Joint Soviet-Mongolian Paleontological Expedition*] 19: 5–119. [In Russian.]

Barsbold, R. 1983b. [On the "bird" features in the structure of carnivorous dinosaurs: Fossil Reptiles of Mongolia]. [*Transactions, Joint Soviet-Mongolian Paleontological Expedition*] 24: 96–103. [In Russian.]

Barsbold, R. 1986. Raubdinosaurier Oviraptoren. In E. I. Vorob'eva (ed.), *Herpetologische Untersuchungen in der Mongolischen Volksrepublik,* pp. 210–223. Moscow: Akad. Nauk SSSR Inst. Evoliutsionnoi morfologii i ekologii zhivotnykh im. A.M. Severtsova. [In Russian with German summary.]

Barsbold, R. 1997. Oviraptorosauria. In P. J. Currie and K. Padian (eds.), *Encyclopedia of Dinosaurs,* pp. 505–509. San Diego: Academic Press.

Barsbold, R., and H. Osmólska. 1990. Ornithomimosauria. In D. B. Weishampel, P. Dodson, and H. Osmólska (eds.), *The Dinosauria,* pp. 225–244. Berkeley: University of California Press.

Barsbold, R., H. Osmólska, M. Watabe, P. J. Currie, and K. Tsogtbaatar. 2000b. A new oviraptorosaur (Dinosauria, Theropoda) from Mongolia: The first dinosaur with a pygostyle. *Acta Palaeontologica Polonica* 45: 97–106.

Barsbold, R., P. J. Currie, N. P. Myhrvold, H. Osmólska, K. Tsogtbaatar, and M. Watabe. 2000a. A pygostyle from a non-avian theropod. *Nature* 403: 155–156.

Bryant, H. N., and Russell, A. P. 1993. The occurrence of clavicles within Dinosauria: Implications for the homology of avian furcula and the utility of negative evidence. *Journal of Vertebrate Paleontology* 13: 171–184.

Burnham, D. A., and Zhou Z. 1999. Comparing the furcula in dinosaurs and birds. *Journal of Vertebrate Paleontology* 19 (3 suppl.): 34A.

Burnham, D. A., K. L. Derstler, P. J. Currie, R. T. Bakker, Zhou Z., and J. H. Ostrom. 2000. Remarkable new birdlike dinosaur (Theropoda: Maniraptora) from the Upper Cretaceous of Montana. *University of Kansas Paleontological Contributions,* no. 13: 1–14.

Carpenter, K. 2002. Forelimb biomechanics of nonavian theropod dinosaurs in predation. *Senckenbergiana Lethaea* 82: 59–76.

Chiappe, L. M. 1996. Late Cretaceous birds of southern South America: Anatomy and systematics of Enantiornithes and *Patagopteryx deferrariisi. Münchner Geowissenschaften Abhandlungen, Reihe A,* 30: 203–244.

Chiappe, L. M., and J. O. Calvo. 1994. *Neuqenornis volans,* a new Upper Cretaceous bird (Enantiornithes: Avisauridae) from Patagonia, Argentina. *Journal of Vertebrate Paleontology* 14: 230–246.

Chiappe, L. M., and A. Lacasa-Ruiz. 2002. *Noguerornis gonzalezi* (Aves: Ornithothoraces) from the Early Cretaceous of Spain. In L. M. Chiappe and L. M. Witmer (eds.), *Mesozoic Birds: Above the Heads of Dinosaurs,* pp. 230–239. Berkeley: University of California Press.

Chiappe, L. M., Ji S.-A., Ji Q., and M. A. Norell. 1999: Anatomy and systematics of the Confuciusornithidae (Theropoda: Aves) from the late Mesozoic of Northeastern China. *Bulletin of the American Museum of Natural History* 242: 1–89.

Chure, D. J., and J. H. Madsen. 1996. On the presence of furculae in some non-maniraptoran theropods. *Journal of Vertebrate Paleontology* 16: 573–577.

Clark, J. M., M. A. Norell, and L. M. Chiappe. 1999. An oviraptorid skeleton from the Late Cretaceous of Ukhaa Tolgod, Mongolia, preserved in an avian-like brooding position over an oviraptorid nest. *American Museum Novitates,* no. 3265: 1–36.

Clark, J. M., M. A. Norell, and R. Barsbold. 2001. Two new oviraptorids (Theropoda: Oviraptorosauria), Upper Cretaceous Djadokhta Formation, Ukhaa Tolgod, Mongolia. *Journal of Vertebrate Paleontology* 21: 209–213.

Cracraft, J. 1971. Caenagnathiformes: Cretaceous birds convergent in jaw mechanism to dicynodont reptiles. *Journal of Paleontology* 45: 805–809.

Currie, P. J. 2000. Theropods from the Cretaceous of Mongolia. In M. J. Benton, M. A. Shishkin, D. M. Unwin, and E. N. Kurochkin (eds.), *The Age of Dinosaurs in Russia and Mongolia,* pp. 434–455. Cambridge: Cambridge University Press.

De Beer, G. 1956. The evolution of Ratites. *Bulletin of the British Museum (Natural History) Zoology* 4 (2): 59–70.

Elźanowski, A. 1999. A comparison of the jaw skeleton in theropods and birds, with a description of the palate in the Oviraptoridae. In Olson, S. L. (ed.), *Avian Paleontology at the Close of the 20th Century: Proceedings of the 4th International Meeting of the Society of Avian Paleontology and Evolution,* pp. 311–323. Smithsonian Contributions to Paleobiology, no. 89. Washington, D.C.: Smithsonian Institution Press.

Feduccia, A., and H. B. Tordoff. 1979. Feathers of *Archaeopteryx:* Asymmetric vanes indicate aerodynamic function. *Science* 203: 1021–1022.

Gauthier, J. 1986. Saurischian monophyly and the origin of birds. In K. Padian (ed.), *The Origin of Birds and the Evolution of Flight,* pp. 1–55. Memoirs of the California Academy of Sciences, no. 8. San Francisco: California Academy of Sciences.

Gauthier, J., and K. Padian. 1985. Phylogenetic functional and aerodynamic analyses of the origin of birds and their flight. In M. K. Hecht, J. H. Ostrom, G. Voihl, and P. Wellnhofer (eds.), *The Beginning of Birds,* pp. 185–197. Eichstätt: Freunde des Jura-Museums.

Holtz, T. R., Jr. 1998. A new phylogeny of the carnivorous dinosaurs. B. P. Pérez-Moreno, T. Holtz Jr., J. L. Sanz, and J. Moratalla (eds.), *Gaia: Aspects of Theropod Paleobiology,* vol. 15, pp. 5–61. Lisbon: Museu Nacional de História Natural.

Holtz, T. R., Jr. 2001. Arctometatarsalia revised: The problem of homoplasy in reconstructing theropod phylogeny. In J. Gauthier and L. F. Gall (eds.), *New Perspectives on the Origin and Early Evolution of Birds: Proceedings of the International Symposium in Honor of John H. Ostrom,* pp. 99–122. New Haven, Conn.: Peabody Museum of Natural History. Yale University.

Huxley, T. H. 1867. On the classification of birds; and on the taxonomic value of the modifications of the cranial bones in that class. *Proceedings of the Zoological Society of London* 1867: 415–472.

Hwang, S. H., M. A. Norell, Ji Q., and Gao K. 2002. New specimens of *Microraptor zhaoianus* (Theropoda: Dromaeosauridae) from Northeastern China. *American Museum Novitates,* no. 3381: 1–44.

Jenkins, F. A. 1993. The evolution of the avian shoulder joint. *American Journal of Science* 293A: 253–267.

Ji Q., P. J. Currie, M. A. Norell, and Ji S.-A. 1998. Two feathered dinosaurs from northeastern China. *Nature* 393: 753–761.

Ji Q., M. A. Norell, Gao K., Ji S., and D. Ren. 2001. The distribution of integumentary structures in a feathered dinosaur. *Nature* 410: 1084–1088.

Ji Q., Ji S.-A., You H.-L., Zhang J.-P., Yuan C.-X., Ji X.-X., Li J.-L., and Li Y.-X. 2002. Discovery of an avialae bird—*Shenzhouraptor sinensis* gen. et sp. nov. from China. *Geological Bulletin of China* 21 (7): 363–369.

Kurzanov, S. M. 1983. [*Avimimus* and the problem of the origin of birds]. *Trudy Sovmestnaia Sovetsko-Mongol'skaia Paleontologicheskaia Ekspeditsiia* 24: 104–109. [In Russian.]

Lowe, P. R. 1928. Studies and observations bearing on the phylogeny of the Ostrich and its allies. *Proceedings of the Zoological Society of London,* pp. 185–247.

Lowe, P. R. 1935. On the relationship of the Struthiones to the Dinosaurs and to the rest of the avian class, with special reference to the position of *Archaeopteryx. Proceedings of the Zoological Society of London* (13) 5: 398–432.

Lü J.-C. 2000. Oviraptorosaurs compared to birds. In Shi L.-Q. and Zhang

F.-C. (eds.), *Fifth International Meeting of the Society of Avian Paleontology and Evolution and the Symposium on Jehol Biota. Vertebrata PalAsiatica* 38 (suppl.): 18.

Lü J.-C. 2003. A new oviraptorosaurid (Theropoda: Oviraptorosauria) from the Late Cretaceous of Southern China. *Journal of Vertebrate Paleontology* 22: 871–875.

Lü J.-C., Dong Z. M., Y. Azuma, R. Barsbold, and Y. Tomida. 2002. Oviraptorosaurs compared with birds. In Zhou Z. and Zhang F. (eds.), *Proceedings of the 5th Symposium of the Society of Avian Paleontology and Evolution,* pp. 175–189. Beijing: Science Press.

Madsen, J. H., Jr. 1976. Allosaurus fragilis: *A Revised Osteology.* Utah Geological and Mineral Survey Bulletin, no. 109. Salt Lake City: Utah Geological and Mineral Survey, Utah Department of Natural Resources.

Makovicky, P. J., and P. J. Currie. 1998. The presence of a furcula in tyrannosaurid theropods, and its phylogenetic and functional implications. *Journal of Vertebrate Paleontology* 18: 143–149.

Makovicky, P., and H.-D. Sues. 1998. Anatomy and phylogenetic relationships of the theropod dinosaur *Microvenator celer* from the Lower Cretaceous of Montana. *American Museum Novitates,* no. 3240: 1–27.

Martin L. D. 1995. A new skeletal model of *Archaeopteryx. Archaeopteryx* 13: 33–40.

Martin L. D. 1997. The difference between dinosaurs and birds as applied to *Mononykus.* In D. L. Wolberg, E. Stump, and G. Rosenberg (eds.), *Dinofest International: Proceedings of a Symposium Held at Arizona State University,* pp. 337–343. Philadelphia: Academy of Natural Sciences.

Martin, L. D. 1983. The origin and early radiation of birds. In A. H. Brush and G. A. Clark Jr. (eds.), *Perspectives in Ornithology,* pp. 291–353. Cambridge: Cambridge University Press.

Martin, L. D. 1991. Mesozoic birds and the origin of birds. In H. P. Schultze and L. Trueb (eds.), *Origins of the Higher Groups of Tetrapods,* pp. 485–539. Ithaca, N.Y.: Cornell University Press.

Martin, L. D., Zhou Z.-H., Hou L.-H., and A. Feduccia. 1998. *Confuciusornis sanctus* compared to *Archaeopteryx lithographica. Naturwissenschaften* 85: 286–289.

Maryańska, T., H. Osmólska, and M. Wolsan. 2002. Avialan status for Oviraptorosauria. *Acta Palaeontologica Polonica* 47: 97–116.

Nicholls, E. L., and A. P. Russell. 1981. A new specimen of *Struthiomimus altus* from Alberta, with comments on the classificatory characters of Upper Cretaceous ornithomimids. *Canadian Journal of Earth Sciences* 18: 518–526.

Norell, M. A., and P. J. Makovicky. 1999. Important features of the dromaeosaurid skeleton II: Information from newly collected specimens of *Velociraptor mongoliensis. American Museum Novitates,* no. 3282: 1–45.

Norell, M. A., P. Makovicky, and J. M. Clark. 1997. A *Velociraptor* wishbone. *Nature* 389: 447.

Norell, M. A., J. M. Clark, and P. J. Makovicky. 2001. Phylogenetic relationships among coelurosaurian theropods. In J. Gauthier and L. F. Gall (eds.), *New Perspectives on the Origin and Early Evolution of Birds: Proceedings of the International Symposium in Honor of John H. Ostrom,* pp. 29–67. New Haven, Conn.: Peabody Museum of Natural History, Yale University.

Novas, F. E. 1996. Alvarezsauridae, Cretaceous basal birds from Patagonia and Mongolia. *Memoirs of the Queensland Museum* 39 (3): 675–702.

Novas, F. E., and P. F. Puerta. 1997. New evidence concerning avian origins from the Late Cretaceous of Patagonia. *Nature* 387: 390–392.

Olshevsky, G. 1991. A revision of the parainfraclass Archosauria Cope, 1869, excluding the advanced crocodylia. *Mesozoic Meanderings* 2: 1–196. [Privately published.]

Olson, S. L. 1973. Evolution of the Rails of the South Atlantic Islands (Aves: Rallidae). *Smithsonian Contributions to Zoology* 152: 1–53.

Olson, S. L., and A. Feduccia. 1979. Flight capability and the pectoral girdle of *Archaeopteryx*. *Nature* 278: 247–248.

Osborn, H. F. 1924. Three new Theropoda, *Protoceratops* Zone, central Mongolia. *American Museum Novitates,* no. 144: 1–12.

Osmólska, H. 1976. New light on the skull anatomy and systematic position of *Oviraptor*. *Nature* 262: 683–684.

Osmólska, H., E. Roniewicz, and R. Barsbold. 1972. A new dinosaur, *Gallimimus bullatus* n. gen., n. sp. (Ornithomimidae) from the Upper Cretaceous of Mongolia. *Palaeontologica Polonica* 27: 103–143.

Ostrom, J. H. 1972. Description of the *Archaeopteryx* specimen in the Teyler Museum, Haarlem. *Proceedings of the Koninklijke Nederlandse Akademie van Wetenschappen* B: 289–305.

Ostrom, J. H. 1974. The pectoral girdle and forelimb function of *Deinonychus* (Reptilia: Saurischia): A correction. *Postilla* 165: 1–11.

Ostrom, J. H. 1976a. Some hypothetical anatomical stages in the evolution of avian flight. In Olson, S. L. (ed.), *Collected Papers in Avian Paleontology Honoring the 90th Birthday of Alexander Wetmore,* pp. 1–21. Smithsonian Contributions to Paleobiology, no. 27. Washington, D.C.: Smithsonian Institution Press.

Ostrom, J. H. 1976b. The origin of birds. *Annual Review of Earth and Planetary Sciences* 3: 55–57.

Ostrom, J. H. 1991. The question of the origin of birds. In H.-P. Schultze and L. Trueb (eds.), *Origins of the Higher Groups of Tetrapods,* pp. 467–484. Ithaca, N.Y.: Cornell University Press.

Ostrom, J. H. 1997. How Bird Flight Might Have Come About. In *Dinofest International Proceedings,* pp. 301–310. Philadelphia: Academy of Natural Sciences.

Ostrom, J. H., S. O. Poore, and G. E. Goslow. 1999. Humeral rotation and wrist supination: Important functional complex for the evolution of powered flight in birds? In S. Olson (ed.), *Avian Paleontology at the Close of the 20th Century: Proceedings of the 4th International Meeting of the Society of Avian Paleontology and Evolution,* pp. 301–309. Smithsonian Contribution to Paleobiology, no. 89. Washington, D.C.: Smithsonian Institution Press.

Padian, K. 1997. Pectoral girdle. In P. J. Currie and K. Padian (eds.), *Encyclopedia of Dinosaurs,* pp. 530–536. San Diego: Academic Press.

Padian, K., and L. M. Chiappe. 1998. The origin and early evolution of birds. *Biological Reviews* 73: 1–42.

Parker, T. J. 1882. On the skeleton of *Notornis mantelli. Transactions and Proceedings of the New Zealand Institute* 14: 245–258.

Paul, G. S. 1988. *Predatory Dinosaurs of the World: A Complete Illustrated Guide.* New York: Simon and Schuster.

Petronievics, B., and A. S. Woodward. 1917. On the pectoral and pelvic arches of the British Museum specimen of *Archaeopteryx. Proceed-*

ings of the General Meetings for Scientific Business of the Zoological Society of London 1–6.

Pycraft, W. P. 1901. On the morphology and phylogeny of the Palaeognathae (Ratitae and Crypturi) and Neognathae (Carinatae). *Transactions of the Zoological Society of London* 15: 151–290.

Sanz, J. L., and A. D. Buscalioni. 1992. A new bird from the early Cretaceous of Las Hoyas, Spain, and the early radiation of birds. *Palaeontology* 35: 829–845.

Sanz, J. L., L. M. Chiappe, B. P. Pérez-Moreno, A. D. Buscalioni, J. J. Moratalla, F. Ortega, and F. J. Poyato-Ariza. 1996. An Early Cretaceous bird from Spain and its implications for the evolution of avian flight. *Nature* 382: 442–445.

Sereno, P. C. 1994. The pectoral girdle and forelimb of the basal theropod *Herrerasaurus ischigualastensis. Journal of Vertebrate Paleontology* 13: 425–450.

Sereno, P. C. 1999. The evolution of dinosaurs. *Science* 284: 2137–2147.

Sternberg, R. M. 1940. A toothless bird from the Cretaceous of Alberta. *Journal of Paleontology* 14: 81–85.

Tarsitano, S., and M. K. Hecht. 1980. A reconsideration of the reptilian relationships of *Archaeopteryx. Zoological Journal of the Linnean Society* 69: 149–182.

Thulborn, R. A. 1984. The avian relationships of *Archaeopteryx*, and the origin of birds. *Zoological Journal of the Linnean Society* 82: 119–158.

Wellnhofer, P. 1992. A new specimen of *Archaeopteryx* from the Solnhofen limestone. *Proceedings of the II International Symposium of Avialian Paleontology* 1988: 3–23.

Wellnhofer, P. 1993. Das siebte exemplar von *Archaeopteryx* aus den Solnhofener Schichten. *Archaeopteryx* 11: 1–47.

Xu X., Wang X.-L., and Wu X.-C. 1999. A dromaeosaurid dinosaur with a filamentous integument from the Yixian Formation of China. *Nature* 401: 262–266.

Xu X., M. A. Norell, Wang X.-L., P. J. Makovicky, and Wu X.-C. 2002. A basal troodontid from the Early Cretaceous of China. *Nature* 415: 780–784.

Zhou Z. H. and J. O. Farlow. 2001. Flight capability and habits of *Confuciusornis*. In J. Gauthier and L. F. Gall (eds.), *New Perspectives on the Origin and Early Evolution of Birds: Proceedings of the International Symposium in Honor of John H. Ostrom*, pp. 237–254. New Haven, Conn.: Peabody Museum Natural History, Yale University.

Zhou Z. H. and L. D. Martin. 1999, Feathered Dinosaur or Bird? A New Look at the Hand of *Archaeopteryx*. In S. L. Olson (ed.), *Avian Paleontology at the Close of the 20th Century: Proceedings of the 4th International Meeting of the Society of Avian Paleontology and Evolution. Smithsonian Contributions to Paleobiology* 89: 289–293.

Zhou Z.-H. and Zhang F.-C. 2002. A long-tailed, seed-eating bird from the Early Cretaceous of China. *Nature* 418:405–409.

Zhou Z.-H. and Wang X.-L. 2000. A new species of *Caudipteryx* from the Yixian Formation of Liaoning, Northeast China. *Vertebrata PalAsiatica* 38: 111–127.

Zhou Z.-H., Wang X.-L, Zhang F.-C., and Xu X. 2000. Important features of *Caudipteryx*: Evidence from two nearly complete new specimens. *Vertebrata PalAsiatica* 38: 241–254.

III. Theropods as Living Animals

14. Sexual Dimorphism in the Early Jurassic Theropod Dinosaur *Dilophosaurus* and a Comparison with Other Related Forms

Robert Gay

Abstract

Although the Early Jurassic theropod *Dilophosaurus* is not known from an abundance of specimens, a careful analysis of the known material, along with recently rediscovered material, allows us to compare this animal to other related animals, such as *Coelophysis bauri* and *Syntarsus* (=*Coelophysis*) *rhodesiensis*. Similar patterns of bimodal adult variation can be seen within these two taxa, as well as the expected ontonogenic and individual differences between specimens. The characteristic bimodal plots of adult specimens, which have been interpreted as sexual dimorphism, are not seen in *Dilophosaurus*, suggesting that the variation is likely the result of individual variation and ontogeny alone, and not sexual dimorphism.

Introduction

Although *Dilophosaurus* is one of the earliest known Jurassic carnivores, it is also one of the least well understood early theropods. Because only a few specimens are known, research is difficult. Furthermore, with all the known material (with the possible exception of *Dilophosaurus sinensis*, from China) coming from the Kayenta Formation near Tuba City, Arizona, on the Navajo Reservation, collecting new material is difficult. The private landowners outside of Tuba City, where the original specimens were found, will not permit further prospecting and excavation. Despite this, some work has been done on sexual dimorphism and speciation in the Kayenta specimens.

Paul (1988) has proposed sexual dimorphism in *Dilophosaurus* but does not provide data to support this position. Pickering (1995) suggests that one of the specimens represents a different species, *D. "breedorum."* Because of the uncertain validity of this name resulting from publication practices, and the lack of significant morphological differences between the purported type of *D. "breedorum"* and *D. wetherilli*, this taxon is considered an invalid species.

To determine whether *Dilophosaurus* exhibits sexual dimorphism, a study was undertaken of two other related primitive theropods, *Coelophysis bauri* and *Coelophysis rhodesiensis*. *C. rhodesiensis* was originally named *Syntarsus rhodesiensis* by Raath (1969). However, Paul (1988) and Downs (2000) noted many similarities in the skull, limbs, and pelvis and synonymized the two genera. Both Colbert (1989, 1990) and Raath (1990) argued for sexual dimorphism in each of these animals. Numerous specimens of *Coelophysis bauri* were examined, as well as a cast of part of *Coelophysis rhodesiensis*. This independent review not only confirmed Colbert's findings but also mirrored, with *C. bauri*, the results Raath achieved with *C. rhodesiensis*, as discussed below. The reason that these two taxa are such good comparisons is twofold. First, both are represented by fairly large sample sizes, which makes identifying trends easier. Secondly, Paul (1988), Rowe and Gauthier (1990), and Sereno (1999) propose close relationships between *C. rhodesiensis*, *C. bauri*, and *Dilophosaurus*. Because both species of *Coelophysis* exhibit sexual dimorphism, they provide good models for recognizing dimorphism in the closely related but less well represented *Dilophosaurus*.

Institutional Abbreviations. AMNH, American Museum of Natural History, New York. MNA, Museum of Northern Arizona, Flagstaff, Ariz. NMMNH, New Mexico Museum of Natural History, Albuquerque, N.M. UCMP, University of California Museum of Paleontology, Berkeley.

Sexual Dimorphism in *Coelophysis*

Coelophysis bauri is probably the best-known early theropod dinosaur from anywhere in the world, owing to the exceptionally large number of specimens (estimated between 500 [Paul 1988] and over 1,000 [Schwartz and Gillette 1994]) and to the exceptional preservation of some of the specimens. Some individuals are complete and entirely articulated. Although there is some debate over the cause for the mass death assemblage (Schwartz and Gillette 1994), the *Coelophysis* quarry at Ghost Ranch one of the most remarkable and important early dinosaur sites in the world.

Colbert (1989, 1990) noted two distinct morphs of adult *Coelophysis bauri*, a "robust" morph and a "gracile" morph. He differentiated them on a suite of morphological characters. One morph had a

TABLE 14.1

Coelophysis bauri Skull Measurements

Specimen number	Skull length (cm)	Orbit length (cm)
NMMNH L-4207(B)	19.3	2
MNA V3139	20	2.7
AMNH 7223	20.2	3.1
MNA V3322	13.3	2.9

larger skull, a longer neck, a smaller forelimb, and fused sacral spines. The other morph was characterized by a smaller skull, a shorter neck, larger forelimbs, and free sacral spines. Although verification of fused sacral spines and neck length was not possible in some specimens, the other two characteristics showed up consistently. The lengths of the humerus, radius, and metacarpal III, when plotted, showed a bimodal distribution for adult animals, mimicking the results from earlier studies. Skull material was more commonly preserved, and plots from this data show a bimodal distribution as well (Table 14.1).

As can be seen in Table 14.1, the skulls fall into two distinct categories. Note that NMMNH L-4207 is the designation for the *Coelophysis* block at the NMMNH. The specimens in the block have not yet been properly labeled or catalogued. The block contains on its surface at least two unprepared individuals, which have not been assigned numbers. For the purpose of identification, the letter "A" denotes the incomplete leg elements, and the letter "B" denotes the partial skull.

Another pattern appears in the data on femur length versus tibia length. The data show the same pattern found by Raath (1990) in *Coelophysis rhodesiensis* for measurements of trochanters. Figure 14.1A shows the bimodal distribution of the hindlimb lengths in *Coelophysis bauri*. The immature individuals are represented by NMMNH L-4207(A), AMNH 2704, MNA V3325, V3318(A), V3318(B), and V1960(B) (MNA designations "A" and "B" denote different individuals in a block with one specimen number). The smaller mature morph is represented by MNA V3320 and V3322, and the larger mature individuals are AMNH 7332 and MNA V1960(A). The data for these specimens are given in Table 14.2.

Coelophysis rhodesiensis is known only from the Lower Jurassic of southern Africa. Raath (1990) proposed that *C. rhodesiensis* shows sexual dimorphism, having the same pattern seen in *C. bauri* for femur/tibia length distribution and for correlation between iliofemoralis (lesser) trochanter width versus femoral head breadth. Raath's data plots with two discrete adult modes for *C. rhodesiensis*, with a tight cluster of individuals at the lower (smaller size) end and two distinct groups at the upper end. This same pattern was found in *C. bauri* (compare Fig. 14.1A and 14.1B). Presumably, this marked division between the smaller specimens and the larger, differentiated ones indicates the age of sexual maturity (Raath 1990).

TABLE 14.2

Coelophysis bauri Leg Measurements (cm), Ratios, and Length Reconstructions

Specimen number	Femur (F)	Tibia (T)	F/T ratio	Mt-III	F/Mt-III ratio	Total leg length
NMMNH L-4207(A)	12.3	13.3	0.9	6.4	1.9	32.1
AMNH 2704	12.2	11.3	0.9	6.4	1.9	29.9
AMNH 7332	19	19.9	1	12.1	1.6	51
V1 960(A)	19.0	20.5	0.9	12.2	1.5	51.7
V1 960(B)	12.8	13.5	0.9	5	2.5	31.3
V3318(A)	12.5	13.4	0.9	8.3	1.5	34.2
V3318(B)	12.8	13.9	0.9	6.7	1.9	37.6
V3320	16.9	18	0.9	12.7	1.3	47.6
V3322	15.5	19.7	0.8	9.1	1.7	44.3
V3325	13.6	13.6	1	4.8	2.8	32
Ratio averages			0.9		1.9	

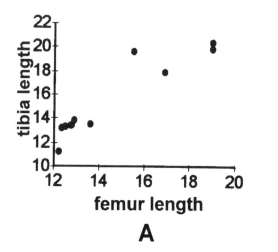

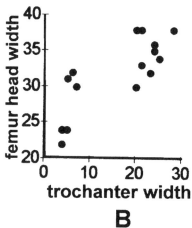

Figure 14.1. Coelophysis *bauri and* C. rhodesiensis *compared, showing bimodal adult distribution. (A)* C. bauri *tibia length vs. femur length (in cm). (B)* C. rhodesiensis *femur head width vs. trochanter width (in mm). (After Raath 1990.)*

Dimorphism in *Dilophosaurus*

As noted above, only a single species, *Dilophosaurus wetherilli* (Welles 1984), is recognized, and any differences among specimens may reflect the presence of more than one morph. The presence of two morphs in *Coelophysis bauri* and *C. rhodesiensis* as noted above are taken to mean sexual dimorphism. With these examples of dimorphism in mind, the various specimens of *Dilophosaurus* can be examined for different morphs, which may indicate sexual dimorphism.

UCMP 37302. This specimen is the holotype and one of the most complete specimens of this taxon. Along with UCMP 37303, it provides the clearest view on the osteology of *Dilophosaurus* (Welles 1984). Welles (1954) published the lengths of the limb bones and total limb lengths, which were verified by me.

TABLE 14.3

Dilophosaurus wetherilli Leg Measurements (cm),
Ratios, and Length Reconstructions (cm)

Specimen number	Femur(F)	Tibia(T)	F/T ratio	Mt-III	F/Mt-III ratio	Total leg length
MNA P1.109	28.1	27.4	0.9	14.6	0.5	70.1
UCMP	54	53	0.9	29	0.5	136
UCMP 77270a	57.5	56	0.9	30.4	0.5	143.9
UCMP 77270b	60.5	58.5	0.9	32.0	0.5	151.0
MNA P1.3145	57.4	55.9	0.9	30.3	0.5	143.6
MNA P1.160/161	54.5	53.0	0.9	28.8	0.5	136.4

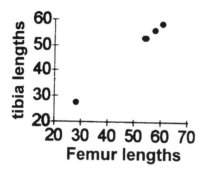

The femur measures 54 cm, the tibia 53 cm, and metatarsal III 29 cm. The ratio for tibia length to femur length is 0.98. The metatarsal III to femur ratio is 0.54. The humerus measures 28 cm in length, the radius 17.5 cm, and metacarpal III 10 cm. The radius to humerus ratio is 0.62. The metacarpal III to humerus ratio is 0.36. The cranial crests are not well preserved in this specimen, showing only the basal-most portion. The femur data are plotted in Figure 14.2, and all ratios and reconstructions are listed in Table 14.3.

UCMP 77270(A,B). This specimen was discovered in 1964 and was originally thought to be a new genus (Welles 1984). However, Welles (1970) later concluded that it was the same taxon as UCMP 37303. This specimen was the first to clearly show the paired crests characteristic of *Dilophosaurus.* Slightly larger than the previous specimen, this animal is not the largest specimen of *Dilophosaurus.*

The femoral lengths are 57.5 cm and 60.5 cm, and the tibial lengths are 56 and 58.5 cm. Because the femoral lengths are significantly different, these specimens may represent two different animals, and they have been plotted as such.

MNA P1.3145. The specimen is the proximal end of a femur; the remainder was estimated from the other specimens of *Dilophosaurus.* Because this partial specimen is large, it is assumed to be an adult, and scaling is not likely to be affected by ontonogenic differences. This restored femur was then used to estimate the rest of the leg. Extrapolat-

Figure 14.2. Dilophosaurus femoral data showing monomodal plot. Tibia lengths vs. femur lengths (in cm).

ing from femoral width versus femur length, the estimated femur length is 57.4 cm. Femur length versus tibia length and metacarpal III length suggest a tibia length of 55.9 cm and a metacarpal III length of 30.3 cm. The entire leg length is estimated to have been about 143.6 cm, or 1.43 meters.

Rock Head Specimens. Recently described by Gay (2002), MNA P1.160 and P1.161 are from MNA locality 219-0, in the Kayenta Formation, near Rock Head. P1.160/161 represent a complete right femur broken in half by a spiraling fracture. Additional specimens are known from this site, including a juvenile femur, MNA P1.109, measuring 28.1 cm. Several other bones, including several phalanges, a complete tooth, several vertebrae, two partial pelvic girdles, and some other partial limb material, are from at least three specimens of *Dilophosaurus*, in various stages of growth (Gay 2002).

Using the average of the ratios for the UCMP specimens, and the femur head width ratio, the femur of MNA P1.160/161 is estimated to have measured 54.5 cm. This is well within the range of variation shown so far for *Dilophosaurus* (Fig. 14.2). A reconstruction of the rest of the leg, based on the ratios mentioned above, computes to a tibia length of 53 cm and a metacarpal III length of 28.8 cm. This yields a leg length of 136.4 cm, or 1.36 meters.

The juvenile femur, MNA P1.109, measures 28.1 cm in length. Using the same ratios as before gives a tibia length of 27.4 cm and a metacarpal III length of 14.6 cm. The total overall leg length of this animal would have been 70.1 cm.

Results. By plotting femur and tibia lengths from these specimens on a graph (Fig. 14.2), some preliminary conclusions can be drawn. The most obvious being that the only measurable difference within the genus *Dilophosaurus* is age. A straight line can be drawn from MNA P1.109 all the way through to MNA P1.3145. It should be noted, however, that the pelves in UCMP 37302 and the largest Rock Head specimen are not fused, indicating that these individuals were not fully grown. This conclusion is also reinforced by the unfused neural arches present on both specimens. The lack of fusion coincides with their placement in Figure 14.2 as young individuals.

Conclusions

This study is only in its preliminary stages. The recovery, preparation, and description of new specimens of *Dilophosaurus*, as well as consideration of other patterns of variation within theropods, will open more avenues of research. Although the data currently indicate that there is no sexual dimorphism in the postcranial skeleton of *Dilophosaurus*, sexual dimorphism cannot be ruled out in the cranial crests. Without additional data, however, the best explanation for the data is ontonogenic variation within the species.

Acknowledgments. I would like to thank Dave Elliott for his help with this chapter. I also extend my thanks to Dave and Janet Gillette, at the Museum of Northern Arizona, for allowing me access to their collections, and to Jane Mason and Pat Holroyd, at UCMP, for allowing me to study the original material of *Dilophosaurus*. I would also like to thank Pete Reser and Spencer Lucas from the New Mexico Museum of Natural History and Science for access to their specimens. Special thanks go to Kenneth Carpenter, as well as to Gregory Paul and an anonymous reviewer, for their thoughtful critiques of the manuscript.

References Cited

Colbert, E. H. 1989. *The Triassic Dinosaur* Coelophysis. Museum of Northern Arizona Bulletin, no. 57. Flagstaff: Museum of Northern Arizona Press.

Colbert, E. H. 1990. Variation in *Coelophysis bauri*. In K. Carpenter and P. J. Currie (eds.), *Dinosaur Systematics: Approaches and Perspectives,* pp. 81–89. Cambridge: Cambridge University Press.

Downs, A. 2000. *Coelophysis bauri* and *Syntarsus rhodesiensis* compared, with comments on the preparation and preservation of fossils from the Ghost Ranch *Coelophysis* Quarry. In S. G. Lucas and A. B. Heckert (eds.), *Dinosaurs of New Mexico,* pp. 33–37. New Mexico Museum of History and Science Bulletin, no. 17. Albuquerque: New Mexico Museum of Natural History and Science.

Gay, R. 2002. New specimens of *Dilophosaurus wetherilli* (Dinosauria: Theropoda) from the early Jurassic Kayenta Formation of northern Arizona. In R. D. McCord and D. Boaz (eds.), *Western Association of Vertebrate Paleontologists and Southwest Paleontological Symposium Proceedings 2001,* pp. 19–23. Mesa Southwest Museum Bulletin, no. 8. Mesa, Ariz.: Mesa Southwest Museum.

Paul, G. S. 1988. *Predatory Dinosaurs of the World: A Complete Illustrated Guide.* New York: Simon and Schuster.

Pickering, S. 1995. A fractal scaling in dinosaurology project. In *Archosauromorpha: Cladistics and Osteologies,* p. 70. Capitola, Calif.: Private publication.

Raath, M. A. 1990. Morphological variation in small theropods and its meaning in systematics: Evidence from *Syntarsus rhodesiensis.* In K. Carpenter and P. J. Currie (eds.), *Dinosaur Systematics: Approaches and Perspectives,* pp. 91–105. Cambridge: Cambridge University Press.

Rowe, T., and J. A. Gauthier. 1990. Ceratosauria. In D. B. Weishampel, P. Dodson, and H. Osmólska (eds.), *The Dinosauria,* pp. 151–168. Berkeley: University of California Press.

Schwartz, H., and D. Gillette. 1994. Geology and taphonomy of the *Coelophysis* Quarry, Upper Triassic Chinle Formation, Ghost Ranch, New Mexico. *Journal of Paleontology* 68: 1118–1130.

Sereno, P. 1999. A rationale for dinosaurian taxonomy. *Journal of Vertebrate Paleontology* 19: 788–790.

Welles, S. P. 1954. A new Jurassic dinosaur from the Kayenta Formation of Arizona. *Geological Society of America Bulletin* 65: 591–598.

Welles, S. P. 1970. *Dilophosaurus* (Reptilia, Saurischia): A new name for a dinosaur. *Journal of Paleontology* 44: 989.

Welles, S. P. 1984. *Dilophosaurus wetherilli* (Dinosauria, Theropoda): Osteology and comparisons. *Palaeontographica Abt. A* 185: 85–180.

15. Sexual Selection and Sexual Dimorphism in Theropods

RALPH E. MOLNAR

Abstract

Theropod dinosaur skulls exhibit features that are probably related to sexual selection. These dinosaurs were unusual among large carnivorous tetrapods in that they had various horns, crests, and rugosities. These structures were probably used to advertise quality to prospective mates, rather than as weapons. Sexual dimorphism has recently been reported in theropods, particularly in *Tyrannosaurus rex*. *T. rex* reportedly exhibits so-called reversed sexual dimorphism in size in which the bones of females are larger than those of males. Reversed sexual dimorphism is intimately related to flight in (monogamous) birds and hence is not analogous to that proposed for theropods. A good case can be made for sexual dimorphism in *Coelophysis*, *Syntarsus*, and *Allosaurus*. Such dimorphism may have been present in other theropods, but the evidence is weak. The aspects of theropod paleobiology discussed here suggest that theropods differed from modern large carnivorous tetrapods in unappreciated ways. Thus they have the potential of enhancing our understanding of modern ecological processes and the scope of response of different lineages to environmental conditions. Theropods were not simply big bipedal lions or crocodiles.

Introduction

Theropod dinosaurs are often thought of as the greatest of hunters. But like all other animals, individual theropods did more than hunt, kill, and feed. These activities provided them with energy to survive, but they not only survived but also reproduced. And since, like most tetrapods, they reproduced sexually, they also must have found and courted mates. They presumably had some strategy to ensure that their offspring had as good a chance as possible of growing and reproducing in their turn. Animals that failed at these activities became extinct just like those that failed at hunting, and the existence of (non-avian) theropods for over 160 million years shows that they reproduced successfully.

Courting and raising offspring are as obvious among modern animals—either in the wild or on television documentaries—as hunting and eating. However, little is known about the reproductive activities of fossil forms both because reproductive structures, unlike locomotory and feeding structures, are not prominent in fossil skeletons and because even among living animals, reproductive aspects have not received much attention until the last twenty to thirty years. Reconstructing the lives of extinct animals requires guidelines: how do we know that a particular structure was used in courting and not some other activity? To some extent, these guidelines are provided by the theory of evolution by natural selection. Selection is manifested in three forms: survival selection, sexual selection, and reproductive selection. For extinct animals, features associated with survival selection—for example, those involved in locomotion or feeding—have often been discussed; although issues like the ability to recover from injury are rarely noted. Features involved in sexual and reproductive selection are less often considered. These different aspects of selection may come into conflict—so the interactions between them must be known if we are to understand an animal's life strategy.

Institutional Abbreviations. AMNH, American Museum of Natural History, New York; CM, Carnegie Museum of Natural History, Pittsburgh, Penn.; DINO, Dinosaur National Monument, Vernal, Utah; FMNH, Field Museum of Natural History, Chicago; MOR, Museum of the Rockies, Bozeman, Mont.; PIN, Palaeontological Institute of the Russian Academy of Sciences, Moscow; RTMP, Royal Tyrrell Museum of Palaeontology, Drumheller, Alberta; SDSM, Museum of Geology, South Dakota School of Mines and Technology, Rapid City; UCMP, University of California Museum of Paleontology, Berkeley; USNM, United States National Museum of Natural History (now National Museum of Natural History), Washington, D.C.; UUVP, Museum of Paleontology, University of Utah, Salt Lake City.

Signals and Cranial Ornament

Theropod dinosaurs were unusual among large carnivorous tetrapods in having various kinds of cranial horns, crests, and rugosities (Tables 15.1, 15.2). The presence of these cranial ornaments was pointed out early by Osborn (1903) and Gilmore (1920). Davitashvili (1961) and Bakker (1986) later interpreted them as weapons, and Hopson (1977) and Molnar (1977) interpreted them as signaling devices. Such ornaments are absent in carnivorous mammals, phororha-

TABLE 15.1

Theropod Genera With and Without Cranial Ornament

In all other theropods the skull is too poorly known to
establish whether or not ornament was present.

Ornament present	Ornament possibly present	Ornament absent
Acrocanthosaurus	Afrovenator	Abelisaurus
Albertosaurus	Angaturama	Alectrosaurus
Alioramus	Aublysodon	Anserimimus
Allosaurus	Chilantaisaurus	Archaeornithoides
Baryonyx	Coelophysis	Avimimus
Carcharodontosaurus	Labocania	Caudipteryx
Carnotaurus	Maleevosaurus	Compsognathus
Ceratosaurus	Neovenator	Conchoraptor
Cristatusaurus	Ornitholestes	Deinonychus
Cryolophosaurus	Tarbosaurus	Dromaeosaurus
Daspletosaurus		Dromiceiomimus
Dilophosaurus		Eoraptor
Giganotosaurus		Erlikosaurus
Gorgosaurus		Gallimimus
Indosaurus		Garudimimus
Irritator		Herrerasaurus
Majungatholus		Indosuchus
Monolophosaurus		Ingenia
"Oviraptor"		Nanotyrannus
Proceratosaurus		Procompsognathus
Suchomimus		Saltopus
Syntarsus		Saurornithoides
Tyrannosaurus		Shanshanosaurus
Yangchuanosaurus		Shuvosaurus
Zupaysaurus		Shuvuuia
		Sinornthoides
		Sinornithomimus
		Sinornithosaurus
		Sinosauropteryx
		Sinraptor
		Struthiomimus
		Torvosaurus
		Velociraptor

TABLE 15.2

Kinds of Cranial Ornament Found among Theropods
Illustrated in Figures 15.1 and 15.2

	Structure	Examples	Age
1	Low, longitudinal median nasal rugosity	*Baryonyx, Suchomimus*	Early Cretaceous
2	High, longitudinal median nasal ridge	*Monolophosaurus, Cristatusaurus* (?)	Middle Jurassic to Early Cretaceous
3	Elongate paired nasal rugosity or ridge	*Yangchuanosaurus*	Middle or Late Jurassic
4	Single median nasal horn	*Proceratosaurus, Ceratosaurus*	Middle to Late Jurassic
5	Paired paramedian series of nasal projections	*Alioramus*	Late Cretaceous
6	Paired paramedian vertical nasal plates (crests)	*Dilophosaurus, Zupaysaurus*	Late Triassic to Early Jurassic
7	Premaxillo-naso-frontal casque	*"Oviraptor" mongoliensis*	Late Cretaceous
8	High, upright median nasolachrymal crest	*Cryolophosaurus*	Early Jurassic
9	Inflated lachrymal horns	*Allosaurus, Albertosaurus*	Late Jurassic to Late Cretaceous
10	Median frontoparietal crest	*Irritator*	Early Cretaceous
11	Median frontal horn	*Majungatholus*	Late Cretaceous
12	Postorbital rugosities (cornual processes) and horns	*Carnotaurus, Tyrannosaurus*	Late Cretaceous

The classification in this table refers to ornament, not taxa, and a few taxa may exhibit more than one type of cranial ornament.

choid and diatrymiform birds, therapsids, varanids, and even crocodilians, with the exception of a single species in each of the latter three (*Tetraceratops insignis, Varanus priscus*, and *Ceratosuchus burdoshi*). This difference implies that the lifestyles of theropods were different from those of all other large land-dwelling carnivores.

In modern mammals, horns (here including antlers and tusks) are often thought to function in defense against predators. However, in males they are usually used in contests against rivals and only secondarily in deterring predators (Andersson 1994; Geist 1966). The functions and evolution of horns in large herbivorous mammals are now reasonably well understood (e.g., Geist 1978; Janis 1982). Since theropods were often the largest predators of their time, we may be skeptical that their horns and other such structures served to deter other predators. Furthermore the structures appear in a wide variety of forms (Table 15.2; Figs. 15.1, 15.2), more so than among horned mammals.

In his study on sexual selection, Andersson (1994) lists four functions for cranial horns in adult male large (herbivorous) mammals: (1) weapons against predators; (2) weapons against rival males; (3) indicators of strength and fighting ability in display to rival males; and (4)

Figure 15.1. The classes of cranial ornament of Table 15.2 illustrated. The numerals designate class, the letters designate taxon. The skulls are not to scale; they are drawn with approximately equal distances from the tip of the premaxilla to the quadrate condyles. 1: A low, longitudinal median nasal rugosity occurs in (A) Baryonyx walkeri (here shown with an elongate snout, as suggested by Sereno et al. [1998]) and (B) Suchomimus tenerensis. 2: A high, longitudinal median nasal ridge occurs in (C) Monolophosaurus jiangi. 3: Elongate paired nasal rugosities or ridges occur in (D) Yangchuanosaurus magnum. 4: A single median nasal horn occurs in (E) Ceratosaurus nasicornis. 5: A paired paramedian series of nasal projections occurs in (F) Alioramus remotus. 6: Paired paramedian vertical nasal plates (crests) occur in (G) Dilophosaurus wetherilli. Unlike the other ornament figured here, which seems to be completely or almost completely preserved, the crests of D. wetherilli are obviously incomplete. The figure shows the extent of the crest as preserved in UCMP 37302 (holotype), heavy line, and UCMP 77274, lighter line. 7: A premaxillo-naso-frontal casque reminiscent of that of the cassowary occurs in (H) "Oviraptor" mongoliensis and (I) Oviraptor ?philoceratops. 8: A high, upright median nasolachrymal crest occurs in (J) Cryolophosaurus ellioti. Redrawn from the literature, except for G, in which the body of the skull is from Welles 1984, and the crests from photos made in 1971.

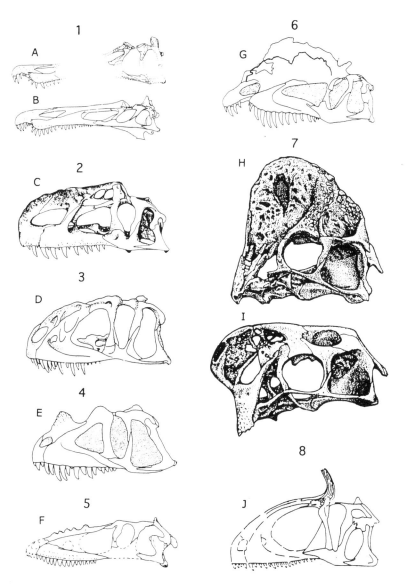

indicators of vigor and quality, assessed by females in mate choice. He also gives reasons why these are the most likely functions. I shall argue that in theropods the signaling functions are less plausible than the weapon functions, for three reasons. First, theropods were often the largest contemporaneous predators, with other predators species often absent. These theropods possessed formidable offensive weapons in their teeth and claws. These structures seem to provide adequate weaponry for crocodilians, varanids, and large mammalian predators. Second, the ornaments generally seem structurally unsuitable as weapons. For example, the crests of *Dilophosaurus* are large, but they are only a few millimeters thick; and the horn-bearing lachrymals of *Allosaurus* are thin walled and hollow (pers. obs.; cf. Gilmore 1920), as is the longitudinal median crest of *Monolophosaurus* (Zhao and Currie 1993).

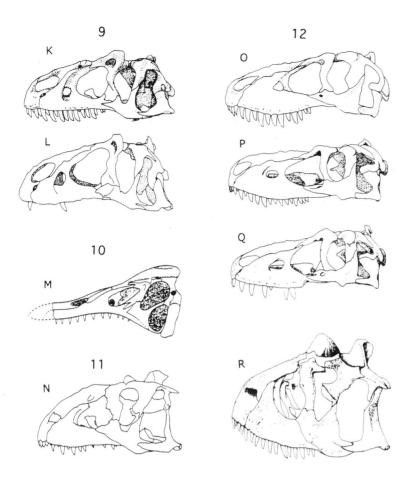

Clearly, these structures are too delicate to serve as weapons. Third, horns are deployed in striking blows with the head, and, as argued below, theropod cranial structures are usually not structurally suited to delivering such blows. In large mammalian herbivores, blows are usually delivered by tapering structures—horns or tusks—that concentrate the force of the blow onto a point and so increase the pressure applied to the opponent and the probability of inflicting damage (Fig. 15.3). The horns of *Carnotaurus*, and perhaps the postorbital rugosities of some tyrannosaurids, are tapered (or edged) and solid, and so are suited to concentrate force. These structures could have been effective weapons. But these are the only examples among theropods, so most theropod cranial ornaments probably served not as weapons but as indicators of strength to rivals or of quality to potential mates.

The process of courtship and mating has several aspects: species recognition, mate recognition, (often) contest for access to mates, female choice, and actual mating. Other aspects, such as sperm competition, are insufficiently documented even for living animals to be reasonably extrapolated to extinct taxa. The key process here is probably mate choice by females (Andersson 1994; Cronin 1991; Gould and Gould 1989; Smith 1993). This undoubtedly occurred among thero-

Figure 15.2. Continuation of the classes of cranial ornament of Table 15.2. 9: Inflated lachrymal horns occur in Allosaurus fragilis *(K) composite reconstruction after Madsen (1976) and (L) AMNH 600. 10: A median frontoparietal crest occurs in (M)* Irritator challengeri. *11: A median frontal horn occurs in (N)* Majungatholus atopus. *12: Postorbital rugosities occur in (O)* Tarbosaurus bataar, *(P)* Daspletosaurus *sp., and (Q)* Daspletosaurus torosus *and horns in (R)* Carnotaurus sastrei. *Redrawn from the literature.*

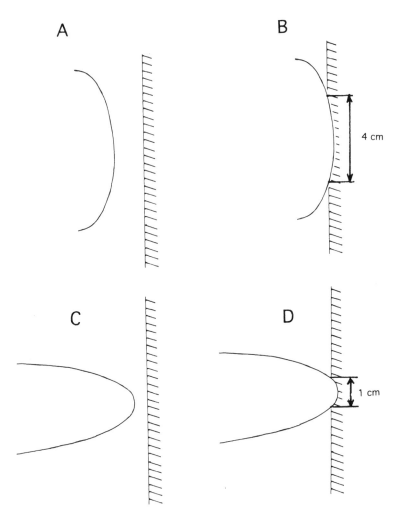

A   B

C   D

Figure 15.3. Relationship of form
of structure delivering impact to
damage suffered by surface
impacted. A broadly rounded
form (A) contacts a flat surface
over a relatively broad area
(B), whereas a more conical (or
horn-like) form (C) delivers the
same force to much more
restricted area (D), thus applying
a greater pressure. Because it is
the amount of pressure that
imposes stress on the target
surface, the horn-like form will
create more damage. Specifically
in this figure if the horn-like form
contacts an area 1 cm in diameter
and the broader structure an area
4 cm in diameter, both delivering
1 N of force, the horn-like
structure will deliver a pressure
of about 1.2 N/cm² (assuming
that both structures are circular
in cross-section). The broader
structure contacts an area of 12.6
square cm, and hence delivers just
over 1/12 N per square cm,
approximately 1/14, or 7.2
percent, as much. This principle
of the force concentration by
narrow forms is well known to
women who wear stiletto heels.

pods just as among most or all other tetrapods. The basis of female
choice has been contentious, but there is now evidence that females
tend to choose by assessing the "quality" of the genotype of their
potential mate (Cronin 1991; Gould and Gould 1989; Zahavi et al.
1997). This quality seems specifically to consist of either good survivor-
ship (the "good genes" hypothesis) or a high degree of attractiveness to
females (the "good taste" hypothesis). Quality is assessed by attention
to signals (Zahavi et al. 1997). A good exposition of this process in
finches is given in by Birkhead (2003, chap. 9). The approach pioneered
by Zahavi to understanding sexual selection seems to make the least
questionable assumptions regarding evolutionary mechanisms. It does
not draw on vague concepts of the "good of the species" or group
selection; instead it involves evolutionary benefit to (i.e., reproductive
success of) individuals or genes. Thus it will be used here as a guide to
understanding the possible function of cranial ornament in theropods.

Certain obvious markings or anatomical structures have generally
been ascribed to signaling the sex or species membership of an individual. Zahavi argues that although they often do serve this purpose,

they are by-products, or exaptations, of their evolution (Zahavi et al. 1997). The structures have actually evolved to signal health, ability to survive, and so on. Thus, the same structures serve for species and mate recognition and also proclaim quality. If we are to recognize such features, it is necessary to specify what properties they must possess. Fundamentally, these properties arise from the nature of the perceptual apparatus of the animals. In tetrapods, all of the "long-distance" sense organs—used for mate recognition—are located in the head. Mammals generally tend to use smell for recognition and hearing for communication: threats are communicated with loud calls, not puffs of scent. However, odor (or taste) is used by male artiodactyls to assess female sexual state in the process of flehmen. Lepidosauromorphs and archosauromorphs tend to use sight more than smell; hence the bright colors often seen on the heads and necks of lizards and birds. Thus, we might expect to find unusual and striking cranial colors or structures in theropod and other dinosaurs. Structures are preserved in the fossil record, but because colors must be inferred, we can reconstruct only in a limited way the role of sight in mate choice for extinct animals. That cranial ornaments could have been used as visual signals among theropods is consistent with the inference that theropods had keen vision, as evidenced by their large orbits and well-developed optic lobes of the brain.

We assume that the signals did not just appear, that they have some history of evolutionary development. This assumption is incorporated in the theory of honest signals (part of "handicap theory"). Current understanding emphasizes that signals must accurately reflect the quality of the signaler (Zahavi et al. 1997). This accuracy is achieved by having "costly" signals, structures whose development incurs sufficient metabolic energy expenditure that "low-quality," for example, sick or parasitized, individuals either cannot produce them or produce them in a way that accurately signals their condition. Mammalian horns fall into this category, even though many also serve as weapons.

If the cranial structures of theropods are honest signals, they should be (1) prominent, (2) variable, and (3) reasonably expensive to grow. Some of the structures are quite prominent, particularly forms 5, 6, and 7 in Table 15.2; others, for example, forms 1 and 2, are less so (Fig. 15.1). The subdued character of the structures may be misleading, however, in that they may have been made much more prominent by integumentary superstructures (in the manner of rhino horns) that have not been preserved. Tyrannosaurids have rugose nasals (Carr 1999; Hurum and Sabath 2003; Molnar 1991) that may have supported a keratinous or integumentary crest. These possible structures will not be further considered here because their existence remains to be established. However, that the integumentary dorsal "frills" of *Sinosauropteryx* (Chen et al. 1998) and *Sinornithosaurus* (Ji et al. 2002) extend well onto the head indicates that such soft structures are not out of the question and that, in these two cases, they were present in forms lacking bony ornament. Such hypothetical features cannot be used as evidence; nevertheless, all the bony structures are, to a greater or lesser degree, prominent.

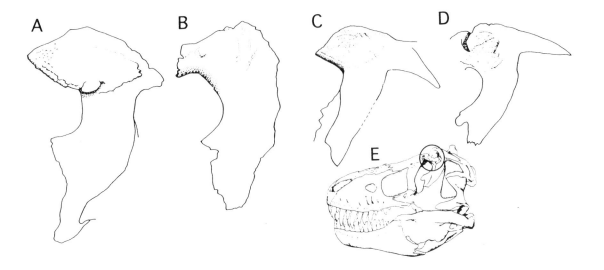

Figure 15.4. Variation in form of the postorbital rugosity in Tyrannosaurus rex. (A) MOR 008; (B) LACM 23844 (reversed); (C) SDSM 12047; (D) AMNH 5027; and (E) position of postorbital rugosity on skull. LACM 23844 lacks the rugosity. A–D original; E after Osborn 1912.

In herbivorous dinosaurs, variation is seen in the crests of hadrosaurids (Dodson 1975) and the horns of *Triceratops* (Ostrom and Wellnhofer 1986), among others. In theropods, variation in the postorbital rugosities (cornual processes) has been documented in *Tyrannosaurus rex* (Fig. 15.4; Molnar 1991, 2001a) and described in *Albertosaurus, Daspletosaurus*, and *Gorgosaurus* as well (Currie 2003). Variation can be seen in the lachrymal horns of *Allosaurus fragilis*: AMNH 600, AMNH 666 (Osborn 1912), USNM 4734 (Gilmore 1920), and DINO 2560 (Glut 1997, p. 106; previously UUVP 6000) all show minor differences in the profile and prominence of the lachrymal horns (Fig. 15.5). These differences do not seem to be due to postmortem distortion or incompleteness of the specimens. Smith (1998) analyzed morphometrically certain features of *Allosaurus,* including the form of the lachrymals. He concluded that the variation in lachrymal form was due to allometry and individual variation, which is consistent with what is to be expected if these structures were used for signaling. Among other theropod species, either there are too few well-preserved skulls to assess variation, or, as for *Coelophysis bauri,* the skulls remain largely undescribed. In some, such as *Dilophosaurus,* the preserved cranial ornament is incomplete (Fig. 15.1G). Because the structures were bony, the metabolic cost of growing them is likely to have been relatively high; that is, growing them would have been more expensive than not growing them.

Many theropods are known from only one or a few good specimens of skulls. *A. fragilis,* however, is known from several, including the skulls just mentioned. So we can be sure that the postulated display structures, the lachrymal horns, do meet all three conditions in specimens of a single taxon, and we may presume that this is true for other theropods as well, until evidence to the contrary emerges.

It is reasonable to assume that, like other amniotes, theropods did not reach sexual maturity until after some period of growth. Thus cranial ornament involved in sexual selection would be expected to be

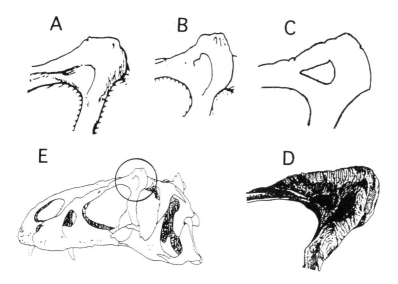

Figure 15.5. Variation in form of the lachrymal horn in Allosaurus fragilis. (A) AMNH 666; (B) AMNH 600; (C) DINO 2560; (D) USMN 4734; and (E) position of lachrymal horn on skull. A, B, and E after Osborn 1912; C, original; D after Gilmore 1920.

absent in immature individuals or to appear only in incipient form. Currie (2003) and Carr (1999) report that the latter is the case in *Gorgosaurus libratus*.

If these structures served as displays to rivals, there should be evidence that such displays were necessary, that is, that intraspecific conflict did occur. Possible tooth punctures in *T. rex* were reported by Molnar (1991) and Larson (2001); and more extensive, and hence more convincing, damage—including tooth drag marks—have been reported in *Sinraptor dongi* (Tanke and Currie 1995), a theropod without prominent bony cranial display structures. Since we cannot yet confidently sex theropods in most cases, we cannot be certain that this conflict was between males, and we cannot determine whether both sexes, or only one, bore ornament. In *T. rex*, the postorbital rugosity ranges from very prominent to almost non-existent (Molnar 1991; 2001a). In almost all cases, the sample of theropod skulls is small, usually fewer than ten, often fewer than five per taxon. Thus, the occurrence of dimorphism in the prominence, or the even possession, of cranial ornament is not out of the question. In those large herbivorous mammals (ungulates) in which both sexes have horns, male horns seem to be adapted as both signals and weapons for contests between rival males, whereas female horns seem to function chiefly as weapons against predators (Packer 1983). If theropods were not dimorphic in cranial ornament, the ornament in females may have functioned in threat displays against other females or males or against competitors of other species.

The appearance of the cranial ornament in general seems not to be due simply to allometric effects, because the ancestral taxa (herrerasaurids and *Eoraptor*) lacked cranial ornament and early forms sometimes had more strongly developed structures (e.g., *Dilophosaurus*) than later, larger forms (*Ceratosaurus*). Instead, some early variant was presumably selected for rugosities or subdued projections. Zahavi et al.

TABLE 15.3

Evidence that Theropod Cranial Ornament Functioned as Signals in Sexual Selection

Expectation	Evidence	Reference
Ornament prominent	Inspection	Figs. 15.1, 15.2
Ornament exhibits variation between individuals	Tyrannosaurid postorbital rugosities; lachrymal horns of *Allosaurus*	Figs. 15.4, 15.5; Currie 2003; Molnar 1991, 2001a
Ornament expensive to grow	None	None
Ornament appears or is most prominent in adult individuals	Postorbital rugosities appear or are most prominent in mature *Gorgosaurus libratus*	Carr 1999; Currie 2003
Intraspecific conflict	Possible tooth puncture marks in *Tyrannosaurus rex*; tooth drag marks in *Sinraptor dongi*—assumed to be inflicted by conspecifics	Larson 2001; Molnar 1991; Tanke and Currie 1995
Ornament sexually dimorphic	None	None
Ornament not due to allometric effects on skulls of phylogenetic size increase	Older forms sometimes having more strongly developed ornament than larger, later taxa	See text

(1997) have proposed for mammals that such structures originated as signals of the direction of gaze and later evolved to become more prominent signals for intimidation, or weapons. This proposal is consistent with the evolution of ceratosaurs: Late Triassic and Early Jurassic forms had thin parasagittal crests, but Cretaceous forms (*Carnotaurus*) had horns structurally suitable to have been weapons. (Forster [1999] recently disputed the existence of the Ceratosauria, but even for her consensus tree this argument still holds.) As can be seen in Table 15.2, Late Triassic–Early Jurassic theropods tended to have paired parasagittal ridges or crests along the snout (forms 3 and 6; Fig. 15.1) that, although retained in *Yangchuanosaurus*, were lost or replaced by a variety of ornament in most later taxa. For increased clarity these arguments in favor of the signaling function of the ornament and the evidence for them are summarized in Table 15.3.

Injury may be inflicted during intraspecific combat; in fact, injury or the threat of injury is the purpose of such conflict. The ability to recover from injury is presumably an important factor in evolution, although it generally receives little attention. Depressed fractures and other puncture wounds in theropod bones probably indicate biting during combat (Tanke and Currie 1995). Intraspecific combat is a significant source of injury in *Caiman crocodilus* (Gorzula 1978), *Crocodylus niloticus* (Cott 1961), and probably *Crocodylus porosus* (Webb and Manolis 1989). Christopher Kofron (pers. comm., 1993) noted that during a study of *C. niloticus,* in which liparoscopy was conducted on animals in the wild, none of the individuals contracted infections from the incisions, in spite of promptly returning to waterways that would seem to be rather septic environments. Together with comments by Brazaitis (1981), this study suggests that the crocodilian immune system is quite effective in suppressing or preventing infection in wounds. This is not the case in large (herbivorous) mammals in the wild, at least regarding infected bites. The injuries to crocodiles observed by Cott had healed without obvious infection. A similar observation of resistance to infection in wounds has been made for birds (Kavanau 1987). This indication of effective immune activity in lineages that represent both plesiomorphic and derived archosaurs suggests that this may have been true of theropods in general.

Molnar (2001b) revealed substantial evidence for injury in theropods (see Rothschild and Tanke, this volume) but no indication of anything other than localized (if that) resultant infections, although there is evidence of more extensive infections in other dinosaurs (e.g., Swinton 1934). This evidence for limited infection is consistent with effective immune activity in theropods. Such activity would have been important in theropod evolution in reducing the effects of certain selective factors—those associated with infection—and hence in amplifying their exposure to other factors, such as sexual selection and hunting efficiency. There is no indication of whether strong powers of healing extended to resistance to disease; in birds, they apparently did not (see Kavanau 1987).

Cranial ornament is generally found on large, rather than small, theropods (Table 15.1), but there are exceptions. It is also likely that other, noncranial structures in theropods functioned as signals, including the dorsal "frill" of *Sinosauropteryx*, the tail fan of *Caudipteryx*—both quite small forms without osseous cranial ornament—as well as the medial dorsal row of osteoderms of *Ceratosaurus* and the "sails" of the spinosaurs. If we can learn to interpret these signals—and find enough specimens—they could provide information regarding the state of general health of individual theropods. Such information would be useful in proposing and assessing hypotheses of the evolution of various theropod (or other fossil) lineages.

The possibility that dinosaurs in general, like modern large mammals, made more extensive use of signaling structures than generally realized is suggested by the following observations. Stegosaurs had dorsal rows of prominent plates; hadrosaurs had crests, and some also had prominent dorsal ridges along the vertebral column; some ankylosaurs had elongate lateral spines and spikes; ceratopians had horns and frills; the sauropods *Amargasaurus* and *Agustinia* (Bonaparte 1999) had dorsal frills or rows of spikes; and other sauropods had keratinous knobs and spikelets along their backs (Czerkas 1992). Signaling behavior has previously been suggested by Hopson (1977).

Sexual Dimorphism: Reversed and Otherwise

There has recently been considerable interest in the occurrence of sexual dimorphism in theropods, particularly *Tyrannosaurus rex*. This taxon reportedly exhibits so-called reversed sexual dimorphism in size, in which females are larger than males. In contrast, in living large terrestrial carnivores the male is often the larger (as in oras, crocodilians, and lions). Reversed sexual dimorphism had previously been proposed for *Syntarsus rhodesiensis* (Raath 1990) and *Tyrannosaurus rex* (Carpenter 1990). In these cases, there are two questions to be answered: (1) Why does sexual dimorphism occur? and (2) Why is the female the larger or more robust; that is, what is the reason for reversed sexual dimorphism? Although the basis of sexual dimorphism in large terrestrial tetrapods seems now generally understood (Jarman 1983; Andersson 1994), the reasons for reversed sexual dimorphism have often been less obvious. Given two sexes, there are only three possible size relationships between them: the female can be the larger than, smaller than, or the same size as the male. Given that the set of possibilities is restricted, that breeding systems are often adapted to environmental conditions, and that there is a large suite of such conditions, different causes can be expected to produce the same results. In other words, dimorphism may well have different causes in different animals (Andersson 1994). Even so, reversed sexual dimorphism, particularly in forms with indeterminate growth, generally seems to result from a fecundity advantage conferred by large size, as suggested by Darwin (1871). In large mammals, it is the males that accrue the most advantage from large size, in contests over access to mates (also as proposed by Darwin). Many other factors probably also contribute to dimorphism (Andersson 1994). Sexual dimorphism in theropods will be compared with that in large, herbivorous mammals and in birds not because these are thought to represent close analogs but because the mammals at least are reasonably well understood and because the differences between theropods and mammals and birds are informative.

The term "sexual dimorphism" connotes a difference in form, but, as here, it has also been used to mean a difference in overall size, a difference that is more accurately termed "sexual size difference" (Fitch 1981). Using the same term for both differences in form and differences in size may conflate different phenomena. Dimorphism, a difference specifically in form, presumably has some genetic basis. If a feature is present in one sex and absent in the other, the genes for this feature are generally assumed to be present or activated only in the sex in which it is manifested. This may also be the case for features that appear in both sexes or that differ substantially in size between the sexes, for example, the jaws of stag beetles. However, in this case the genes would presumably be those controlling the growth of the feature, not its expression.

But sexual difference in overall body size (or weight) may have a different ontogenetic origin: sexual size difference can conceivably be nongenetic in origin. Consider a male and female that mature under conditions in which food is readily available but not unlimited. Assume that they can continue to grow for some substantial period of their lives after reaching sexual maturity and that intrinsic growth rates are the same in individuals of both sexes. As they approach sexual maturity, the male and female may grow at the same rate, but having reached sexual maturity, the female may choose to reproduce. Reproduction can involve an internal shifting of resources away from growth and toward reproduction. The resources that would otherwise have gone into increasing the size of the female go instead into producing offspring. As the male continues to grow, the growth of the female would lag behind until reproduction was completed. If this scenario happened repeatedly in a population, the females would be smaller on average than the males—but not because they possessed an explicit genetic "instruction" for being smaller. (They may, however, have a genetic proclivity for timing the start of breeding.) This difference in internal allocation of resources in the female may account for differences in apparent growth rates of females and males after the approach to maturity, but would not obtain in cases in which growth rates of juveniles differ (as reported by Magnusson et al. [1989], but not seen by Chabreck and Joanen [1979], in *Alligator mississippiensis*). Alternatively the female might increase her feeding rate to accommodate both growth and reproduction simultaneously. If the population density was near the carrying capacity of the environment, this increase might not be possible, and the female either would not reproduce or would cease growing during the reproductive period. In short, it is possible that even without an explicit genetic contribution, females would be smaller than males simply because of an allocation of physiological resources to reproduction rather than to growth in an environment with limited food.

In this model, for a female predator to grow larger than the male, either she would have to have significantly greater hunting success, or the costs of reproduction would have to be negligible (which seems unlikely). Neither situation seems to occur among modern large terrestrial predators, indicating that in some unrecognized way, the ecology of the large, terrestrial predators of the Mesozoic differed from that of today. Another alternative is that one of the features of the model is violated. Theropods might not have continued to grow after reaching reproductive maturity, for example (cf. Chinsamy 1990). Thus, whereas sexual dimorphism (in form) may well have a genetic basis, sexual size difference need not.

Reversed sexual dimorphism is common in two unrelated avian groups, waders and raptors, and is found among several more (Jehl and Murray 1986). As previously mentioned, sexual size difference in which the male is the larger is commonly the result of sexual selection favoring larger males (Selander 1972). In groups in which the sex roles are reversed, for example, phalaropes, reversed sexual dimorphism is thought to result from the same action of sexual selection. The problem with birds has been to explain reversed sexual dimorphism in taxa that do not exhibit reversed sex roles (Mueller 1990), such as raptors and waders. Because sex roles cannot be observed in theropods, two questions must be answered: (1) Were the sex roles reversed? and (2) What is the origin of the sexual size difference? But before these questions can be answered, the existence of sexual size difference, rather than just individual variation, must be confirmed. This requires a sufficiently large sample, preferably from a single population; otherwise the possibility that the supposed difference is actually due to other kinds of variation is difficult to exclude (cf. Horner and Lessem 1993; Paul 1988). Examples of fossil tetrapods in which dimorphism or sexual size difference has been confirmed and identifications of the sexes proposed are given in Chapman et al. 1997.

The possibility that theropods exhibited reversed sexual dimorphism was first proposed by Raath (1990), for *Syntarsus rhodesiensis*. A deposit on the Chitake River, in the central Zambezi Valley of northern Zimbabwe, yielded remains of more than thirty individuals. These were studied by Raath and compared with specimens from two other localities. In addition to variations that he related to maturity, Raath found dimorphism in the forelimb, sacrum, and pelvis (Raath, pers. comm., 1999), but the dimorphism was clearest in the form of the femur, which had gracile and robust forms (Fig. 15.6A; Raath 1990). This he plausibly interpreted as evidence of sexual dimorphism. Almost all the remains in the Chitake River deposit belonged to *S. rhodesiensis*, and thus the deposit was interpreted as the remains of a group of *Syntarsus* associated during life: it included juveniles as well as adults. The remains were all generally similar in form, and dimorphism appeared only in the larger specimens. The material was interpreted as indicating that monomorphic juveniles matured into dimorphic adults. However, the robust morph was more common than the gracile, at least twelve versus seven individuals. At least five juveniles were also present; although more than thirty individuals were represented in the deposit, information on only twenty-four (Raath 1990, fig. 7.10) or twenty-six (table 7.3) was presented. Raath's argument assumes that the dimorphism is not due merely to sampling error, although he was aware that given the small sample size this is just what it might have been. Since the Chitake River deposit was interpreted as the result of a catastrophic event, he speculated that the dominant robust morphs might have been females that were "reluctant to desert distressed young" and so were killed. In this case, the sample would have been biased for sex ratio, although not because of simple sampling error.

Although females often defend their young, the plausibility of this scenario is not beyond question. Female kangaroos with young that have emerged from the pouch are known to abandon them under deteriorated environmental conditions, such as drought (Tyndale-Biscoe 1973); and large African and North American ungulates may abandon young attacked by predators. However, the recent discovery of three oviraptorid skeletons associated with nests, as well as the original discovery of *Oviraptor mongoliensis* also associated with a nest (Clarke et al. 1999; Dong and Currie 1996; Osborn 1924), indicates that theropod parents did sometimes perish with their offspring. This does not indicate that the parent was the mother: in these theropods, as in ostriches, males may have guarded the young.

Were Raath's sample unbiased, it would indicate, rather than the usual sex ratio of 1:1, an unusual ratio of about 2:1. This ratio could result from a breeding system in which a few males bred with many females, as in harem-keeping species. So in this scenario, too, the abundant morph would have been female. Thus, two mutually exclu-

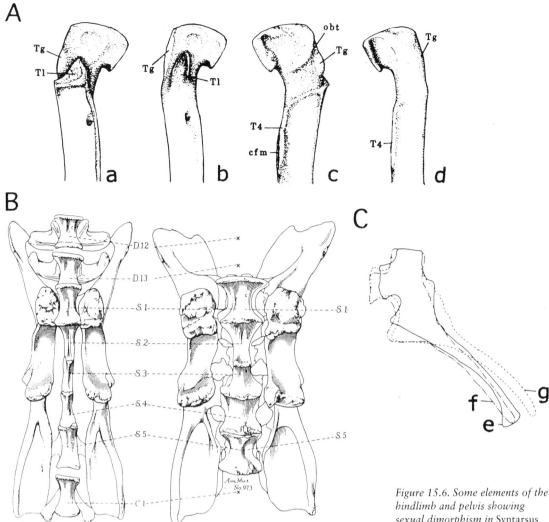

sive scenarios were presented, and both imply reversed sexual dimorphism. In addition, Raath pointed out that such sexual size difference is found among living raptors. Raath did not link his proposed reversed sexual dimorphism in *Syntarsus* to that in raptors as, for example, being the plesiomorphic condition for theropods.

Chinsamy (1990), in a histological examination of the femur of *S. rhodesiensis,* found that perimedullar erosion occurred only in femora assigned to the robust morph. This erosion was thought to result from the mobilization of calcium or phosphates from the bone, which is known to occur during egg production in birds (Bloom et al. 1941). Unfortunately it is not clear how many specimens Chinsamy examined, and he does caution that the erosion may be due to other causes. But this provides the best evidence in favor of Raath's interpretation—even though a larger sample is desirable.

The claim of reversed sexual dimorphism in *S. rhodesiensis* was

Figure 15.6. Some elements of the hindlimb and pelvis showing sexual dimorphism in Syntarsus rhodesiensis *and reportedly in* Tyrannosaurus rex. *(A) The proximal femur of S. rhodesiensis in anterior (a, b) and posterior (c, d) views (after Raath 1990). The robust morph is seen in a and c, and the gracile morph in b and d. (B) The difference in pelvic and sacral form in T. rex as originally figured by Osborn (1916). Left, AMNH 5027; right, CM 9380 (previously AMNH 973). (C) The differences in ischial form of T. rex found by Carpenter (1990): (e) CM 9380; (f) RTMP 81.61; and (g) AMNH 5027. Abbreviations: cfm— insertion for M. caudofemoralis; obt—obturator ridge; T4—fourth trochanter; Tg—greater trochanter; Tl—lesser trochanter.*

presented as a speculation. Since the claim for this phenomenon in *T. rex* has been presented as more than speculation (Larson 1994, 1995, 2001), examining Raath's scenarios in greater detail is worthwhile. The "reluctant mothers" scenario requires no further comment here, but the "harem" scenario does. Sexual size difference has been linked with polygyny (Andersson 1994), which may involve harems. In all living harem-keeping tetrapods that have been studied, intrasexual competition among males for access to mates is considerable. In such circumstances, there is corresponding selection for large size in males (hence the size difference). As a result, males are slightly (as in alcelaphines and oryx) to substantially (as in large kangaroos, African elephants, and particularly sea elephants) larger than females: instances in which the females are larger are unknown. Thus, this scenario of Raath posits a situation unlike any among modern tetrapods, in that some unknown process counteracted the effects of selection for large size among the males.

Reversed sexual dimorphism does occur among non-avian tetrapods, especially snakes and turtles (Brochu 2003); however, its occurrence among birds has received the most attention. Because explaining reversed sexual dimorphism in taxa in which the sex roles are reversed is not a problem, interest has focused on those birds that exhibit reversed sexual dimorphism but lack reversed sex roles. These birds tend to be monogamous. Many hypotheses have been proposed, but there is now evidence that reversed sexual dimorphism in waders is due to sexual selection (Figuerola 1999). Females preferentially mate with agile males, and the smaller males are the more agile. This may also apply to raptors, for many of which aerobatics are a prominent feature of courtship, as well as to other flying birds that show reversed size difference (Jehl and Murray 1986). Reversed sexual dimorphism may be intimately related to flight in (monogamous) birds and, hence, their situation is not analogous to that proposed for theropods. Brochu (2003) believes that reversed sexual dimorphism is related to the pursuit of prey birds in flight. Although the pursuit of prey is a different determinant than the one presented here, it also implies that reversed sexual dimorphism in birds is causally related to flight. Thus inferences drawn from these birds regarding the breeding system of (nonflying) theropods (e.g., Larson 1994, 1995) are premature.

Neither of Raath's scenarios, nor the evidence of Chinsamy, unequivocally indicates that the larger morph was female. The only conclusion that can be confidently reached from the literature is that *S. rhodesiensis* exhibited sexual dimorphism.

Carpenter (1990), in the same volume in which Raath 1990 appeared, also referred to reversed sexual dimorphism among theropods. Unlike Raath, Carpenter studied the large derived theropod *T. rex* and, also unlike Raath, made only brief mention of dimorphism. Carpenter illustrated impressive variation among the specimens. He suggested that a robust and gracile morph were present, as in *Syntarsus*. Among the varying features, he noted orientation of the distal ischial shaft. In two of the specimens he examined, this shaft diverged more from the sacral vertebrae than in the third specimen (Fig. 15.6C). These differences, he reported, came from robust individuals, so he suggested that

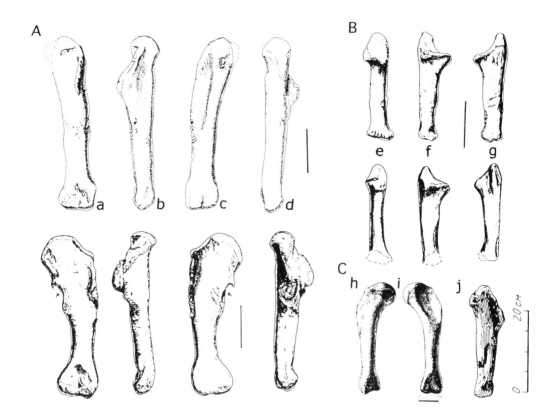

these might have been female, with the orientation of the shaft being an adaptation for the easier passage of eggs. This is a plausible suggestion based on anatomy rather than taphonomic association, but it is based on a sample of only three; and, furthermore, Brochu (2003, p. 128–129) maintains that "hip-related sexual dimorphism has not been demonstrated in any egg-laying amniote." Two years later, Carpenter (1992) noted differences in the proportions of the maxilla, lachrymal, dentary, and teeth. These, also, were based on examination of only three specimens, all that were accessible at the time. Carpenter and Smith (2001) reported and figured differences in humeral form in *Tyrannosaurus rex* (Fig. 15.7A) and *Tarbosaurus bataar* (Fig. 15.7C)). They also reported a difference in the robustness of the ulna of *T. rex*, between FMNH PR2081 and MOR 555 (Fig. 15.7B). Carpenter and Smith did not explicitly relate this difference to sexual dimorphism, although if they are correct in their assessment that the humeri did exhibit dimorphism, differences in ulnar form may be related.

Some years later, there were reports that *T. rex* did indeed have robust and gracile morphs and that the former was female (Larson 1994, 1995; Anonymous 1995). Larson argued that robust and gracile morphs were discernible in the material of *T. rex*. These morphs were interpreted as evidence for sexual dimorphism. He further reported that Eberhard Frey had found that the anterior-most chevron is on the second caudal in male crocodilians and on the third in females. In males, this chevron is the attachment for the retractor of the intromit-

Figure 15.7. Some elements of the forelimb reportedly showing sexual dimorphism in Tyrannosaurus rex *(A–B) and* Tarbosaurus bataar *(C). (A) The left humerus of MOR 555 above and the right (reversed) of FMNH PR2081 below (*T. rex*):*
(a) anterior; (b) lateral;
*(c) posterior; and (d) medial views. Scale bars = 10 cm. The arrows indicate the pathological region of the humerus of FMNH PR2081. (B) The ulnae of FMNH PR2081 above and MOR 555 (reversed) below (*T. rex*):*
*(e) anterior; (f) lateral; and (g) medial views. Scale bar = 10 cm. (C) The humeri of PIN 552-1 in posterior (h) and anterior (i) views and of PIN 552-2 (j) in medial view. PIN 552-1 is interpreted as the gracile form and PIN 552-2 as the robust. Scale bar of PIN 552-1 = 5 cm, of PIN 552-2 = 20 cm. (*T. rex *figures from Carpenter and Smith 2001;* T. bataar *from Maleev 1974.)*

tent organ, and in females the anterior-most chevron is smaller than the succeeding ones. This pattern may provide a skeletal indicator of sex; however, Brochu (2003) was unable to verify these observations. Even if the observations are correct, Larson was able to check this condition only for a single skeleton of *T. rex*, MOR-555. This skeleton reportedly showed that the gracile morph had the "masculine" pattern of chevrons, but this report was also disputed by Brochu, who argued that the reported "first" chevron was actually the second, the first not having been found until later. Larson (1994) reported that examination of Asian troodontids showed that they also exhibited robust and gracile morphs, and the gracile individuals exhibited the "masculine" chevron pattern. This makes an interesting hypothesis, but further information—and evidence—are desirable. The work on the Asian troodontids, for example, has yet to be published.

Presumably Larson based his evidence for robust and gracile morphs in *Tyrannosaurus* on proportions of individual elements. The measurements given (Larson 1994: table 2), however, are simply lengths and so cannot support the existence of two morphs differing in proportion. The differences in proportions reported (Larson 1994) pertain mainly to cranial elements (Fig. 15.8). No information regarding which element derives from which specimen is given; however, those attributed to the robust morph are from FMNH PR2081 (Carpenter, pers. comm., 2003). The skull of this specimen is crushed and distorted (Fig. 15.8D) and hence inappropriate for this purpose without compensation for the effects of preservation. Only reported differences in the width of the pelvis and humeral (and possibly ulnar) form remain. The only two examples of humeri (FMNH PR2081, RTMP 81.6.1) from the robust morph are pathological, and thus the robust form of this element may result from either pathological alteration or dimorphism (Brochu 2003). The robust ulna also derives from FMNH PR2081. Because the humerus of this specimen is pathological, whether the robustness of the ulna is actually evidence for dimorphism or some result of the antebrachial pathology (or individual variation) is not clear.

Osborn (1916) illustrated two pelves of *T. rex* that differed substantially in breadth, that of AMNH 5027 being narrow and that of AMNH 973 (now CM 9380) broad (Fig. 15.6B). Brochu (2003) noted that differences in breadth of the sacrum—presumably equivalent, for purposes of this argument, to differences in breadth of the whole pelvis (cf. Osborn 1916, fig. 20)—were also seen in FMNH PR2081 (broad) and MOR 555 (narrow). Larson (2001) stated that the narrow pelvis of AMNH 5027 is not crushed; however, Osborn (1916, p. 767) states that it is "naturally crushed and deformed." As with the cranial elements, there is no indication that the pelvis of MOR 555 is uncrushed and undistorted (cf. Molnar 2001a). If the robust humeri were not pathological, the variation in ulnar form would support sexual dimorphism, as would the differences in pelvic width, if the narrower pelves were not crushed. In the absence of such information on postmortem alterations, the only evidence of sex are the three ischia studied by Carpenter and the chevron pattern of a single specimen (MOR 555). Neither of these provides unambiguous evidence for the existence of

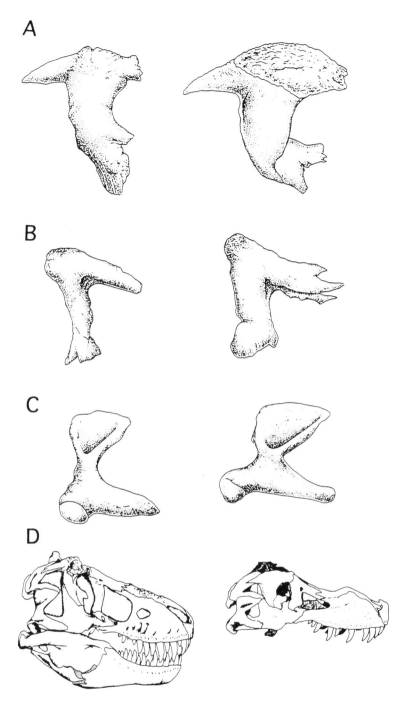

A

B

C

D

Figure 15.8. Some cranial elements reportedly showing sexual dimorphism in Tyrannosaurus rex. Those on the left are from the "gracile form" (specimen numbers not given for individual elements; skull is AMNH 5027); on the right are those of the "robust form" (all FMNH PR2081). (A) postorbital; (B) lachrymal; and (C) quadrato-jugal. As can be seen from the skulls (D), that of FMNH PR2081 has been substantially crushed dorsoventrally. It is also sheared to the left (Brochu 2003, figs. 5, 6), as is that of AMNH 5027 (Osborn 1912, figs. 3, 4), although to a lesser degree. The issue is whether this postmortem deformation of FMNH PR2081 accounts for the differences in form and robustness of its elements from those of the "gracile form." (Individual elements from Larson 1994; AMNH 5027 from Osborn 1912; FMNH PR2081 from Brochu 2003.)

two morphs as opposed to individual variation, so this interpretation is currently without compelling published support (unlike the situation with *Syntarsus* and *Coelophysis*, for which Raath and Colbert have made descriptions, measurements, and photographs available).

It was not claimed for either *Syntarsus* or *Tyrannosaurus* that one sex was actually significantly bigger than the other; what was claimed

Sexual Selection and Sexual Dimorphism in Theropods • 303

was that one sex had more robust elements (transversely broader in the case of the pelvis) than the other. In the only case for which supporting data have been provided, *Syntarsus,* only the femur is involved. For purposes of interpretation, the dimorphism has been equated with the differences in body mass found in sexually dimorphic birds, that is, with sexual size difference; but that it actually implies a comparable difference in body mass between the morphs is not clear. So, strictly speaking, the dimorphism is not reversed sexual dimorphism as found in birds, which does involve body size and mass. It may be rash to assume that the factors producing a difference in body mass would also produce the observed difference in femoral form, and it remains to be shown that theropods exhibited dimorphism in body mass. Furthermore, since Osborn, only Molnar (1991) has published stratigraphic data on the occurrences of *T. rex,* and these data are not adequate to demonstrate that the specimens examined actually derive from a single population. In contrast, the Chitake River specimens of *S. rhodesiensis* (and those of *Coelophysis,* which is discussed next) are a single population. If the specimens derive from a single population, then the variants represent organisms living together at the same time, and the variation supports the interpretation of sexual dimorphism. If the specimens are not known to be from a single population, the possibility remains that the variants may represent geographic or temporal variation (Schwartz 1995)—although for a sufficiently large sample, drawing from a single population is probably unnecessary. Thus, the conclusion that the variation seen in *T. rex* is sexual dimorphism, much less reversed sexual dimorphism, seems premature: it may be both, but this cannot be demonstrated without further work.

However, these are not the only reported cases of sexual dimorphism among theropods: Colbert (1989, 1990) reported dimorphism in *Coelophysis bauri.* The dimorphism involves proportional differences in the head and neck and in the forelimb, as well as fusion of the sacral neural spines versus its absence. Colbert (1989) remarked that the limb material was generally quite well preserved. Smith's (1997, p. 42) report of dimorphism in cranial ornament in *Coelophysis bauri* was not based on personal observation, and he recommends treating the report with caution (Smith, pers. comm., 1999). Rowe and Gauthier (1990) also discussed dimorphism in plesiomorphic ceratosaurs (*Coelophysis* and *Syntarsus*) and noted dimorphism in forelimb length; development of the deltopectoral crest, olecranon, and epicondyles; breadth of the humeral head; and form of the proximal femur. Rowe (1989) also reported dimorphism in *Syntarsus kayentakatae* but gave few details. He did comment that robust individuals were "distinguished by hypertrophied muscle attachments" (Rowe 1989, p. 125). None of these workers speculated which morph represented which sex. Gay (2001, this volume), in a rare report of a negative conclusion, was unable to find evidence for dimorphism in *Dilophosaurus wetherilli.*

Sexual dimorphism in *Coelophysis* and *Syntarsus* seems clear. Both are known from relatively large numbers of specimens, enough to be confident that we are not just seeing individual variation. Both reports plausibly concern single populations. The only other theropod known from a relatively large number of specimens is *Allosaurus fragilis*. A brief note on this animal by Thulborn (1994) reports that dimorphism is present in the proportions of the premaxilla and maxilla, but not the dentary. The upper jaw elements of adults were grouped as gracile or robust, the former having sixteen to seventeen maxillary teeth, the latter fourteen to fifteen. Thulborn identified this as reversed sexual dimorphism, presumably following Raath, Carpenter, and Larson. Both forms were approximately equally abundant in the samples (taken from Madsen 1976) of thirty-six maxillae and forty-four premaxillae. Thulborn noted that the robust morph may have been able to take different (presumably larger) prey than the gracile morph, and this ability might reflect the higher costs of reproduction for females than for males. He noted that immature elements were all gracile, as is the case in *Syntarsus* (Raath 1990). In both *Allosaurus* and *Syntarsus*, the robust morph deviates in its growth trajectory from the juvenile condition. This deviation is unusual in that in sexual dimorphism, the male, not the female, usually deviates from the juvenile condition (Gould and Gould 1989).

So we may conclude that sexual dimorphism was present in basal ceratosaurs and (probably) basal tetanurans (Tables 15.4, 15.5). In spite of the work cited on *Tyrannosaurus* (and Molnar 1991), sexual dimorphism in that taxon is not yet established beyond doubt. Determining whether the dimorphism was reversed is more problematic and requires sexing individual specimens, which is not completely reliable, even for human skeletal remains (Schwartz 1995). Exceptional discoveries, like that of eggs preserved in place in the body cavity in *Sinosauropteryx* (Chen et al. 1998), cannot be relied upon for identification. However, reasonably consistent differences in large samples that can be functionally related to sex, as with Carpenter's work on ischia, may be appropriate (see also Chapman et al. 1997; Brochu 2003).

The dimorphism of theropods seems unlike that of other large tetrapods, especially large terrestrial carnivores. Simply arguing that theropods exhibited reversed sexual dimorphism because that condition is common is not sufficient (Larson 1994, 1995). After all, the largest single taxon of animals has six legs and wings (Evans 1975; Lewin 1991), but this does not mean that tyrannosaurs are derived coleopterans. Brochu (2003) argues that among living reptiles, reversed sexual dimorphism probably occurred only twice. With respect to arguing its possible occurrence among theropods, this reversal is the significant factor, not how much those lineages in which it did happen have diversified. In view of this scarcity, there are several aspects requiring further work:

1. Proportional differences between the morphs in derived theropods need to be documented.
2. Postmortem distortion needs to be taken into account, either to demonstrate that there was none or to compensate for its effects.
3. The specific nature and degree of the dimorphism as it is manifested in body size and mass needs to be ascertained.
4. The deviation of the female growth trajectory, rather than that of the male, from the juvenile condition requires explanation.

TABLE 15.4
Reported Instances of Sexual Dimorphism in Theropods

Expectation	Evidence	Reference
Syntarsus rhodesiensis	Form of forelimb, sacrum, pelvis, femur	Raath 1990
Syntarsus rhodesiensis	Perimedullar erosion in femur	Chinsamy 1990
Tyrannosaurus rex	Orientation of distal shaft of ischium	Carpenter 1990
Tyrannosaurus rex	Proportions of maxilla, lachrymal, dentary, teeth	Carpenter 1992
Tyrannosaurus rex	Form and proportions of humerus and possibly ulna	Carpenter and Smith 2001
Tyrannosaurus rex	Form and position of anterior-most chevron	Larson 1994
"Asian troodontids"	Position of anterior-most chevron	Larson 1994
Coelophysis bauri	Proportions of head and neck, and of forelimb, fusion of sacral neural spines	Colbert 1989, 1990
Coelophysis, Syntarsus	Length of forelimb, development of deltopectoral crest, olecranon, and humeral epicondyles, breadth of humeral head, form of proximal femur	Rowe and Gauthier 1990
Syntarsus kayentakatae	Development of muscular attachments	Rowe 1989
Allosaurus fragilis	Proportions of premaxilla and maxilla	Thulborn 1994

TABLE 15.5
Skeletal Structures Exhibiting Sexual Dimorphism in Theropods

Taxon	Anatomical region	Reference
Syntarsus rhodesiensis	Forelimb, pelvis, hindlimb	Raath 1990
Syntarsus rhodesiensis	Hindlimb	Chinsamy 1990
Tyrannosaurus rex	Pelvis	Carpenter 1990
Tyrannosaurus rex	Head	Carpenter 1992
Tyrannosaurus rex	Forelimb	Carpenter and Smith 2001
Tyrannosaurus rex	Tail	Larson 1994
"Asian troodontids"	Tail	Larson 1994
Coelophysis bauri	Head, neck, forelimb, pelvis	Colbert 1989, 1990
Coelophysis, Syntarsus	Forelimb, hindlimb	Rowe and Gauthier 1990
Syntarsus kayentakatae	?	Rowe 1989
Allosaurus fragilis	Head	Thulborn 1994

5. Reversed sexual dimorphism in monogamous birds seems related to flight and hence is not analogous to that proposed for theropods; therefore, reversed sexual dimorphism in theropods requires independent explanation.

6. For the harem scenario, the male is expected to be the more robust. Therefore, if this scenario is supported, reversed dimorphism requires explanation.

7. Larger samples are desirable, not just finding more specimens but also making more data available on known specimens.

If reversed sexual dimorphism is present in theropods, it may well have resulted from greater reproductive success of larger individuals. This may, in turn, have been the selective factor for the evolution of the robust morph, which may then—as Thulborn suggested—have been able to obtain more or larger prey to support both physiological and reproductive costs. Rowe's observation of hypertrophied muscle attachments is consistent with this interpretation. Strictly speaking, this hypertrophy seems not to have been a dimorphism of general body size (sexual size difference), in the sense of linear dimensions—for no evidence has been forthcoming of a significant difference in size among adults—but rather in limb and jaw proportions and muscular development. The advantage to the female would have accrued from the enhanced capacity for obtaining prey as well as enhanced ability to provision the young, following from greater muscular development. If general body mass were involved, benefits of proportionately increased energy reserves with greater mass (Calder 1984; Reiss 1989) would have accrued. On the other hand, a fecundity advantage with greater size generally occurs in forms with indeterminate growth (Andersson 1994); however, the only theropod for which data is available, *Troodon,* apparently had determinate growth (Chinsamy 1990; Varricchio 1993, 1997).

It is incidental, but interesting, that the large majority of modern birds lack a penis, although it has been retained or a new intromittent organ evolved in some (Briskie and Montgomerie 1997). If dimorphism in the proximal chevrons occurred in theropods, this suggests that the penis was lost at about the same time that the avian lineage developed flight: if it did not occur, we could ascertain when in the evolution of the theropod lineage the penis was lost.

If the sexual dimorphism was not reversed, the driving factor presumably would have been male contests in which the cranial ornament played a role. Since intraspecific combat likely occurred in at least some theropods, superiority in those contests may have driven the evolution of the robust morph. These scenarios are speculative but should, at least in part, be susceptible to empirical test.

Conclusion

Features probably related to sexual selection can be seen or inferred in theropods. Theropod dinosaurs were unusual among large carnivorous tetrapods in the possession of various cranial horns, crests, and rugosities. The ornaments (generally) seem structurally unsuitable for weapons. Because the ancestral taxa lacked cranial ornament and early forms sometimes had more strongly developed structures than more derived, later, larger forms, the appearance of the cranial ornament in general seems not to be due simply to allometric effects. Theropod cranial ornaments probably served as signaling structures in displays to rivals and as advertisements of quality to potential mates.

In raptors (and probably waders), reversed sexual dimorphism seems to be the result of female selection for aerobatic ability (or, perhaps, hunting skill). This implies that reversed sexual dimorphism is intimately related to flight in (monogamous) birds and hence not analogous to that proposed for theropods. In neither *Syntarsus* nor *Tyrannosaurus* was it claimed that one sex was actually significantly bigger than the other; rather it was claimed that one sex had more robust skeletal elements than the other. This is neither a difference in overall form nor a sexual size difference, but rather a dimorphism of bone robustness, a kind apparently unknown in modern tetrapods. In the only case for which supporting data have been provided, *Syntarsus*, only the femur is involved. Sexual dimorphism was present in both *Coelophysis* and *Syntarsus*. It is also reliably reported in *Allosaurus fragilis*.

The theropod paleobiology discussed here suggests that theropods may well have differed from modern large carnivores in previously unappreciated ways. Furthermore, they can enhance our understanding of modern ecological processes and the scope of response of different lineages to environmental conditions. Therefore, theropods should not be viewed simply as big bipedal lions or crocodiles.

Acknowledgments. I am grateful to Mike Raath for his comments on this manuscript and discussion of this subject, as well as for elucidating the situation in *Syntarsus rhodesiensis*. Tony Thulborn, David Smith, Frank Seebacher, and Kenneth Carpenter contributed useful information and discussion on aspects of the material discussed here. You Hailu and an anonymous reviewer also provided significant assistance. This essay was inspired in various ways by the work of Christine Janis, Peter Larson, and Mike Raath.

References Cited

Andersson, M. 1994. *Sexual Selection.* Princeton, N.J.: Princeton University Press.

Anonymous (Tokyo Broadcasting System). 1995. *The T. rex World Exposition Official Guide Book.* Tokyo: Tokyo Broadcasting System. [In Japanese.]

Bakker, R. T. 1986. *The Dinosaur Heresies.* New York: William Morrow and Co.

Birkhead, T. 2003. *A Brand-New Bird.* New York: Basic Books.

Bloom, W., M. A. Bloom, and F. C. McLean. 1941. Calcification and ossification: Medullary bone changes in the reproductive cycle of female pigeons. *Anatomical Record* 81: 443–475.

Bonaparte, J. F. 1999. An armoured sauropod from the Aptian of northern Patagonia, Argentina. In Y. Tomida, T. H. Rich, and P. Vickers-Rich (eds.), *Proceedings of the Second Gondwanan Dinosaur Symposium,* pp. 1–12. National Science Museum Monographs, no. 15. Tokyo: National Science Museum.

Brazaitis, P. 1981. Maxillary regeneration in a marsh crocodile, *Crocodylus palustris*. *Journal of Herpetology* 15: 360–362.

Briskie, J. V., and R. Montgomerie. 1997. Sexual selection and the intromittent organ of birds. *Journal of Avian Biology* 28: 73–86.

Brochu, C. A. 2003. Osteology of *Tyrannosaurus rex*: Insights from a nearly complete skeleton and high-resolution computed tomographic analysis of the skull. *Journal of Vertebrate Paleontology, Memoir* 7: 1–138.

Calder, W. A., III. 1984. *Size, Function, and Life History.* Cambridge: Harvard University Press.

Carpenter, K. 1990. Variation in *Tyrannosaurus rex*. In K. Carpenter and P. J. Currie (eds.), *Dinosaur Systematics: Approaches and Perspectives*, pp. 141–145. Cambridge: Cambridge University Press.

Carpenter, K. 1992. Tyrannosaurids (Dinosauria) of Asia and North America. In N. J. Mateer and Chen P.-J. (eds.), *Aspects of Nonmarine Cretaceous Geology*, pp. 250–268. Beijing: China Ocean Press.

Carpenter, K., and M. Smith. 2001. Forelimb osteology and biomechanics of *Tyrannosaurus rex*. In D. H. Tanke and K. Carpenter (eds.), *Mesozoic Vertebrate Life*, pp. 90–116. Bloomington: Indiana University Press.

Carr, T. D. 1999. Craniofacial ontogeny in Tyrannosauridae (Dinosauria, Coelurosauria). *Journal of Vertebrate Paleontology* 19: 497–520.

Chabreck, R. H., and T. Joanen. 1979. Growth rates of American alligators in Louisiana. *Herpetologica* 35: 51–57.

Chapman, R. E., D. B. Weishampel, G. Hunt, and D. Rasskin-Gutman. 1997. Sexual dimorphism in dinosaurs. In D. L. Wolberg, E. Stump, and G. D. Rosenberg (eds.), *Dinofest International: Proceedings of a Symposium Held at Arizona State University*, pp. 83–93. Philadelphia: Academy of Natural Sciences.

Chen P., Dong Z., and Zhen S. 1998. An exceptionally well-preserved theropod dinosaur from the Yixian Formation of China. *Nature* 391: 147–152.

Chinsamy, A. 1990. Physiological implications of the bone histology of *Syntarsus rhodesiensis* (Saurischia: Theropoda). *Palaeontologia Africana* 27: 77–82.

Clarke, J. M., M. A. Norell, and L. M. Chiappe. 1999. An oviraptorid skeleton from the late Cretaceous of Ukhaa Tolgod, Mongolia, preserved in an avian-like brooding position over an oviraptorid nest. *American Museum Novitates*, no. 3265: 1–36.

Colbert, E. H. 1989. *The Triassic Dinosaur* Coelophysis. Museum of Northern Arizona Bulletin, no. 57. Flagstaff: Museum of Northern Arizona Press.

Colbert, E. H. 1990. Variation in *Coelophysis bauri*. In K. Carpenter and P. J. Currie (eds.), *Dinosaur Systematics: Approaches and Perspectives*, pp. 81–90. Cambridge: Cambridge University Press.

Cott, H. B. 1961. Scientific results of an inquiry into the ecology and economic status of the Nile Crocodile (*Crocodilus niloticus*) in Uganda and Northern Rhodesia. *Transactions of the Zoological Society of London* 29: 211–357.

Cronin, H. 1991. *The Ant and the Peacock.* Cambridge: Cambridge University Press.

Currie, P. J. 2003. Cranial anatomy of tyrannosaurid dinosaurs from the Late Cretaceous of Alberta, Canada. *Acta Palaeontologica Polonica* 48: 191–226.

Czerkas, S. A. 1992. Discovery of dermal spines reveals a new look for sauropod dinosaurs. *Geology* 20: 1068–1070.

Darwin, C. R. 1871. *The Descent of Man, and Selection in Relation to Sex.* London: Thomas Murray.

Davitashvili, L. Sh. 1961. *Teoriia polovogo otbora.* Moscow: Akademia Nauk.

Dodson, P. 1975. Taxonomic implications of relative growth in lambeosaurine hadrosaurs. *Systematic Zoology* 24: 37–54.

Dong Z. and P. J. Currie. 1996. On the discovery of an oviraptorid skeleton on a nest of eggs at Bayan Mandahu, Inner Mongolia, People's Republic of China. *Canadian Journal of Earth Sciences* 33: 631–636.

Evans, G. 1975. *The Life of Beetles.* London: George Allen & Unwin.

Figuerola, J. 1999. A comparative study on the evolution of reversed size dimorphism in monogamous waders. *Biological Journal of the Linnean Society* 67: 1–18.

Fitch, H. S. 1981. *Sexual Size Differences in Reptiles.* University of Kansas Museum of Natural History, Miscellaneous Publications, no. 70. Lawrence: University of Kansas, Museum of Natural History.

Forster, C. A. 1999. Gondwanan dinosaur evolution and biogeographic analysis. *Journal of African Earth Sciences* 28: 169–185.

Gay, R. J. 2001. Evidence for sexual dimorphism in the Early Jurassic theropod dinosaur, *Dilophosaurus* and a comparison with other related forms [abstract]. *Journal of Vertebrate Paleontology* 21 (3 suppl.): 53A.

Geist, V. 1966. The evolution of horn-like organs. *Behaviour* 27: 175–214.

Geist, V. 1978. On weapons, combat and ecology. In L. Krames, P. Pliner, and T. Alloway (eds.), *Aggression, Dominance, and Individual Spacing,* pp. 1–30. New York: Plenum Press.

Gilmore, C. W. 1920. *Osteology of the Carnivorous Dinosauria in the United States National Museum, with Special Reference to the Genera* Antrodemus (Allosaurus) *and* Ceratosaurus. U.S. National Museum Bulletin no. 110. Washington, D.C.: Government Printing Office.

Glut, D. F. 1997. *Dinosaurs: The Encyclopedia.* Jefferson, N.C.: McFarland & Co.

Gorzula, S. J. 1978. An ecological study of *Caiman crocodilus crocodilus* inhabiting savanna lagoons in the Venezuelan Guayana. *Oecologia* 34: 21–34.

Gould, J. L., and C. G. Gould. 1989. *Sexual Selection.* New York: HPLHP.

Hopson, J. A. 1977. Relative brain size and behavior in archosaurian reptiles. *Annual Review of Ecology and Systematics* 8: 429–448.

Horner, J. R., and D. Lessem. 1993. *The Complete* T. rex. New York: Simon and Schuster.

Hurum, J. H., and K. Sabath. 2003. Giant theropod dinosaurs from Asia and North America: Skulls of *Tarbosaurus bataar* and *Tyrannosaurus rex* compared. *Acta Palaeontologica Polonica* 48: 161–190.

Janis, C. 1982. Evolution of horns in ungulates: ecology and paleoecology. *Biological Reviews* 57: 261–318.

Jarman, P. 1983. Mating system and sexual dimorphism in large, terrestrial, mammalian herbivores. *Biological Reviews* 58: 485–520.

Jehl, J. R., Jr., and B. G. Murray Jr. 1986. The evolution of normal and reverse sexual size dimorphism in shorebirds and other birds. *Current Ornithology* 3: 1–86.

Ji Q., Ji S., Yuan C.-X., and Ji X.-X. 2002. Restudy on a small dromaeosaurid dinosaur with feathers over its entire body. *Earth Science Frontiers* 9: 57–63. [In Chinese.]

Kavanau, J. L. 1987. *Lovebirds, Cockatiels, Budgerigars: Behavior and Evolution.* Los Angeles: Science Software Systems.

Larson, P. L. 1994. *Tyrannosaurus* sex. In G. D. Rosenberg and D. L. Wolberg (eds.), *Dino Fest: Proceedings of a Conference for the General Public, March 24, 1994,* pp. 139–155. Special Publication, The Paleontological Society, no. 7. Indianapolis: Geology Department, Indiana University–Purdue University Indianapolis, and Knoxville, Tenn.: Paleontological Society.

Larson, P. L. 1995. To sex a *rex. Australia Nature* 25 (2): 46–53.

Larson, P. L. 2001. Paleopathologies in *Tyrannosaurus rex. Dino Press* 5: 26–35. [In Japanese.]

Lewin, R. 1991. Too many insects, and not enough entomologists. *New Scientist* 129 (1751): 33. [The pagination of this magazine differs in regional editions, this is the North American pagination.]

Madsen, J. H., Jr. 1976. Allosaurus fragilis: *A Revised Osteology.* Utah Geological and Mineral Survey Bulletin, no. 109. Salt Lake City: Utah Geological and Mineral Survey, Utah Department of Natural Resources.

Magnusson, W. E., K. A. Vliet, A. C. Pooley, and R. Whitaker. 1989. Reproduction. In L. Dow and C. Craig (eds.), *Crocodiles and Alligators,* pp. 118–135. Sydney: Weldon Owen.

Molnar, R. E. 1977. Analogies in the evolution of combat and display structures in ornithopods and ungulates. *Evolutionary Theory* 3: 165–190.

Molnar, R. E. 1991. The cranial morphology of *Tyrannosaurus rex. Palaeontographica Abt. A* 217: 137–176.

Molnar, R. E. 2001a. Paleobiology of large carnivorous dinosaurs, or sex and violence among the theropods. *Dino Press* 4: 85–91. [In Japanese.]

Molnar, R. E. 2001b. Theropod paleopathology: A literature survey. In D. H. Tanke and K. Carpenter (eds.), *Mesozoic Vertebrate Life,* pp. 337–363. Bloomington: Indiana University Press.

Mueller, H. C. 1990. The evolution of reversed sexual dimorphism in size in monogamous species of birds. *Biological Reviews* 65: 553–585.

Osborn, H. F. 1903. The skull of *Creosaurus. Bulletin of the American Museum of Natural History* 19: 697–701.

Osborn, H. F. 1912. Crania of *Tyrannosaurus* and *Allosaurus. Memoirs of the American Museum of Natural History,* n.s., 1: 1–30.

Osborn, H. F. 1916. Skeletal adaptations of *Ornitholestes, Struthiomimus, Tyrannosaurus. Bulletin of the American Museum of Natural History* 35: 733–771.

Osborn, H. F. 1924. Three new Theropoda, *Protoceratops* Zone, central Mongolia. *American Museum Novitates,* no. 144: 1–12.

Ostrom, J. H., and P. Wellnhofer. 1986. The Munich specimen of *Triceratops* with a revision of the genus. *Zitteliana* 14: 111–158.

Packer, C. 1983. Sexual dimorphism: the horns of African antelopes. *Science* 221: 1191–1193.

Paul, G. S. 1988. *Predatory Dinosaurs of the World: A Complete Illustrated Guide.* New York: Simon and Schuster.

Raath, M. A. 1990. Morphological variation in small theropods and its meaning in systematics: Evidence from *Syntarsus rhodesiensis.* In K. Carpenter and P. J. Currie (eds.), *Dinosaur Systematics: Approaches and Perspectives,* pp. 91–105. Cambridge: Cambridge University Press.

Reiss, M. J. 1989. *The Allometry of Growth and Reproduction.* Cambridge: Cambridge University Press.

Rowe, T. 1989. A new species of the theropod dinosaur *Syntarsus* from the Early Jurassic Kayenta Formation of Arizona. *Journal of Vertebrate Paleontology* 9: 125–136.

Rowe, T., and J. A. Gauthier. 1990. Ceratosauria. In D. B. Weishampel, P. Dodson, and H. Osmólska (eds.), *The Dinosauria*, pp. 151–168. Berkeley: University of California Press.

Schwartz, J. H. 1995. *Skeleton Keys*. Oxford: Oxford University Press.

Selander, R. K. 1972. Sexual selection and dimorphism in birds. In B. Campbell (ed.), *Sexual Selection and the Descent of Man 1871–1971*, pp. 180–230. Chicago: Aldine.

Smith, D. K. 1997. Cranial allometry in *Coelophysis:* Preliminary results. In B. Anderson, D. Boaz, and R. D. McCord (eds.), *Southwest Paleontological Symposium Proceedings*, vol. 1, pp. 41–48. Mesa, Ariz.: Southwest Paleontological Society.

Smith, D. K. 1998. A morphometric analysis of *Allosaurus*. *Journal of Vertebrate Paleontology* 18: 126–142.

Smith, M. F. 1993. *Female Choices*. Ithaca, N.Y.: Cornell University Press.

Swinton, W. E. 1934. *The Dinosaurs*. London: Thomas Murby & Co.

Tanke, D. H., and P. J. Currie. 1995. Intraspecific fighting behavior inferred from toothmark trauma on skulls and teeth of large carnosaurs (Dinosauria). *Journal of Vertebrate Paleontology* 15 (3 suppl.): 55A.

Thulborn, R. A. 1994. Dimorphic jaw bones in the theropod dinosaur *Allosaurus fragilis* [abstract]. In *Abstracts & Programme, Australasian Palaeontological Convention 94*, p. 54 Sydney: Association of Australian Palaeontologists.

Tyndale-Biscoe, H. 1973. *Life of Marsupials*. London: Edward Arnold.

Varricchio, D. J. 1993. Bone microstructure of the Upper Cretaceous theropod dinosaur *Troodon formosus*. *Journal of Vertebrate Paleontology* 13: 99–104.

Varricchio, D. J. 1997. Growth and embryology. In P. J. Currie and K. Padian (eds.), *Encyclopedia of Dinosaurs*, pp. 282–288. San Diego: Academic Press.

Webb, G., and C. Manolis. 1989. *Crocodiles of Australia*. Sydney: Reed Books.

Zahavi, A., A. Zahavi, N. Zahavi-Ely, and M. P. Ely. 1997. *The Handicap Principle*. Oxford: Oxford University Press.

Zhao X.-J. and P. J. Currie. 1993. A large crested theropod from the Jurassic of Xinjiang, People's Republic of China. *Canadian Journal of Earth Sciences* 30: 2027–2036.

16. An Unusual Multi-Individual Tyrannosaurid Bonebed in the Two Medicine Formation (Late Cretaceous, Campanian) of Montana (USA)

Philip J. Currie, David Trexler, Eva B. Koppelhus, Kelly Wicks, and Nate Murphy

Abstract

A bonebed in north-central Montana includes the remains of three or more individuals of the tyrannosaurid *Daspletosaurus* intermingled with the skeletons of at least five hadrosaurs. The species of *Daspletosaurus* cannot be identified with certainty at this time but appears to be different from *Daspletosaurus torosus* of the Oldman Formation of southern Alberta. The bonebed occurs in mudstone and shows little evidence of current sorting. It provides the first evidence that *Daspletosaurus* may have been a gregarious animal.

Introduction

In 1997, a field survey in Teton County, Montana, by Timescale Adventures staff and participants revealed a large amount of dinosaur bone, including vertebrae and a partial lower jaw of a tyrannosaurid. A grid was set up, and surface fragments were assigned grid coordinates and collected before excavation began. All of the remains initially identified belonged to an adult tyrannosaurid, but as excavation progressed, hadrosaurid fossils were found as well. Work is still progressing on the quarry. As of 2002, more than 1,400 individual specimens have been collected. These specimens range from fragments to complete and well-preserved bones.

Thus far, an area of over 100 m² has been excavated, and the quarry has produced evidence of at least three tyrannosaurs and five hadrosaurs. Dozens of shed tyrannosaurid teeth have been found in the bone-bearing layer, and conspicuous tooth marks are evident on several of the hadrosaur bones. Some of the larger bones are broken or distorted in such a manner as to suggest that large animals stepped on them.

Geology and Deposition

The Two Medicine Formation in the research locality (Trexler 2001) is slightly more than 600 m thick (Gulbrandsen 1992) and was deposited between 80.5 and 74.5 Ma (Rogers et al. 1993). The formation is postulated to represent rapid deposition that occurred on a broad, flat upper coastal plain to the west of the Cretaceous Interior Seaway. The Laramide Orogeny was ongoing at that time, and the region was also experiencing increased volcanic activity. Rapid erosion of the uplands, large volumes of volcanic ash, and moderate sinking of the crust are thought to have provided the conditions necessary for the rapid buildup of the sediments that became the Two Medicine Formation.

The TA 1997.002 site is located in the upper portion of the Two Medicine Formation in Teton County, Montana. No exposed section occurs in the area, and the exposure at the site occupies an area less than 70 m by 50 m. Stratigraphic placement within the formation is therefore difficult, although interpolation from well logs and contour maps places a nearby fossil locality at 336 m above the base of the formation (Gulbrandsen 1992). By measuring the difference in elevation between the two places, and allowing for regional dip, the site is tentatively placed 385 m above the base of the formation.

The bone-bearing layer is up to 1 m thick, and iron-rich horizontal bedding planes occur at various intervals of vertical distance through this facies. The specimens vary in preservational quality and range from scattered, isolated fragments to articulated sections of skeletons. The bones are preserved in fine-grained, poorly sorted, muddy siltstone. Nodules identified as redeposited caliche, transported pebbles, and mud clasts are present but are primarily confined to the upper portions of the facies. The sequence thus possibly represents a series of recurrent flooding events. Some specimens in the quarry are almost vertical in orientation, suggesting rapid burial.

The fossils from the quarry have several types of preservation. Some of the bones are rounded bone "pebbles," and others are complete and well preserved. Most are fractured, and many are shattered to a degree more severe than in other nearby fossil localities. No more than 25 cm of overburden was present anywhere in the quarry. The locality has changed little in over 100 years, according to reports from descendants of the original homestead family. Because of shallow burial, the fossils have been subjected to freeze and thaw cycles for a long time, thus contributing to the shattering. Mountain-building processes in the area have broken and separated portions of some bones by several centimeters.

The surfaces of virtually all specimens have been etched by recent plant root systems. However, a significant but variable amount of bone surface weathering also appears to have occurred before burial. The rounded, smooth edges of some bones suggest that they might have been stream-carried bed load. However, most of these rounded fragments came from an area of less than 2 m². Varricchio (2001) suggested that such occurrences might represent tyrannosaurid stomach contents rather than water-worn bone pebbles.

The hadrosaurid material is, in general, more scattered and more weathered than the tyrannosaurid remains. Several of the hadrosaur bones are tooth marked, presumably by feeding tyrannosaurids (Fiorillo 1991). This assumption is supported by the numerous shed (rootless) tyrannosaur teeth recovered from the quarry. Hadrosaurid remains have been found above and below the tyrannosaurid fossils but are usually intermingled. The intermingling suggests that the remains of the two types of animal were deposited concurrently, although there may have been a significant time lag between death and burial.

The hadrosaurid remains appear to belong to a single species within the Lambeosaurinae. They represent at least two juveniles, one sub-adult, and two adults. The smallest juvenile is estimated at approximately 2 m in length, and the largest was over 9 m long. The hadrosaurid fossils are largely unprepared at this time, and taxonomic placement below the level of subfamily is not possible.

All of the identifiable tyrannosaurid remains from the Two Medicine quarry can be referred to *Daspletosaurus*. Russell (1970) described *Daspletosaurus torosus* on the basis of a specimen from the Oldman Formation (Eberth et al. 2001) of Dinosaur Provincial Park. A new species of this genus was reported from the Two Medicine Formation by Horner et al. (1992), but the species has not been formally described. The new material from the Two Medicine Formation likely represents the same species, although further study and description of *Daspletosaurus* will be necessary before the species can be determined with certainty.

Institutional Abbreviations. AMNH, American Museum of Natural History, New York; FMNH, Field Museum of Natural History, Chicago; NMC, Canadian Museum of Nature, Ottawa, Ontario; TA, Timescale Adventures Research and Interpretive Center, Bynum, Mont.; TMP, Royal Tyrrell Museum of Palaeontology, Drumheller, Alberta.

Description

Over 1,400 elements and fragments have been collected from the site, although only the bones studied during the preparation of this

TABLE 16.1

Some of the Tyrannosaurid Material from Bonebed TA.1997.002,
Teton County, Montana

Specimen no.	Identification	Grid no.	Size class
TA.1997.002.1435	premaxilla, left	8-m/n	large
TA.1997.002.064	premaxilla, right	8-Y	medium
TA.1997.002.423	maxilla, left	1-K	large
TA.1997.002.487	maxilla, right	2-B	large
TA.1997.002.682	maxilla, left	7-N	small
TA.1997.002.1436	maxilla, right	8-A	small
TA.1997.002.168	nasals, fused	2-C	large
TA.1997.002.388	lacrimal, right	1-F	large
TA.1997.002.563	lacrimal, left	7-T	large
TA.1997.002.1308	quadratojugal, right	8-A	small
TA.1997.002.1384	jugal, left	—	large
TA.1997.002.899	quadrate, left	8-S	large
TA.1997.002.834	quadrate, right	8-M/N	large
TA.1997.002.1282	dentary, right	8-F	small
TA.1997.002.302	dentary, right	8-T	large
TA.1997.002.057	dentary, left	2C,2D	large
TA.1997.002.140	dentary, left	2C	large
TA.1997.002.390	surangular, right	8-J	large
TA.1997.002.710	furcula	7N	medium
TA.1997.002.781	ilium, left	7-S	large
TA.1997.002.1437	ilium, left	8-B/G	
TA.1997.002.1440	ilium, left	7-W, 8-C	medium
TA.1997.002.1239	ischium, left	8-B	small
TA.1997.002.1428	pubis, left	8-B	small
TA.1997.002.232	metatarsal II, left	2-L/M	small
TA.1997.002.787	metatarsal III, left	8-N	small
TA.1997.002.163	metatarsal III, right	2-B	large
TA.1997.002.316	metatarsal ?IV, left	8-E	medium
TA.1997.002.496	metatarsal ?IV, right	1-Q	large
TA.1997.002.200	?metatarsal fragment	2-Q	medium
TA.1997.002.385	manual I-2	8-I	large
TA.1997.002.838	pedal ungual ?IV-5	8R	?
TA.1997.002.516	pedal phalanx ?II-1	8E	?
TA.1997.002.321	pedal phalanx III-2	2G	small
TA.1997.002.223	pedal phalanx ?IV-4	2P	small
TA.1997.002.350	metacarpal II	1P	?medium
TA.1997.002.2	pedal phalanx ?III-4	1E	?small
TA.1997.002.395	manual phalanx ?II-2	7J	?medium
TA.1997.002.264	pedal phalanx	2K	large
TA.1997.002.318	pedal phalanx	2L	small
TA.1997.002.71	pedal phalanx	3p	medium
TA.1997.002.648	pedal phalanx IV-1	7-N	large

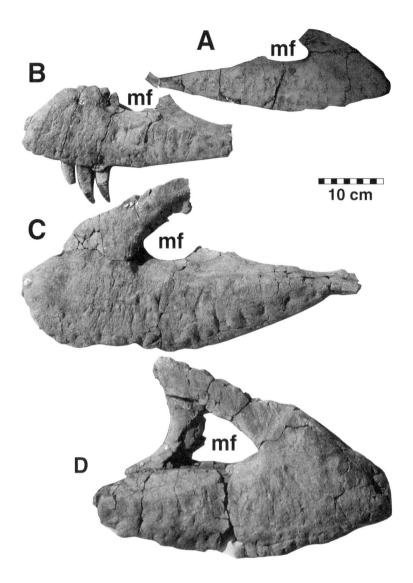

Figure 16.1. Daspletosaurus *sp.*
maxillae in lateral views.
(A) TA.1997.002.1436;
(B) TA.1997.002.682;
(C) TA.1997.002.423; and
(D) TA.1997.002.487.
Abbreviation: mf—maxillary
fenestra.

paper are listed in Table 16.1. The mapping grid was designed to allow the quarry to be extended in any direction.

A large (67 mm mediolaterally) left premaxilla (TA.1997.002.1435) has been recovered with four empty alveoli. A fragmentary right premaxilla (TA.1997.002.064) is 60 mm across and is from a slightly smaller individual.

Four maxillae have been recovered from the quarry, and they seem to represent the right (TA.1997.002.1436, Fig. 16.1A) and left (TA.1997.002.682, Fig. 16.1B) sides of one or two small individuals, and the left (TA.1997.002.423, Fig. 16.1C) and right (TA1997.002.487, Fig. 16.1D) sides of a large individual. The tooth row of TA.1997.002.423 is 460 mm long, and that of TA.1997.002.682 is 310 mm. Both sizes of individuals have sixteen alveoli. Characteristic of *Daspletosaurus*, the maxillary fenestra is large and triangular (Fig. 16.1D) in the large

10 cm

Figure 16.2. Daspletosaurus *sp. lacrimals in lateral view. (A) TA.1997.002.388 and (B) TA.1997.002.563. Arrows point to pneumatic foramina.*

animal, in which the promaxillary fenestra has shifted far enough forward that it cannot be seen in lateral aspect. The maxillary fenestra is relatively smaller in each of the smaller maxillae, and its anterior margin is posterior to the anterior edge of the antorbital fossa. These differences are ontogenetic and have also been observed in other tyrannosaurines (Currie 2003b). As in *Albertosaurus sarcophagus,* the ventral surface of the dental shelf has deep pits for reception of the tips of at least twelve of the dentary teeth. The pits are also conspicuous in at least one of the smaller specimens (TA.1997.002.682), so this is not an age-related character.

The fused nasals of the large individual are partially preserved (TA.1997.002.168). The convex dorsal surface of the middle region is only moderately rugose in comparison with that of other tyrannosaurids. Posteriorly, the region between the lacrimals and prefrontals is relatively narrow (7 cm) as in all specimens of *Daspletosaurus.*

Both the right (TA.1997.002.388, Fig. 16.2A) and left (TA.1997.002.563, Fig. 16.2B) lacrimals of the large individual were recovered and are typical for a tyrannosaurine like *Daspletosaurus.* The lacrimal horn has developed into a longitudinal crest, and the pneumatopore is relatively small (22 mm high). Five centimeters in front of the pneumatopore of the right lacrimal, there is a second, smaller opening in the bottom of the anterodorsal process (Fig. 16.2A). Posterodorsally, there is a suture for the contact with the postorbital, so the frontal slot would not have opened directly onto the orbital rim in this individual. The right lacrimal is 27 cm high and 25.5 mm in anteroposterior length and is almost complete.

An isolated left postorbital (Fig. 16.3) has a low, C-shaped ridge posterodorsal to the orbit as in most tyrannosaurines, including specimens of *Daspletosaurus* from the Dinosaur Park (FMNH PR365, TMP

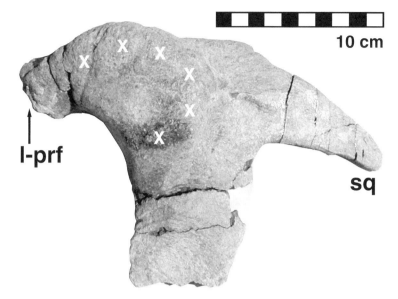

Figure 16.3. Daspletosaurus *sp. left postorbital (TA.1997.002.1383) in lateral aspect. White crosses indicate the crest of the C-shaped ridge. Abbreviations: l-prf—contact surface for lacrimal and prefrontal; sq—squamosal process.*

Figure 16.4. Daspletosaurus *sp. left jugal (TA.1997.002.1384) in lateral view. Abbreviations: mx— maxillary suture; or—orbital margin; pn—margin of pneumatopore in posteroventral corner of the antorbital fossa; po—ventral limit of postorbital suture; qj—quadratojugal suture.*

85.62.1) and Two Medicine formations (MOR 590) of Alberta and Montana. In *Daspletosaurus torosus* (NMC 8506, TMP 2001.36.1) of the Oldman Formation of Alberta, the C-shaped ridge rises ventrally into a prominent tubercle that protrudes 3 cm from the lateral surface of the postorbital bar. Although the ventral end of the postorbital is incomplete, the orbital margin is vertical, and there is no evidence of the suborbital bar characteristic of other tyrannosaurid genera.

A single left jugal (Fig. 16.4) of a large individual has been recovered. As in other specimens of *Daspletosaurus*, the pneumatic opening is relatively large and is separated from the antorbital fossa of the lacrimal by a ridge on the posterodorsal margin of the jugal. It is also typical in that the suborbital bar is relatively low, the postorbital bar is anteroposteriorly broad and shallowly concave laterally, the jugal cornua (Carr 1999) is prominent, and the anterior margin of the quadratojugal contact is vertical rather than tapering.

10 cm

Figure 16.5. Daspletosaurus *sp. right quadratojugal (TA.1997.002.1308) in lateral view. Abbreviations: j—jugal process; qf—quadrate fenestra; sq—squamosal contact.*

TA.1997.002.1308 (Fig. 16.5) is a well-preserved, right quadratojugal from the small individual. Like another juvenile *Daspletosaurus* (TMP 94.143.1), the quadrate and jugal rami meet in an acute angle. In contrast with the condition in *Albertosaurus* and *Gorgosaurus*, the anterior margin of the squamosal process is relatively straight.

A left (TA.1997.002.899) and a right (TA.1997.002.834) quadrate are from the large individual, and both are 232 mm high. The condyle of the left quadrate is 110 mm across, and the quadratojugal fenestra is 52 mm high. The two quadrates are asymmetric in that the large pneumatopore entering the bone between the medial condyle and the base of the pterygoid ala is taller but narrower on the right quadrate.

Two right dentaries (TA.1997.002.1282, TA.1997.002.302) have been recovered (Fig. 16.6), and they represent animals of different sizes. Two pieces of a partial left dentary (TA.1997.002.057) represent a large individual. Another dentary fragment (TA.1997.002.140) is slightly thinner (56 mm) at the equivalent position of TA.1997.002.302 (62 mm), suggesting that it may represent a third animal. The smaller jaw (TA.1997.002.1282) has some erupted teeth in position (Fig. 16.6A), and both of the complete jaws have germ teeth in their seventeen

10 cm

alveoli. The tooth row of the small jaw is 290 mm long, and that of the larger specimen is 42 percent longer (412 mm). The larger jaw is relatively deeper (minimum jaw depth is 27 percent the length of the tooth row) than the smaller (22%) and is also labiolingually relatively wider (16% versus 12%, when compared with tooth row length). These differences are expected because tyrannosaurid jaws are known to increase in relative height and width as they grow larger (Carr 1999; Currie and Dong 2001; Currie 2003a).

TA.1997.002.390 is a right surangular (Fig. 16.7) from a large individual. Most of the bone is preserved, and it is 430 mm long. When articulated with the right dentary (TA.1997.002.302), it gives a lower jaw length of 890 mm. The anterior surangular foramen is narrow and slit-like, and the posterior surangular fenestra is large (75 × 69 mm) and triangular. At the back of the mandibular shelf above the posterior margin of the fenestra, there is a low bump, which is pierced antero-dorsally by what seems to be a pneumatic fenestra.

Numerous vertebrae and ribs have been recovered from the quarry, although tyrannosaurid vertebrae cannot be identified to generic level at this time. An almost complete furcula (TA.1997.002.710, Fig. 16.8) is 155 mm across but probably would have been 15–20 mm wider in the living animal. The distal end of the right side expands into a 17-mm-high epicleideal facet as in other specimens of *Daspletosaurus* (NMC 8506 and 11315) for which this bone is known (Makovicky and Currie 1998). In other tyrannosaurids (*Albertosaurus, Gorgosaurus*), the epi-cleideal facets do not expand to the same degree, and the interclavicular angles are sharper.

Three left ilia have been identified, but none are fully prepared. TA.1997.002.781 is large (1085 mm long, 370 mm between the top of the acetabulum and the dorsal margin, 500 in postacetabular length). This specimen may represent a considerably larger animal than what is represented by the cranial material in the quarry. However, if this

Figure 16.6. Daspletosaurus sp. right dentaries in lateral view. (A) TA.1997.002.1282 and (B) TA.1997.002.302.

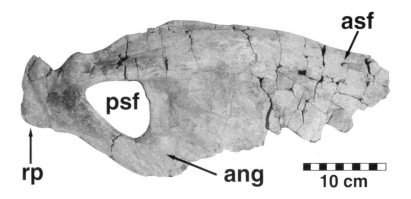

Figure 16.7. Daspletosaurus *sp. right surangular (TA.1997.002.390) in lateral view. Abbreviations: ang—suture for angular; asf—anterior surangular foramen; psf— posterior surangular fenestra; rp—retroarticular process.*

specimen does represent a larger individual, then the remains collected would include a fourth tyrannosaur, and so far, no other evidence of a fourth individual has been identified. TA.1997.002.1437 is part of an articulated left pelvis from the smallest animal. This ilium is only 680 mm in length (320 mm in postacetabular length), or roughly one-half the length of the large ilium. A third ilium, which has a postacetabular length of 420 mm, TA.1997.002.1440, is only partially exposed at present, but is intermediate in size between the other two. A left pubis (TA.1997.002.1438) and a left ischium (TA.1997.002.1239) were discovered in partial articulation with the ilium from the smaller individual.

Numerous metatarsals have been collected, but few are complete. The distal end of a pathologic, left metatarsal II (TA.1997.002.232) is about 58 mm across and is presumably from the small individual. It articulates with the distal end of a left metatarsal III (TA.1997.002.787), whose distal end is 25 mm across. A large right metatarsal III (TA.1997.002.163) is 530 mm long, with a 102-mm-wide distal end. TA.1997.002.316 is a left metatarsal IV that is 458 mm long. Comparison with other tyrannosaurs (Currie 2003a) suggests that this came from a large animal, estimated to be approximately 7 m long. The distal end is 58 mm across, which compares in size with the metatarsal II (TA.1997.002.232). TA.1997.002.496 is a right metatarsal IV from a much larger individual, and the distal end is 69 mm across.

Discussion and Conclusions

The presence of articulated sections of skeletons, the lack of sorting and stream-induced orientation, and the fact that the specimens can be assigned to individuals of discrete body sizes suggest that the tyrannosaurids and hadrosaurs may have died, decomposed, and been buried together at approximately the same time. Even the rounded bone fragments seem to be tyrannosaurid stomach contents rather than channel lag deposits. Tooth-marked hadrosaurid bones and shed tyrannosaurid teeth suggest that the tyrannosaurids were feeding on the hadrosaurs.

The theropod material from the bonebed can be attributed to at

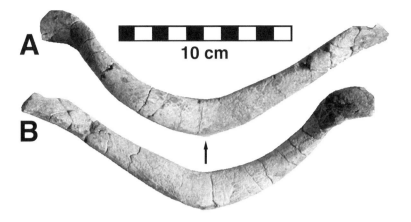

Figure 16.8. Daspletosaurus *sp.* furcula (TA.1997.002.710) in ventral (A) and dorsal (B) views. Arrow marks the position of the hypocleidium.

least three individuals of *Daspletosaurus*. This species was not the only tyrannosaurid species living in this area at that time, as *Gorgosaurus* is known from a site only 3 km away. These genera are both recovered from Dinosaur Provincial Park (Eberth et al. 2001) and were clearly contemporaries. The presence of multiple individuals of only one tyrannosaurid genus (*Daspletosaurus*) at the same site might be coincidental, but this is improbable. Association of large and small *Daspletosaurus* specimens in the same quarry, the abnormally high occurrence of tyrannosaur bones in the bonebed, and depositional factors such as partial articulation, co-mingling with the hadrosaurid remains, and the fact that all are found at the same level suggest that these animals were together in the same area, perhaps even interacting, at the time of their deaths. At present, there is no taphonomic information to indicate the cause of death for any of the animals. The multiple hadrosaurid individuals associated with the tyrannosaurs still need to be studied, but the identifiable specimens can be assigned to the subfamily Lambeosaurinae. Ample evidence already exists for herding behavior in hadrosaurs (Horner 1997).

This discovery represents the first report of possible gregarious behavior for *Daspletosaurus*. Such behavior has already been suggested for the tyrannosaurids *Albertosaurus* (Currie 2000) and *Tyrannosaurus* (Larson 1997). During Late Cretaceous times, some species of hadrosaurs and ceratopsians were collecting together into large herds (Horner 1997) at certain times of the year. It makes sense that tyrannosaurids adopted gregarious behavior that allowed them to break through the defenses of these herds.

Acknowledgments. The tyrannosaurid bonebed was discovered by Bob Kahn, who noticed small fragments of bone almost immediately after leaving his vehicle in 1997.

References Cited

Carr, T. D. 1999. Craniofacial ontogeny in Tyrannosauridae (Dinosauria, Coelurosauria). *Journal of Vertebrate Paleontology* 19: 497–520.

Currie, P. J. 1998. Possible evidence of gregarious behavior in tyrannosaurids. B. P. Pérez-Moreno, T. Holtz Jr., J. L. Sanz, and J. Moratalla
(eds.), *Gaia: Aspects of Theropod Paleobiology*, vol. 15, pp. 271–277.
Lisbon: Museu Nacional de História Natural.

Currie, P. J. 2003a. Allometric growth in tyrannosaurids (Dinosauria:
Theropoda) from the Upper Cretaceous of North America and Asia.
Canadian Journal of Earth Sciences 40: 651–665.

Currie, P. J. 2003b. Cranial anatomy of tyrannosaurid dinosaurs from the
Late Cretaceous of Alberta, Canada. *Acta Palaeontologica Polonica*
48: 191–226.

Currie, P. J., and Dong Z.-M. 2001. New information on *Shanshanosaurus
huoyanshanensis,* a juvenile tyrannosaurid (Theropoda, Dinosauria)
from the Late Cretaceous of China. *Canadian Journal of Earth Sciences* 38: 1729–1737.

Eberth, D. A., P. J. Currie, D. B. Brinkman, M. J. Ryan, D. R. Braman, J.
D. Gardner, V. D. Lam, D. N. Spivak, and A. G. Neuman. 2001.
Alberta's Dinosaurs and other fossil vertebrates: Judith River and
Edmonton Groups (Campanian-Maastrichtian). In C. L. Hill (ed.),
*Guidebook for the Field Trips, Society of Vertebrate Paleontology 61st
Annual Meeting: Mesozoic and Cenozoic Paleontology in the Western
Plains and Rocky Mountains,* pp. 49–75. Museum of the Rockies,
Occasional Paper 3. Bozeman, Mont.: Museum of the Rockies.

Fiorillo, A. R. 1991. Prey bone utilization by predatory dinosaurs. *Palaeogeography, Palaeoclimatology, Palaeoecology* 88: 157–166.

Gulbrandsen, R. A. 1992. The stratigraphic horizon of the Two Medicine
Formation at the dinosaur site in the McKracken area west of Pendroy.
Unpublished manuscript.

Horner, J. R. 1997. Behavior. In P. J. Currie and K. Padian (eds.), *Encyclopedia of Dinosaurs*, pp. 45–50. San Diego: Academic Press.

Horner, J. R., D. J. Varricchio, and M. B. Goodwin. 1992. Marine transgressions and the evolution of Cretaceous dinosaurs. *Nature* 358: 59–
61.

Larson, P. L. 1997. The king's new clothes: A fresh look at *Tyrannosaurus
rex*. In D. L. Wolberg, E. Stump, and G. D. Rosenberg (eds.), *Dinofest
International: Proceedings of a Symposium Sponsored by Arizona
State University*, pp. 65–71. Philadelphia: Academy of Natural Sciences.

Makovicky, P. J., and P. J. Currie. 1998. The presence of a furcula in
tyrannosaurid theropods, and its phylogenetic and functional implications. *Journal of Vertebrate Paleontology* 18: 143–149.

Rogers, R. R., C. C. Swisher III, and J. R. Horner. 1993. $^{40}Ar/^{39}Ar$ age and
correlation of the nonmarine Two Medicine Formation (Upper Cretaceous), northwestern Montana, U.S.A. *Canadian Journal of Earth
Sciences* 30: 1066–1075.

Russell, D. A. 1970. *Tyrannosaurs from the Late Cretaceous of Western
Canada*. National Museums of Canada, National Museum of Natural
Sciences, Publications in Palaeontology, no 1. Ottawa.

Trexler, D. 2001. Two Medicine Formation, Montana: Geology and fauna.
In D. H. Tanke and K. Carpenter (eds.), *Mesozoic Vertebrate Life,* pp.
298–309. Bloomington: Indiana University Press.

Varricchio, D. J. 2001. Gut contents from a Cretaceous tyrannosaurid:
Implications for theropod dinosaur digestive tracts. *Journal of Paleontology* 75: 401–406.

17. Evidence for Predator-Prey Relationships
Examples for *Allosaurus* and *Stegosaurus*

Kenneth Carpenter, Frank Sanders,
Lorrie A. McWhinney, and
Lowell Wood

Abstract

Unequivocal evidence of predator-prey relationships in the vertebrate fossil record is rare owing to the vagaries of preservation and the difficulties of interpretation. Occasionally, commutative evidence may be found that strongly implies such a relationship. Several pathological, contemporaneous specimens of the large theropod *Allosaurus* and the large stegosaur *Stegosaurus* from the Upper Jurassic Morrison Formation suggest an antagonistic relationship that is most likely between predator and prey. The specimens include a *Stegosaurus* cervical plate with a bite pattern that matches that of an *Allosaurus* mouth, an *Allosaurus* anterior caudal with a partially healed wound congruent with what a *Stegosaurus* tail-spike puncture would be expected to cause, and *Stegosaurus* spikes with broken tips exhibiting remodeling of the bone, implying that the spikes were broken well before death.

In this study, the spike force levels generated by the tail motion of *Stegosaurus* are estimated and compared to the stress levels likely to fracture the spikes. Tail spikes were susceptible to failure when imposed stresses exceeded their bending strength at impact or when shear or torsion loads exceeded spike strength. Estimates of spike penetration performance below load-induced failure levels at plausible spike-strike speeds and geometries suggest that life-threatening wounds could have been imposed on a fully grown *Allosaurus*, as well as other predators and even conspecifics. The spike-bearing tail of *Stegosaurus* thus appears to have been a formidable weapon for both offensive and defensive purposes and may have contributed to the prolonged, widespread success of the stegosaurid taxon in the presence of the allosaurid one.

Introduction

Imputing behaviors to extinct organisms is difficult, especially when such behavioral hypotheses are untestable. Thus, analogies with the behaviors of living creatures are often cited to support particular hypotheses. When such analogies are not available, inductive reasoning may be presented, although such interpretative endeavors are even more hazard prone unless tested with pertinent models. Mechanical (Farlow et al. 1976) and computer (Myhrvold and Currie 1997) models have been used to test various hypotheses about behavior in dinosaurs, although the applicability of the models, and thus their results, can be open to question (e.g., Carpenter 1998a). Even the currently highly regarded phyletic bracketing must be employed with great caution. For example, Witmer (1995) used phyletic bracketing to suggest that ornithischians did not have cheeks. Whereas it is true that many birds and crocodilians do not have cheeks, their presence in ornithischians has been argued on morphological grounds (Galton 1973), on the basis of features not seen in any extant bird or crocodile. Furthermore, fleshy cheeks are present in the extant condor (Gregory Paul, pers. comm.), thereby negating the use of phyletic bracketing in this instance.

Only when direct evidence involving a specific behavior is preserved in the fossil record can a hypothesis be considered to have been adequately tested (Boucot 1990). For example, predator-prey interaction between two dinosaurs is documented in the famous pair of "fighting dinosaurs" of Mongolia: a *Velociraptor* with the sickle claw of its left foot extended in the neck region of a *Protoceratops,* and its right ulna and radius held within the *Protoceratops*'s beak (Carpenter 1998b). The close association between the two taxa has what Boucot (1990) referred to as a Category 1 behavioral-inference reliability. Because many of Boucot's (1990) categories are specific to invertebrates, a modified classification is presented in Table 17.1 that is specific to dinosaurs.

Some behavioral activities can result in trauma-related injuries on the skeleton. For example, Cott (1961) reports high instances of jaw injuries in extant crocodilians engaged in intraspecific combat, and similar injuries are known from the crocodilian fossil record (Buffetaut 1983). Paleopathological evidence for predation (vs. scavenging) by a theropod is very rare, with only one unequivocal example known: that of a hadrosaur, *Edmontosaurus,* that survived an attack dorsally on its mid-tail by *Tyrannosaurus* (Carpenter 1998b). This example also serves as evidence of Category 4 reliability for predator-prey interaction between two antagonists during the Late Cretaceous. Pathological specimens (described below) that are even older than the *Edmontosaurus* specimen provide evidence of comparable reliability for predator-prey

TABLE 17.1

Behavioral Reliability Categories for the Dinosaurian Record (adapted from Boucot 1990)

Category	Definition	Example	Reference
1	Organisms "frozen" in their interactions	Fighting dinosaurs of Mongolia; parental care of egg clutch by *Oviraptor*	Carpenter 1998b; Clarke, J. et al. 1999
2A	Organisms in close association implying behavior	Ceratopsian bonebed imply herding	Currie and Dodson 1984
2B	Functional morphology by direct comparison of extant organisms	Biomechanics of the caudofemoralis in theropods	Gatsey 1990
3	Evidence of behavior where the identity of the maker can be inferred with certainty	Embryos within eggs indicates that *Megaloolithus* eggs were laid by sauropods	Chiappe et al. 2001
4	Trace evidence of behavior where the identity of the maker is inferred, but not established conclusively	Healed bite mark on the caudals of *Edmontosaurus* by a large theropod (probably *Tyrannosaurus*)	Carpenter 1998b
5	Behavioral interpretation based on phyletic bracketing	Parental care (of sorts) must occur in dinosaurs because of care in extant crocodiles and birds	Horner 1997
6	Behavior for which interpretation is speculative, although reasonable	Thick domes of pachycephalosaurs implies flank butting	Carpenter 1997
7	Behavior so highly speculative as to have little reliability	Tail club of *Euoplocephalus* mimics the head so as to draw predators towards the wrong end	Thulborn 1993

interactions between two Late Jurassic candidate antagonists, the large theropod *Allosaurus* and the somewhat more massive stegosaurid *Stegosaurus*. All of the specimens are from the Upper Jurassic Morrison Formation.

Institutional Abbreviations. DMNH, Denver Museum of Nature and Science (formerly Denver Museum of Natural History); UMNH, Utah Museum of Natural History; UUVP, University of Utah, Vertebrate Paleontology collection.

Paleopathological Evidence

Example 1. Allosaurus *Caudal Vertebra*

Description. A caudal vertebra of *Allosaurus fragilis* (UMNH 10781; formerly UUVP 3811) (Fig. 17.1) was collected from the Cleveland-Lloyd Quarry, Emery County, Utah. The centrum is 11.5 cm long and 11.1 cm wide, suggesting that it is from a large individual of ~11 m total length. The left transverse process has a large hole measuring about 4 × 4.1 cm; its exact shape is difficult to characterize owing to

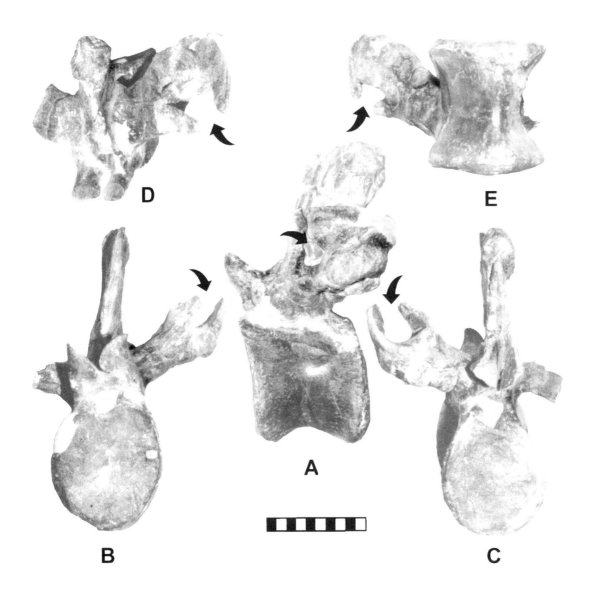

A

B

C

D

E

Figure 17.1. First caudal vertebra of Allosaurus fragilis *(UMNH 10781) showing a possible puncture in the caudal rib (arrow) in lateral (A), anterior (B), posterior (C), dorsal (D), and ventral (E) views. Scale in cm.*

pronounced dorsomedial reflection of the anterior portion of the transverse process. This hole opens into the distal edge of the process in an erupted, three-dimensional manner, generating an isthmus wound of roughly half the maximum diameter of the hole. This displaced process fragment has been frozen in its reflected position by remodeling of bone. The transverse process is extensively remodeled, especially on the ventral side as it attaches to the vertebral body. The surface of this remodeling displays a filigree pattern and a scattering of various-size matrix-filled openings. These openings appear to mark sinus tracts, for example, for the release of pus buildup within the underlying, severely distorted bone. Contralateral to this remodeling, on the ventral side of the transverse process, are several large, solid nodes formed by exuberant periosteal bone.

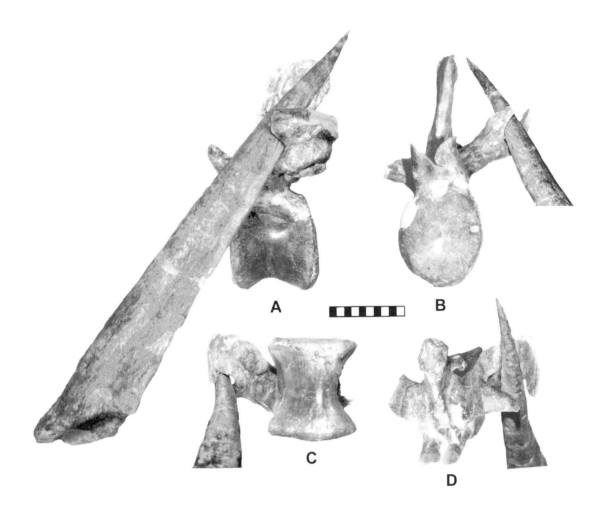

A

B

C

D

Diagnosis. The pathological hole was caused by sharp-object trauma (the imposition of a piercing or penetrating object sufficiently forcefully as to mechanically fail the bone), as attested by a fragment of the transverse process that was reflected dorsomedially while the bone was still fresh (Fig. 17.2A,B,D). In addition, there is a ~0.5-cm-diameter, conical indentation in the transverse process located proximal to the hole itself. It is interpreted as the initial impact hole generated by the spike, which was then deflected sideways ~2 cm distally where it successfully created the main hole in the impacted process. The resulting hole is too large to have been formed by the tooth of any known Jurassic theropod. Furthermore, tooth punctures into bone leave distinct crush zones that record the shape of the tooth (Erickson and Olson 1996; Erickson et al. 1996), a shape that is distinct from the rounded, quite featureless shape of the hole in the *Allosaurus* process. Remodeling of the bone indicates that the wound was not immediately lethal. However, because little bone deposition occurred within the puncture, a substantial fragment of the spike may have broken off and remained lodged in the hole, where it mechanically blockaded bone deposition; or the animal may have expired before full remodeling could occur, for

Figure 17.2. Supporting evidence that the puncture in the Allosaurus *caudal was caused by a* Stegosaurus *spike is seen by how well the spike fits within the pathology in lateral (A), anterior (B), ventral (C), and dorsal (D) views. When the spike pierced the caudal rib, it deflected a piece of bone upward as seen in B and D. Note in C the lobular periosteal reactive bone caused by bone infection and associated drainage sinuses for release of pressure caused by buildup of pus.*

A   B   C

Figure 17.3. Stegosaurus *tail spikes broken in life and showing remodeling of the broken bone surface (A, DMNH 2212; B, USNM 6646) compared to a normal spike (DMNH 1483).*

example, from the systemic effects of wound infection. A more detailed description of the pathology will be presented elsewhere (Carpenter and McWhinney, in prep.).

Example 2. Broken Stegosaurus *Spikes*

Description. Several *Stegosaurus* tail spikes are known that have incomplete distal ends (Fig. 17.3). The truncated surfaces show remodeling of bone, indicating that the spike ends were lost significantly before the corresponding animals expired (McWhinney et al. 2001). The incidence of loss is interestingly high, appearing in 10 percent of the fifty-one specimens examined (though small-number statistics should be interpreted with caution). Although most specimens exhibited loss of at most a quarter of the distal portion of the spike, one specimen showed loss of a third of the estimated spike length.

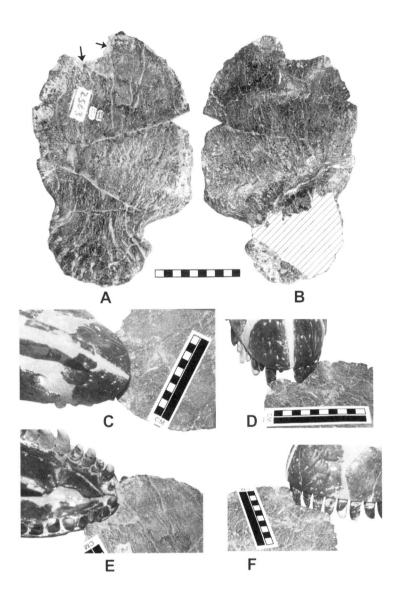

Diagnosis. The loss of the distal tip of the spike supports the hypothesis that the spikes were employed as defensive weaponry to initiate penetrating injuries that were sufficiently vigorous as to fail relatively high-strength bony structures. The remodeled surfaces show filigreed bone, rounding of the broken surfaces, and sinus tracks, all characteristic of post-traumatic chronic osteomyelitis (McWhinney et al. 2001). Estimates of the force levels necessary to fracture the spikes in this manner are presented below.

Example 3. Bitten Stegosaurus *Cervical Plate*

Description. A posterior cervical plate of *Stegosaurus*, which is characterized by the ventrolaterally flaring base (Carpenter and Miles, in prep.), shows a U-shaped notch along one of its distal edges. The plate (UMNH 5443), from the Cleveland-Lloyd Quarry, measures 28

Figure 17.4. Possible bitten cervical plate of Stegosaurus *sp. (UMNH 5543) in lateral (A) and medial (B). Note the matrix still present along the inner edge (arrows) demonstrating that this damage occurred preburial. That this damage most likely was caused by an* Allosaurus *bite is seen in the multiple views of an* Allosaurus *skull in relation to the damage (C–F). Scale in cm.*

cm in maximum height and 17.5 cm in greatest anteroposterior length. The U-shaped notch has a maximum width of 5.7 cm (Fig. 17.4A,B). That this damage preceded burial, collection, and preparation is attested to by the presence of matrix still adhering along the damaged section (Fig. 17.4A, arrows).

Diagnosis. The pattern of notching in this pathology dimensionally matches well the broad muzzle of a large *Allosaurus* (Fig. 17.4C–F), which argues for this animal's being the one that imposed this particular bite-type wound, rather than the more narrowly snouted *Ceratosaurus*, the other large theropod known from the Cleveland-Lloyd Quarry. For a variety of reasons, this plate damage is not likely to have been imposed postmortem via scavenging. First, the plates of *Stegosaurus* are believed to have consisted of skin overlaying bone, with no intervening muscle or viscera. Second, the geometric difficulty involved in accessing the plate when the *Stegosaurus* was lying dead, presumably on its side, with the distal edges of its plates drooping downward. Thus, the probable high ratio of bone to soft tissue and the noncongenial geometry suggest that the plate was likely to be of relatively low nutritive value and accessibility, further decreasing the likelihood of its being a scavenging target.

Interpretation

The examples of damage observed on several specimens of *Allosaurus* and *Stegosaurus*, both contemporaneous in the Morrison Formation, provide strong evidence for interspecific interaction between these two taxa. The large size and near-oval shape of the pathology in the caudal of the *Allosaurus* (Fig. 17.1) matches very well the cross-sectional geometry of a *Stegosaurus* spike (Fig. 17.2). On the basis of an empirically determined best fit, the spike pierced the transverse process from 58° below the horizontal plane, 33° anteriorly from the vertical transverse plane, and 10° laterally from the sagittal vertical plane. The spike apparently did not exit cleanly from the wound, as indicated by the anterior deflection of the transverse process fragment, as well as the tearing of an opening into its distal edge. This deflection of the bone may have resulted from the tip of the spike pivoting anteriorly relative to the vertebra, thus widening the hole. At this moment, the spike functioned as a first-class lever, with the motive force being provided by its inertia and by the *Stegosaurus* tail along the axis of the spike, applied at the spike base. The force applied to the *Allosaurus* transverse process was sufficient to deflect the fragment anteriorly and possibly resulted in the application of a shearing force sufficient to break off the tip of the spike. A fragment of the spike probably lodged in the wound hole because there is no evidence of bone deposition within the hole, though there is such on the outer surfaces. Such a spike fragment may have consisted primarily of the keratinous sheathing because its much lower mechanical strength (compared to the bone core of the spike) would result in its likely failure and fragmentation as the spike penetrated the process.

That *Stegosaurus* tail spikes did break off well before the deaths of the individuals is evidenced by pathological spikes that quite unequivocally exhibit remodeling of fractured tips (McWhinney et al. 2001). As noted above, such damaged spikes occur in approximately 10 percent of the total spike sample, indicating that conditions for nonlethal tip-fracturing forces acting on spikes were not rare. The most probable forces included buckling, torsion, and shear: for example, when a spike underwent an axial impact on high-strength surfaces, such as pelvic structures, it would be subjected to forces down the axis of the spike (buckling); when a spike lodged and twisted about its axis, as it might in an *Allosaurus* caudal, it would be subjected to forceful, effectively rotational motion (torsion); and when a spike forcefully but obliquely struck a high-strength skeletal feature, it would be subjected to shear. Estimates of the stress levels needed to mechanically fail a tail spike in each of these three modes are developed below. Anticipating these results, we note that although spike penetration of soft tissues might sprain the regions where spikes were attached to *Stegosaurus* tails, only spike contact with mineralized tissues having intrinsic strengths comparable to the strength of spike bone is likely to fracture the spike itself in any of these three modes.

Further evidence of *Allosaurus-Stegosaurus* interaction is provided by what we have identified as a bitten cervical plate (Fig. 17.4). The neck of prey animals is a region typically attacked by large extant carnivores because several vulnerable structures occur there, including the trachea, carotids, and spinal cord (because of the relatively short neural spines and reduced muscle and tendon masses associated with comparatively low skull masses and moments). The existence of a bite-notched cervical plate in *Stegosaurus* supports the hypothesis that *Allosaurus* attacks did indeed occur on the neck. These plates, as well as the mosaic of ossicles surfacing the ventral portion of the *Stegosaurus* neck, may have served to defeat attacks aimed at tracheal crushing (Carpenter 1998a).

Stress Estimations and Force Analyses

The punctured *Allosaurus* vertebral transverse process and broken *Stegosaurus* tail spikes raise questions about the forces that the tail of *Stegosaurus* could have generated and transmitted to its tail spikes and about whether the corresponding stress levels were sufficient to mechanically fail the spikes and the transverse process in the manners described above. To answers these questions, we undertook two analyses: the first considers the force levels that may have been generated by the motions of the tail spike (Part 1), and the second considers the stress levels needed to mechanically fail the tail spikes in the three modes sketched above, as well as the stress levels required for the observed penetration of the vertebral transverse process (Part 2).

Part 1. Stegosaurus *Tail-Force Analysis*

Although the forces generated by *Stegosaurus* tail spikes cannot be directly measured, the forces can be analytically estimated within useful bounds. Such an analysis can be used in turn to estimate the amount

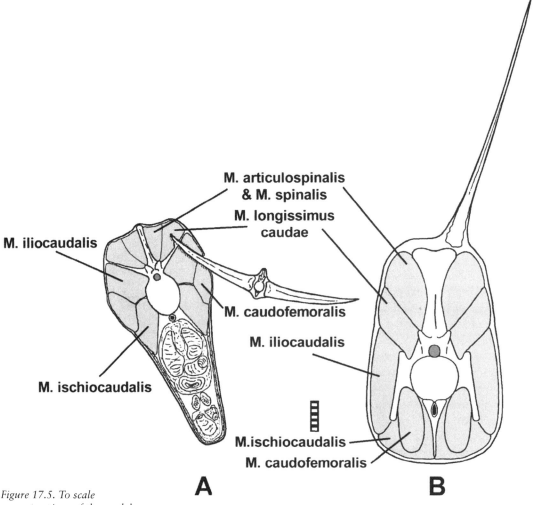

M. articulospinalis
& M. spinalis

M. longissimus
caudae

M. iliocaudalis

M. caudofemoralis

M. iliocaudalis

M. ischiocaudalis

M.ischiocaudalis

M. caudofemoralis

A   B

Figure 17.5. To scale
reconstructions of the caudal
muscles of Allosaurus (A) and
Stegosaurus (B) based on
dissections of an adult Caiman
sp. These illustrations are only
aids to understanding Allosaurus
and Stegosaurus as living
creatures. In A, based on the
position of the puncture in the
caudal rib and angle of the
puncture, the reconstruction
suggests that the M. caudofemoralis, M. iliocaudalis, and M.
longissimus caudae were pierced.
The M. caudofemoralis is shown
at its lateral position just before
extending toward the fourth
trochanter. In B, the muscle
masses used to approximate the
caudal strength in Stegosaurus
are shown. Scale in cm.

and type of damage that a *Stegosaurus* tail spike could have inflicted upon tissue and bone. This analysis is performed in two major parts. In the first part, the maximum striking force and related parameters are computed. In the second part, that force is compared to the force necessary to break two fossilized spikes that were damaged in life. The results of the analysis show that one of the spikes was probably broken as an immediate result of impact with a predator's body and that the second specimen must have been damaged as a result of having lodged in a predator's body (probably its skeleton) prior to being snapped in two as the two animals struggled to disengage themselves.

Two methods were used to calculate the force generated at the tip of the spike, and the results differ by about an order of magnitude. The more detailed approach yields the smaller value. Data for *Stegosaurus* are from a very large mounted skeleton, DMNH 1483. The spike used in the analysis has a length of 57 cm and a mid-shaft circumference of 17 cm (Fig. 17.3C).

TABLE 17.2

Estimated Range of Strength for Various Muscles in the Tail of *Stegosaurus*

Muscle	Cross-sectional area (cm²)	Strength at 39 N/cm²	Strength at 78 N/cm²
M. articulospinalis and M. spinalis	200	7,800 N	15,600 N
M. longissimus caudae	200	7,800 N	15,600 N
M. iliocaudalis	225	8,800 N	17,600 N
M. caudofemoralis	200	7,800 N	15,600 N
M. ischiocaudalis	125	4,900 N	9,800 N
Total	950	37,000 N	74,000 N

Method 1

The first method assumes that the greatest amount of movement in the tail occurs near its base (Carpenter 1998a) and therefore that the proximal caudal muscles are the most important, especially the M. iliocaudalis, but also the M. articulospinalis, M. spinalis, M. caudofemoralis, and M. ischiofemoralis (Fig. 17.5B). This interpretation is supported by the vertically tall, cleat-like caudal ribs on the first ten to eleven caudal vertebrae for the M. iliocaudalis and their minimal development on the postcaudal eleven vertebrae. These tall caudal ribs leave little room for a large M. caudofemoralis, suggesting that this muscle was reduced relative to other dinosaurs in cross-sectional area (see Fig. 17.5B). Nevertheless, the muscle also contributed some force to the lateral movement of the tail by pulling against a statically held hind leg. The inferred force generated by these muscles can be approximated from their cross-sectional area, and the estimates range from 39 to 78 N/cm² (Ikai and Fukunaga 1968), although higher extremes (98 N/cm²) have also been reported (Fick 1910). Using dissection of a modern adult caiman as a guide, the muscles were restored (Fig. 17.5B) to produce the muscle cross-sectional areas given in Table 17.2. Several of the cross-sectional areas are identical (numerical values were rounded to the nearest 5 cm²). Muscle strengths based on the cross-sectional areas are also shown for the conservative force estimates of 4 kg/cm² and 8 kg/cm².

To determine the force at the tip of the spike at the moment of impact with the *Allosaurus* tail, the spike is treated as a slender cone (~10° full-angle), with the force concentrated at its tip. Two assumptions are made about the spike: first, that the keratin covering had a conservative thickness of 3 mm and, second, that *Stegosaurus* had no method of sharpening the tip of the spike the way a cat sharpens its claws. Therefore, in our model, the tip is slightly blunt rather than tapered to a point, with a conservative surface 6 mm in diameter and an area of 28 mm². The level of stress at the spike tip available for penetrating tissues may be estimated from momentum conservation considerations. To move through a distance of 0.5 cm (taken to approximate the total cortical-bone thickness of the reference *Allosaurus* caudal vertebral transverse process) while uniformly decelerating by 10 percent, from a precontact speed V to a speed of $0.9V$, involves 20 percent of the spike's initial kinetic energy of $MV^2/2$, or $0.1MV^2$, where M is the spike's mass. This energy must be equal to a force, F, acting over a distance Dx (0.5 cm in this case): $F = (10^{-1})MV^2/Dx$, or $MV^2/5$ (we have adopted cgs units so that mass is in grams and force in dynes). The tail spike is estimated to have a mean density of 1.9 gm/cm³ (based on the density of modern bone; Currey 2002). From the dimensions given above, its estimated volume is 2,150 cm³, and its mass (sheathed with 3 mm of keratin) is about 4,000 gm. The mass of the entire end of the tail would be that of four spikes plus the connective tissue and the last few vertebrae, for a total of about 17,000 gm. The tail architecture of extant archosaurs supports tail-tip motion through a distance on the order of the animal's length within an interval on the order of 1 second, so that a characteristic maximum speed for the tail tip of a typical fully grown archosaur is on the order of the animal's length over 1 second of time (within, e.g., a factor of ~2, depending on the details of tail architecture). From this we estimate that the maximum speed, V, of an adult stegosaur's tail tip (and thus spike) likely was around 10 m/s, or 10^3 cm/s (roughly 20 mph).

We adopted this as a reference tail-tip speed and scaled our various results from it. The force estimate for slowing a 14,000-gm tail end (Tables 17.3 and 17.4) by a reference 10 percent from a reference speed of 10 m/s over a reference 0.5-cm distance then is $F = MV^2/5 = (14,000)(1,000)^2/5$, or 2.8×10^9 dynes. Applied over the reference spike-tip area of 0.28 cm², this force represents a pressure of 11.2×10^9 dynes/cm², or 11.2×10^4 N/cm² (11,200 bars, ~160,000 psi). This force level is 10–13 times that measured (Erickson et al. 1996, fig. 17.2) as being required to drive a curved conical penetrator—a shape replicated in a *Tyrannosaurus* tooth simulator—through fresh cortical bone 0.25 cm thick. It is more than enough to penetrate successively the two layers of cortical bone of roughly this thickness facing the two sides of the *Allosaurus* transverse process and to generate the observed wound (considering that the diameter of the wound in the *Allosaurus* caudal vertebra transverse process is twice that of the simulated *Tyrannosaurus* tooth-puncture hole in the bovid ilium and that the required penetration force scales linearly with wound diameter).

In summary, a small fraction (~10%) of the maximum kinetic energy of a stegosaurid tail spike moving at the reference speed of 10 m/s likely suffices to generate stress levels at the spike tip that are adequate to penetrate the *Allosaurus* caudal vertebral transverse process to the observed extent. Even if the stegosaur tail tip were moving only half as fast—5 m/s—less than half of the kinetic energy of a single tail spike would suffice to generate the stress levels sufficient to cause the observed wound in the *Allosaurus* caudal vertebral process.

Because the shear strength of living cortical bone is conservatively ~100 MPa (10^4 N/cm²; Currey 2002), the estimated impact force of the mass of even a single spike, much less the mass of the end of the tail, is

Figure 17.6. (opposite page) (A) Maximum lateral motion of the Stegosaurus *tail is constrained by the rigid plate along the dorsal surface, thus dividing the tail into a set of rigid links. The tail cannot move beyond the limits imposed by the plates, otherwise gaps (tears) would occur (arrows) in the skin. To maximize lateral tail motion and maintain large plate size (light dashed lines), the base of the plate tapers (see Carpenter 1998a for further discussion). (B) The base to the system of links is the hips. Articulation angles (θa–e) are relative to the mid-line of the pelvis for proximal segment (a) and relative to succeeding links for distal segment (b–e). The basic model is a set of five rigid links that are determined by the presence of four rigid dorsal plates and a set of tail spikes on the distal segment. Movement is constrained to a single degree of freedom, within the horizontal plane. Muscles apply motive torque at the five points of articulation along the tail. (C) Relative velocity of* Stegosaurus *tail spikes as a function of articulation angle. Maximum impact force occur when velocity is maximized at the half-angle (left or right) of articulation shown in B. (D) Growth in angular velocity (ω) along the* Stegosaurus *tail as a function of time. Adapted from Kreighbaum and Barthels 1985.*

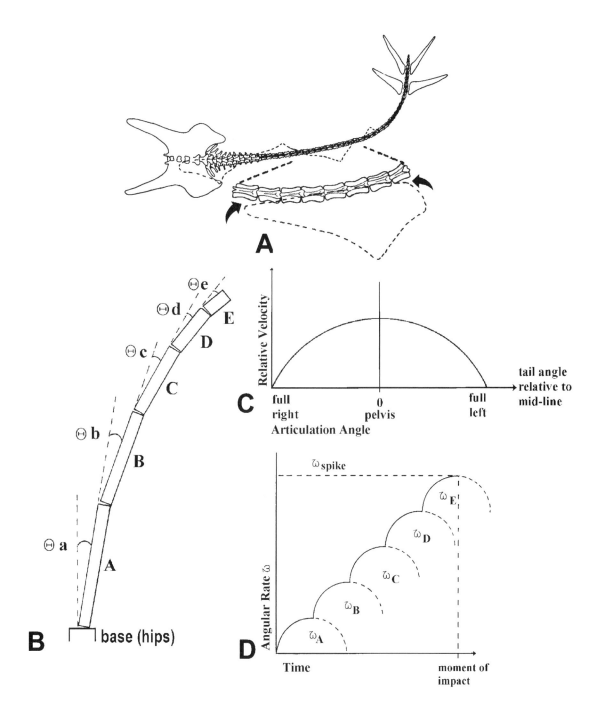

sufficient to break high-strength bone in shear. The impact-force estimate supports the basic reasonableness of our penetration estimates, as
it is hundreds of times the amount necessary to penetrate soft tissues,
and tens of times the minimum needed to crush thoracic cavities (and
fracture gastralia) if applied over non-negligible fractions of their surfaces (Glasstone and Dolan 1977). It is several times the force that will
potentially break the spike. Thus, if projected at reference speed against

likely target tissues, the "working tip" of a stegosaurid tail spike would not only penetrate and tear them, by large margins, but could also be mechanically threatened itself (if it were loaded predominantly in shear when it struck).

Method 2

The second method models the tail as a series of rigid links because of the constraints imposed by the dermal plates over large segments of the tail (Fig. 17.6A). Most movement can occur only where the plates do not overlap (Carpenter 1998a). The basic model (Fig. 17.6B) is a set of five rigid links that are determined by the presence of four rigid dorsal plates and a set of tail spikes on the distal segment. Movement is constrained to a single angle of medial-lateral movement in the horizontal plane. Muscles are assumed to apply motive torque at the five points of articulation along the tail. The maximum articulation angle available to each link is a critical parameter in the calculation of spike impact forces. *Stegosaurus* could have swung the tail laterally on both sides of the body (Fig. 17.6C); but for simplification, analysis is done with half-angles of articulation. Spike velocity would have necessarily dropped to zero at each end of a full-angle swipe (Fig. 17.6C).

Theoretical Analysis of Tail Motion. Each rigid link has rotational movement about its proximal point of articulation. The rotational equations of motion for each link are

$$L = I\omega, \text{ (angular momentum) (1a)}$$
$$T = I\dot{\omega} = |\vec{r} \times \vec{F}|, \text{ (torque) (1b)}$$

and

$$KE = \tfrac{1}{2}I\omega^2 \text{ (kinetic energy), (1c)}$$

where
I = moment of inertia (kg·m²);
ω = the link's rate of rotation (rad/s);
$\vec{r}$ = half-width at the base of each link; and
$\vec{F}$ = muscle force exerted at the base of each link.

For each link,

$$I = \int_{L1}^{L2} x^2 \, \rho dx = x^3 \Big|_{L1}^{L2} \left(\frac{\rho}{3}\right) = \frac{(L2^3 - L1^3)}{3}\rho \ , (2)$$

where ρ is the segment's average mass density per unit length; $L1$ and $L2$ are the distances from its proximal and distal ends, respectively, to the base of the tail; and x is the variable of integration between $L1$ and $L2$ along the link. Plate mass and moment are considered to be negligible compared to the mass and moment of other link tissues and bone.

Note that the angular momentum and kinetic energy (Eqs. 1a, 1c) of a link are measures of its state of motion. But the motive torque, $T = |r \times F|$ (Eq. 1b), is derivable from the tail's physical characteristics and the performance of modern muscle. Torque is generated at the proximal end of each link by muscle contraction. The torque available at the proximal end of each link is

$$T = (A_{\text{cross-section}}) \times (F_{\text{muscle}}) \times \left(\frac{W_{\text{link}}}{2}\right), \quad (3)$$

where $A_{\text{cross-section}}$ is the total area of cross-sectional muscle, F_{muscle} is the force the muscle can exert per unit cross-sectional area, and W_{link} is the width of the link. The factor of $1/2$ in Eq. 3 is required because only half of the muscles at the base of each link (those on the side toward which the tail is swinging) can apply pulling (torquing) force. The other half must elongate and can apply only minor torque that stabilizes, controls, or both.

The impact force generated by a tail spike is related to the spike's velocity during the impact and to the mass of the link to which the spike is attached. Other physical parameters related to the impact, such as maximum pressure applied, are also derivable. The key to deriving all these quantities is to determine the angular rate of movement, ω, of the spike at the moment of impact. As the *Stegosaurus* tail swings laterally, ω should add from one link to the next as shown in Figure 17.3.

The angular rate of the tail spike equals the sum of the maximum angular rates of each link in Figure 17.1:

$$\omega_{\text{spike}} = \omega_A + \omega_B + \omega_C + \omega_D + \omega_E. \quad (4)$$

Each ω must in turn be related to the physically measurable or derivable quantities of I, T, and the angle through which the corresponding link moves, θ. Conversion to these quantities is done mathematically as follows:

$$\omega = \dot{\theta} = \int \dot{\omega} dt = \dot{\omega} t \quad (5a)$$

$$\theta = \iint \dot{\omega} dt = \int \dot{\omega} t = \frac{\dot{\omega} t^2}{2} = \frac{\omega t}{2}. \quad (5b)$$

From Eq. 5b, ω is related to θ and to time, t, as

$$\omega = \frac{2\theta}{t} = \dot{\omega} t. \quad (5c)$$

Rearranging algebraically yields

$$t^2 = \frac{2\theta}{\dot{\omega}}. \quad (5d)$$

Substituting for ω (from Eq. 1b) yields

$$t^2 = \frac{2\theta I}{T} \quad (5e)$$

and

$$t = \sqrt{\frac{2\theta I}{T}}. \quad (5f)$$

Substituting Eq. 5f for t into Eq. 5c yields ω in terms of θ, T, and I:

$$\omega = \sqrt{\frac{2\theta T}{I}}. \quad (5g)$$

Equation 5g can be used to represent the individual terms in Eq. 4 as follows:

$$\omega_{\text{spike}} = \sqrt{\frac{2\theta_A T_A}{I_{\text{tail}}}} + \sqrt{\frac{2\theta_B T_B}{I_{\text{tail-A}}}} + \sqrt{\frac{2\theta_C T_C}{I_{\text{tail-A-B}}}} + \sqrt{\frac{2\theta_D T_D}{I_{\text{tail-A-B-C}}}} + \sqrt{\frac{2\theta_E T_E}{I_{\text{tail-A-B-C-D}}}} \quad (6)$$

TABLE 17.3.

Measured Values for *Stegosaurus stenops* Tail (DMNH 1438)

Link	Length (cm)	Height (cm)	Width (cm)	Plate base length (cm)	Plate height (cm)	Maximum width at base (cm)	Half-angle of articulation (deg, radians)
A	100	50	18	70	75	5	4, 0.070
B	63	32	13	64	64	6	2.3, 0.040
C	41	26	9	34	31	4	2.8, 0.049
D	37	20	7.5	28	24	5	3.1, 0.054
E	69	11.5	7	9*	50*	n/a	1, 0.017

*For Link E, the plate base length and plate height parameters refer to the base diameter and length, respectively, of a spike.

The moments of inertia range from that of the entire tail in the first term to that of the most distal link in the last term.

The value of Eq. 6 for the spike's angular rate is converted into force by first converting it to linear spike velocity, V_{spike}:

$$V_{spike} = (L_{tail} \cdot \omega_{spike}), \quad (7)$$

where L_{tail} is the length of the tail from the proximal end of link A to the distal end of the spike.

Assuming that the spike halts inside the target tissue and bone, the impulse delivered by the spike, P_{spike}, is

$$P_{spike} = (m_E \cdot V_{spike}), \quad (8)$$

where m_E is the mass of the distal tail link. The pressure exerted by the spike on the target is this impulse divided by the surface area of the spike tip:

$$Pressure_{spike} = \frac{P_{spike}}{A_{spike\ tip}} . \quad (9)$$

Measurements and Analysis of Stegosaurus stenops *Tail.* The measurements, taken at the distal end of each link in the *Stegosaurus* tail, included the height and width of the vertebral structures (including transverse processes, chevrons, and neural spines) and the half-angles of articulation (Fig. 17.6B). The articulation half-angles were directly measurable because the tail is mounted nearly at its maximum possible articulation. Dorsal plates were measured for base length, height, and maximum width at the base. Spikes were measured for length and base diameter. Measurements are summarized in Table 17.3.

The cross-sectional area of each link was calculated as the arithmetic mean of the ellipsoidal and parallelogram areas implied by the heights (H) and widths (W) in Table 17.3. Ellipsoidal area is given by

$$A_{ellipsoidal} = \pi \left(\frac{\left(\frac{H}{2}\right)^2 + \left(\frac{W}{2}\right)^2}{2} \right) \quad (10)$$

and parallelogram area by

TABLE 17.4

Summary of Volume and Mass for *Stegosaurus stenops* Tail (DMNH 1438)

Link	Cross-sectional area(cm²)	Bone (ρ = 1.98) to muscle (ρ = 1.0) ratio	Nonplate mass per unit length (gm/cm)	Total cross-section muscle area(cm²)	Link length (cm)	Cross-sectional unit length mass × length(kg)	Plate volume (cm³)	Plate volume × bone density = mass(kg)	Total link mass(kg)	Total mass per unit length (gm/cm)
A	780	1:3	971	585	100	97	4,375	8.7	106	1,060
B	338	1:2	448	225	63	28	4,096	8.1	36	571
C	207	1:1	308	104	41	12.6	703	1.4	14	341
D	127	1:1	189	64	37	7.0	560	1.1	8	216
E	56	1:1	83.4	28	69	5.8	4,240*	8.4*	14	203

*The volume of a single spike (1,060 cm³) and the mass of a single spike are multiplied by four to yield total spike volume (4,240 cm³) and total spike mass (8.4 kg).

TABLE 17.5

Link Torques and Moments of Inertia

Link	Muscle cross-sectional area(cm²)	Force range (half of muscle cross-section multiplied by 39 N/cm² to 78 N/cm² (N)	Link half-width, r (m)	Torque at base of link $\|r \times F\|$ (N·m)
A	585	11,400–22,800	0.09	1,000–2,000
B	225	4,400–8,800	0.065	285–570
C	104	2,000–4,100	0.045	91–182
D	64	1,250–2500	0.0375	47–94
E	28	550–1100	0.035	19–38

TABLE 17.6

Link Moments of Inertia

Link	Unit length density (Table 17.4) (gm/cm)	I_{tail} (kg·m²)	I_{tail-A} (kg·m²)	$I_{tail-A-B}$ (kg·m²)	$I_{tail-A-B-C}$ (kg·m²)	$I_{tail-A-B-C-D}$ (kg·m²)
A	1,060	35	n/a	n/a	n/a	n/a
B	571	63	4.8	n/a	n/a	n/a
C	341	47	9.9	0.78	n/a	n/a
D	216	40	12	2.9	0.36	n/a
E	203	107	44	18	7.7	2.2

$$A_{parallelogram} = \left(\frac{H \cdot W}{2} \right) . (11)$$

Therefore, the mean cross-sectional area is each tail link is estimated to be

$$A_{cross-sectional} = \left(\frac{A_{ellipsoidal} + A_{parallelogram}}{2} \right). (12)$$

Plate volume is estimated to be the area of the equivalent triangle multiplied by $1/3$ the base width:

$$Volume_{plate} = \frac{Width}{3} \left(\frac{Base \cdot Height}{2} \right) = \frac{1}{6} \left(Base \cdot Height \cdot Width \right), (13)$$

and spike volume is derived from the formula for the volume of a cone:

$$Volume_{spike} = \frac{1}{3} \pi \left(\frac{d}{2} \right)^2 L, (14)$$

where d is the base diameter and L is the length. Masses were calculated by multiplying the average tissue and bone density of each object with the volumes. The density of tissue is 1 gm/cm³, and the density of bone is 1.98 gm/cm³. Table 17.4 summarizes these values for volume and mass. Table 17.5 summarizes the calculation of motive torque (Eq. 3) and moment of inertia (Eq. 2) for each link. Based on modern muscle

TABLE 17.7

Link Angular Rates for Equation 6

Link	Torque at base of link sequence (Table 17.5) (N·m)	Cumulative moment of inertia (sum of columns in Table 17.6) (kg·θ·m²)	Half-angle of articulation (Table 17.4) (rad)	ω term (Eq. 6) (rad/s)	ω term (Eq. 6) (deg/s)
A+B+C+D+E	1,000–2,000	290	0.070	0.70–0.98	40–56
B+C+D+E	290–570	71	0.040	0.57–0.80	33–46
C+D+E	91–180	22	0.049	0.64–0.90	37–52
D+E	47–94	8.1	0.054	0.79–1.1	45–63
E	19–38	2.2	0.017	0.54–0.76	31–43

performance, the force exerted by muscles could range between 39 N/cm² and 78 N/cm². This performance range provides lower and upper bounds for this analysis.

Table 17.6 contains individual angular rates and the calculated moment of inertia terms that appear in Eq. 6. Note that the moment of inertia of the distal link is large because of the presence of the four spikes. This large moment of inertia produces a telling effect on the target.

Calculated Maximum Performance of the Tail in Stegosaurus. The sum in Eq. 6 is the sum of the ω values in Table 17.7. The range is 3.2–4.5 rad/s, depending upon the muscle force that can be brought to bear (as noted above). The total tail length is 3.1 m, but the last spike is 39 cm short of the end. Therefore the spike is 2.71 m from the proximal end of link A. From Eq. 7, the spike tip velocity is (assuming a spike radius of 2.71 m from the proximal end of link A):

$$(\omega_{spike} \cdot L_{spike}) = (3.2 - 4.5 \text{rad/s}) \cdot (2.71\text{m}) = 8.7 - 12\text{m/s}. \quad (15)$$

In English units, the spike velocity range is 20–27 mph. This value brackets the estimate of 10 m/s (~20 mph) determined by Method 1, above.

Using a conservative value of 14 kg for the mass of the end of the stegosaur tail (Table 17.4), the impulse delivered to the target is given by Eq. 8:

$$P_{spike} = (14 \text{ kg}) \cdot (8.7 - 12\text{m/s}) = 120 - 170 \frac{\text{kg} \cdot \text{m}}{\text{s}}. \quad (16)$$

If the stopping time in the target tissue and bone is (conservatively) estimated to be about $\frac{1}{3}$ s, then the maximum force, F, exerted on the target (the impulse given in Eq. 16 divided by this interval) is

$$F_{max} = \frac{P_{spike}}{t_{stopping}} = \frac{120 - 170 \frac{\text{kg} \cdot \text{m}}{\text{s}}}{\frac{1}{3}\text{s}} \approx 360 - 510 \text{ newtons}. \quad (17)$$

The minimum calculated impact force of 360 N is more than adequate to damage bone and tissue struck by the *Stegosaurus* spike (see discussion under Method 1).

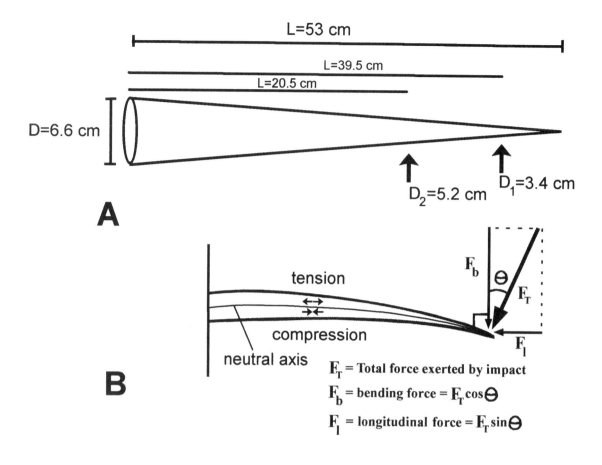

F_T = Total force exerted by impact

F_b = bending force = $F_T\cos\Theta$

F_l = longitudinal force = $F_T\sin\Theta$

Figure 17.7. (A) Schematic illustration of a Stegosaurus *spike and values used to determine breakage at D_1 (based on USNM 6646) and D_2 (DMNH 2818). (B) Schematic of a* Stegosaurus *spike as a non-uniform cantilever with forces applied to it. The total impact force on the spike, F_T, is reduced to two components, one parallel to the spike's longitudinal axis and the second perpendicular to the first. The perpendicular component, F_b, tends to bend the spike through tension and compression on either side of a force-neutral central axis. If these stresses exceed the strength of the bone at some location along the spike, then the bone will break. The impulsive force application causes the bone to behave with only about 80 percent of the strength it would have if the force were applied gradually.*

Both methods of analysis for *Stegosaurus* indicate that the tail spikes could generate more than sufficient force to puncture the *Allosaurus* vertebra.

Part 2. Stegosaurus Spike Breakage Force Analysis

In the analysis of *Stegosaurus* spike pathologies, most specimens showed breakage near their tips, but one showed loss of the distal third (McWhinney et al. 2001). How much force was needed to cause these breaks, and are the results within the range of force the tail is estimated to have generated?

The breakage pattern of all the spikes indicates loading perpendicular to the long axis of the spike (Fig. 17.7). The broken surface is not long and oblique as would be seen when if loading were parallel to the axis (Currey 2002). Breakage could occur when the spike hit a hard object (e.g., bone) at an angle roughly perpendicular to the longitudinal axis of the spike, or when the spike had already penetrated deeply into the target body and torsion was applied by the struggling animal. Regardless, for the spike, the direction of forces is the same.

The spike is treated as a cantilevered beam with conical shape (of length L and base diameter D), as shown in Figure 17.7. (Actual spike cross-sections are slightly flattened, but the circular approximation simplifies the analysis and provides no significant change in the final result.) The impact force, F, causes tension stress along one side and compression stress on the opposite side (Fig. 17.7B). These forces balance along the spike's center axis, or neutral line. The stress level varies along the length of the spike, increasing from the base to the tip as described below. If at any point the stress exceeds the strength of the bone, then the spike will break at that point.

The force vector that tends to bend the spike along its length (and which ultimately may break it), denoted F_b, is somewhat less than the total striking force, F_T. The two components are related by the cosine of the angular difference, θ, between the impact angle of F_T and $90°$ (which is the perpendicular to the spike's long axis):

$$F_b = F_T \cos\theta. \quad (18)$$

Although F_b is expected to be somewhat less than F_T, the swept-back angle of *Stegosaurus* spikes (about $65°$ relative to the tail centerline as shown in Figure 17.5) results in an impact angle of $65°$ for the case in which the attacker's body is parallel to the *Stegosaurus* body and the *Stegosaurus* tail hits with maximum force, just as it crosses the centerline of the *Stegosaurus* hips. This geometry suggests that *Stegosaurus* spikes must have ordinarily been used more for slashing than for spearing or stabbing. For this geometry,

$$\cos\theta = \cos(90°-65°) = \cos(25°) \approx 0.906. \quad (19)$$

Because the cosine of the difference angle ($25°$) is nearly unity, it is apparent that for this geometry F_b and F_T can be considered to be nearly equal. (At the end of this analysis, the factor of 0.906 in Eq. 19 will be taken into account.) In the remainder of this analysis, the force term that tends to bend the spike is thus rendered simply as F and is assumed to be nearly equal to the impact force of the spike. Also note that a spike might not be broken upon impact and might instead become embedded in target tissue or bone. In that case, the wrenching force exerted by the exertions of the two animals is likewise expected to be exerted at roughly right angles to the spike's longitudinal axis and can also be rendered simply as F.

Referring again to Figure 17.7, the diameter, d, of the spike at any length, l, measured from the base is

$$d(l) = D\left(\frac{L-l}{L}\right) = D\left(1-\frac{l}{L}\right), \quad (20)$$

where
D = spike diameter at the base; and
L = total spike length.
The moment of inertia, I, about any point is (from Eq. 2):

$$I = \frac{\pi d^4}{64} = \frac{\pi D^4 (1-\frac{l}{L})^4}{64}. \quad (21)$$

The section modulus, Ω, is (from Eq. 2):

$$\Omega = \frac{2I}{d} = \frac{\pi d^3}{32} = \frac{\pi D^3}{32}\left(1-\frac{l}{L}\right)^3 = \frac{\pi D^3}{32} \bullet \frac{(L-l)^3}{L^3}. \quad (22)$$

TABLE 17.8

Force Necessary to Cause Spike Failure at Observed Lengths for Specimens 1 and 2

The correction in the last column is obtained from Equations 18 and 19.

Specimen number	$\kappa(m^2)$	$S_{critical}$ (Mpa)	$F_{applied} = \kappa S_{critical}$ (N)	$F_{total} = F_{applied} / 0.906$ (N)
1 (static loading)	3.46×10^{-6}	100 (static)	346	382
1 (dynamic loading)	3.46×10^{-6}	80 (dynamic)	276	305
2 (static loading)	2.00×10^{-5}	85 (static)	1700	1880
2 (dynamic loading)	2.00×10^{-5}	68 (dynamic)	1360	1500

Stress, S, is equal to the torque applied at any distance from the tip, $F(L–l)$, divided by the section modulus:

$$S = \frac{F(L-l)}{\Omega}. \quad (23)$$

Substituting for Ω from Eq. 22 yields

$$S = \frac{32FL^3}{\pi D^3 (L-l)^2} = \frac{C}{(L-l)^2}, \quad (24)$$

where C is a constant term, $32FL^3/\pi D^3$. Equation 24 shows that the stress is minimized at the base and increases toward the tip. Dividing the constant term, which has units of force, by the square of the distance from the base, $(L–l)^2$, provides stress (force per unit area).

The spike will fail mechanically (break) at the place where the stress equals or exceeds a critical value, $S_{critical}$ (see Fig. 17.9). This will occur at a corresponding location designated as l_{break} (Fig. 17.9 points A and B):

$$S_{critical} = \frac{32FL^3}{\pi D^3 (L-l_{break})^2}. \quad (25)$$

Solving this equation for force yields

$$F = \frac{\pi D^3 (L-l_{break})^2 S_{critical}}{32L^3} = \kappa S_{critical}, \quad (26)$$

where $\kappa = \pi D^3 (L-l_{break})^2/(32L^3)$ and has units of area. Measurements of unbroken fossilized spikes provide nominal values for D and L of 0.066 m and 0.53 m, respectively. For the two broken spikes, designated as specimens 1 and 2 with corresponding subscripts, the values of l_{break} are 0.395 m and 0.205 m, respectively. The constant κ thus has a unique value for each of these specimens:

$$\kappa_1 = 1.896 \cdot 10^{-4} \cdot (0.53-0.395)^2 = 3.46 \cdot 10^{-6} m^2 \quad (27a)$$
$$\kappa_2 = 1.896 \cdot 10^{-4} \cdot (0.53-0.205)^2 = 2.00 \cdot 10^{-5} m^2. \quad (27b)$$

Finally, the value of $S_{critical}$ must be estimated for each of the two broken specimens. The porosity of specimen 1 at its break point is 11 percent, and that of specimen 2 is 16 percent (porosity is estimated from

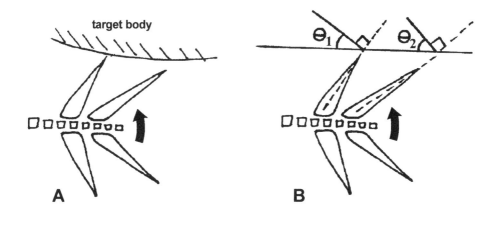

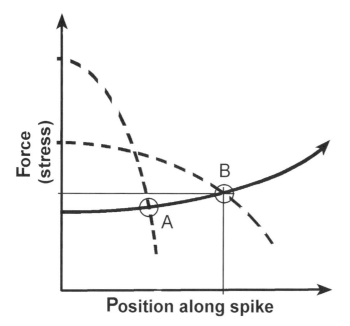

Figure 17.8. (above) The spike impact geometry is idealized from an actual impact. (A) A case in which a predator's body is parallel to the Stegosaurus. The Stegosaurus tail strikes the target body with maximum possible force, just as it crosses the animal's medial axis. (B) The geometry is simplified. The angle θ between the total impact force and the component that may break the spike (see Fig. 17.9) varies depending upon whether a spike is anterior or posterior and upon the exact orientations of the animals' bodies; the angle is on the order of 25°.

the normal spike, DMNH 1483). Based upon the characteristics of modern bone, the static failure stress values for these porosities are as shown in Table 17.8. Failure under dynamic loading (which is more realistic for a *Stegosaurus* spike impact) is expected to occur at about 80 percent of these values, as noted in Table 17.8. In Table 17.8, the necessary applied force to cause mechanical spike failure at the observed lengths is computed from Eqs. 26 and 27.

The total dynamically applied force required to break the first specimen (Figs. 17.3B, 15.7A) at the observed length of 39.5 cm from the base (with the observed 11 percent porosity value at that point) is 305 N. This is well within the minimum available impact force of 360 N computed above. Even the statically applied force that would be required is close to that limit.

Figure 17.9. Bending stresses exerted on spike of length L at the moment of impact as a function of position along the length of the bone. Minimum stress value C (described in text) occurs at the base and increases as the square of the distance from the base. The bone will break at position A or B, which corresponds to l_{break} for the specimens; this is where the stress level equals $S_{critical}$. The curve showing stress required to break the spike is schematic only.

For the second specimen, with the observed value of 16 percent porosity at its break point of only 20.5 cm from the base (Figs. 17.3A, 17.7A), the dynamically applied force at the tip would have to be about 1500 N. This is three times the 510-N maximum impact force limit computed above, implying that the mechanism of direct impact could not have caused this spike to break at the observed location.

The break observed on specimen 1 could have been caused by a slashing impact on a predator's body (fig. 17.8). The required force value of 305 N for dynamic force required to generate the break compares favorably with the minimum computed impact force of 360 N. Therefore, the damage observed in specimen 1 was probably caused by an impact at slightly less than the minimum theoretical force, in a geometry where the impact force was nearly at right angles to the spike's longitudinal axis (fig. 17.9). This further implies that the *Stegosaurus* and its predator were roughly parallel to each other when the impact occurred, and that the *Stegosaurus* tail struck when it was approximately aligned on the medial line of the animal's body.

The break observed on specimen 2, which at 1500 N of dynamic loading required three times the maximum possible computed force of 510 N, could not have been generated by spike impact. The only remaining likely possibility is that this spike struck not with the slashing action of specimen 1 but rather with a stabbing action in which the main force of impact was directed along the spike's longitudinal axis and in which the spike ultimately lodged somewhere in the predator's skeleton. When the two animals twisted apart, the spike would likely have been broken at the high force level we have computed. Evidence that such events did occur is seen by the pierced *Allosaurus* vertebra (Fig. 17.1) and a possible puncture in an *Allosaurus* pubis (Bruce Rothschild, pers. comm.).

Conclusions

Antagonistic interaction between *Allosaurus*, a predator, and *Stegosaurus*, a potential prey animal, has long been hypothesized, but only recently has evidence for this interaction been found. A bitten cervical plate of *Stegosaurus* shows that *Allosaurus* did attack *Stegosaurus* and, in at least one instance, did target the neck. Furthermore, a punctured caudal vertebra of *Allosaurus* and broken tail spikes show that *Stegosaurus* did use its spikes as weapons against *Allosaurus*. These occurrences demonstrate that there were frequent interactions between these two antagonists, probably as predator and prey.

The two broken spikes, specimens 1 and 2, apparently represent two ways in which *Stegosaurus* spikes were used as weapons. The first injury, caused by the slashing-action impact when the predator attempted to attack with its body parallel to the *Stegosaurus*, might be expected to have occurred more commonly. The second type of injury, caused when the spike piercing the tissue of the attacker, would be less likely, owing to the more difficult geometry of impact. But it would potentially cause the most serious injuries for both predator and prey because the spearing action could deeply embed the spike somewhere in the predator's skeleton. The resulting break when the two animals disengaged would have left the predator with a large piece of spike stuck in its body and the *Stegosaurus* with a gaping wound on its tail, because the spike would have broken close to the base. Secondary infections in both animals would likely have ultimately killed both of them, if they survived their immediate encounter.

Both of the force values estimated for breaking the spikes are well within the range of computed values that could be exerted during the impact of a *Stegosaurus* spike into a target. This analysis and the observed breakage *Stegosaurus* spikes indicate that the problem with the use of these defensive weapons was not in generating enough force to penetrate the tissues of a predator but rather in restricting the off-axis impact forces to values small enough to prevent spike breakage, either on impact or subsequent to impact as the spikes were wrenched out of predators' soft tissues or bony parts.

Acknowledgments. We thank Kenneth Stadman, Earth Science Museum, Brigham Young University, Provo, Utah, and Mike Getty and Scott Sampson, Utah Museum of Natural History, Salt Lake City, Utah, for access to the *Allosaurus* specimen. We also thank Thomas Barsch, M.D., Radiology Department, Kaiser Permanente, Denver, Colorado, for discussions on the reactive bone pathology and Bruce Rothschild for discussions about the *Allosaurus* pathology. Finally, special thanks to David Fyhrie, Galateia Kazakia, John Cotton, James Funk, and John Currey for discussions about how to solve the problem of *Stegosaurus* tail spike breakage at different points.

References Cited

Boucot, A. J. 1990. *Evolutionary Paleobiology of Behavior and Coevolution.* Amsterdam: Elsevier Publishers.

Buffetaut, E. 1983. Wounds on the jaw of an Eocene mesosuchian crocodilian as possible evidence for the antiquity of crocodilian intraspecific fighting behaviour. *Paläontologische Zeitschrift* 57: 143–145.

Carpenter, K. 1997. Agonistic behavior in pachycephalosaurs (Ornithischia: Dinosauria): A new look at head-butting behavior. *Contributions to Geology, University of Wyoming* 32: 19–25.

Carpenter, K. 1998a. Armor of *Stegosaurus stenops,* and the taphonomic history of a new specimen from Garden Park, Colorado. In K. Carpenter, D. Chure, and J. I. Kirkland (eds.), *The Morrison Formation: An Interdisciplinary Study. Modern Geology* 23: 127–144.

Carpenter, K. 1998b. Evidence of predatory behavior by carnivorous dinosaurs. In B. P. Pérez-Moreno, T. Holtz Jr., J. L. Sanz, and J. Moratalla (eds.), *Gaia: Aspects of Theropod Paleobiology,* vol. 15: 135–144. Lisbon: Museu Nacional de História Natural.

Chiappe, L. M., L. Salgado, and R. A. Coria. 2001. Embryonic skulls of titanosaur sauropod dinosaurs. *Science* 293: 2444–2446.

Clarke, J. M., M. A. Norell, and L. M. Chiappe. 1999. An oviraptorid skeleton from the late Cretaceous of Ukhaa Tolgod, Mongolia, preserved in an avian-like brooding position over an oviraptorid nest. *American Museum Novitates,* no. 3265: 1–36.

Cott, H. B. 1961. Scientific results of an inquiry into the ecology and economic status of the Nile Crocodile (*Crocodilus niloticus*) in Uganda and Northern Rhodesia. *Transactions of the Zoological Society of London* 29: 211–357.

Currey, J. D. 2002. *Bones: Structure and Mechanics*. Princeton, N.J.: University of Princeton Press.

Currie, P. J., and P. Dodson. 1984. Mass death of a herd of ceratopsian dinosaurs. In W.-E. Reif and F. Westphal (eds.), *Third Symposium on Mesozoic Terrestrial Ecosystems: Short Papers*, pp. 61–66. Tübingen: Attempto Verlag.

Erickson, G. M., and K. H. Olson. 1996. Bite marks attributable to *Tyrannosaurus rex:* Preliminary description and implications. *Journal of Vertebrate Paleontology* 16: 175–178.

Erickson, G. M., S. D. Van Kirk, J. Su, M. E. Levenston, W. E. Caler, and D. R. Carter. 1996. Bite-force estimation for *Tyrannosaurus rex* from tooth-marked bones. *Nature* 382: 706–708.

Farlow, J., C. Thompson, and D. Rosner. 1976. Plates of the dinosaur *Stegosaurus:* Forced convection heat loss fins? *Science* 192: 1123–1125.

Fick, R. 1910. *Handbuch der Anatomie und Meckanik der Gelenke Unter Berücksichtigung der bewegenden Muskeln*. Jena: Fischer.

Galton, P. M. 1973. The cheeks of ornithischian dinosaurs. *Lethaia* 6: 67–89.

Gatsey, S. M. 1990. Caudofemoral musculature and the evolution of theropod locomotion. *Paleobiology* 16: 170–186.

Glasstone, S., and P. J. Dolan. 1977. *The Effects of Nuclear Weapons*. Washington, D.C.: U.S. Department of Defense and U.S. Department of Energy.

Horner, J. R. 1997. Behavior. In P. J. Currie and K. Padian (eds.), *Encyclopedia of Dinosaurs*, pp. 45–50. San Diego: Academic Press.

Ikai, M., and T. Fukunaga. 1968. Calculation of muscle strength per unit cross-sectional area of human muscle by means of ultrasonic measurement. *Internationale Zeitschrift für Angewandte Physiologie Einschlägig Arbeitsphysiologie* 26: 26–32.

Kreighbaum, E., and K. Barthels. 1985. *Biomechanics: A Qualitative Approach for Studying Human Movement*. Minneapolis: Burgess Publishing Co.

McWhinney, L., B. Rothschild, and K. Carpenter. 2001. Posttraumatic chronic osteomyelitis in *Stegosaurus* dermal spikes. In K. Carpenter (ed.), *The Armored Dinosaurs*, pp. 141–156. Bloomington: Indiana University Press.

Myhrvold, N., and P. J. Currie. 1997. Supersonic sauropods? Tail dynamics in the diplodocids. *Paleobiology* 23: 393–409.

Thulborn, T. 1993. Mimicry in ankylosaurid dinosaurs. *Records of the South Australian Museum* 27: 151–158.

Witmer, L. M. 1995. The extant phylogenetic bracket and the importance of reconstructing soft tissues in fossils. In J. J. Thomason (ed.), *Functional Morphology in Vertebrate Paleontology*. Cambridge: Cambridge University Press.

18. Theropod Paleopathology
State-of-the-Art Review

Bruce Rothschild and
Darren H. Tanke

Abstract

Paleopathology and theropods have a long literary association, even preceding Sir Richard Owen's dinosaur appellation. Although the earliest recognition of pathology cannot be further characterized because specimens were lost during World War II, surviving specimens evidence altered gait (derived from study of trackways) and specific pathologies. The latter include injury-related pathologies (fractures, exostoses, stress fractures, bites, infection), congenital pathology (abnormal teeth and thoracic vertebral fusion), and arthritis (gout, but not osteoarthritis). The frequency of activity-related pathologies portrays theropods as "enthusiastic" in their daily activities. Speculation about a possible *Allosaurus* tumor proved premature, and the specimen actually was a humerus with an infected fracture. This catalogue of known theropod paleopathology indicates the rarity of all but tooth and trauma- or injury-related pathology and identifies those areas as fruitful for future study of their population frequency (paleoepidemiology).

Indeterminate Evidence

The history of pathology recognition in theropods is traceable to *Poekilopleuron* in 1838 (Eudes-Deslongchamps 1838), three years prior even to the naming of dinosaurs as such by Sir Richard Owen (Rothschild and Martin 1993). Fusion of chevrons to proximal caudal

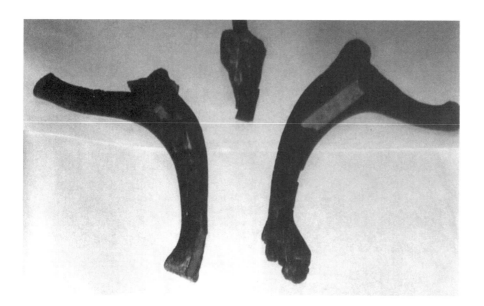

Figure 18.1. Allosaurus *DMNH 2149 thoracic rib fractures, including one (right) with a large callus formation and pseudoarthrosis.*

vertebrae and an abnormal proximal pedal phalangeal surface were reported by Tanke and Rothschild (2002). The former was more likely either congenital or trauma-related. Identifying the cause of the phalangeal surface phenomenon (specimen lost in World War II) requires consideration of infection and perhaps some form of arthritis. Unfortunately, there are two hundred varieties of arthritis, of which at least forty affect bone. Even if the specimen were available, the cause of an abnormality in an isolated joint surface cannot usually be identified (Rothschild and Martin 1993). Similarly, reports of pathologic humeri in *Daspletosaurus* and *Gorgosaurus* (Russell 1970) provide inadequate information to allow identification of their cause (diagnosis) (Tanke and Rothschild 2002).

Indirect Evidence

Footprint and trackway evidence of limping theropods (Dantas et al. 1995; Ishigaki 1986b) suggests injury or arthritis, but the published record does not yet allow frequency assessment. Examples include *Anchisauripus, Eubrontes,* and *Sauroidichnites;* the latter is even given the species name *abnormis* because of an abnormally positioned toe. Gait analysis may provide insights to the significance of or interference caused by limping in theropod activity. The report of Ishigaki (1986a) suggests an injured toe but does not provide the necessary comparative spacing information for analysis of biomechanical effect on gait (Tanke and Rothschild 2002).

Direct Evidence

Injuries. Trackway signs of abnormal gait suggest the likelihood of injuries that will leave direct bony evidence—and such evidence has

TABLE 18.1

Fracture Distribution in Theropods*

Dentary	1
Scapula	3
Humerus	3
Radius	1
Ulna	1
Manus phalanx	3
Neural spine	3
Cervical rib	2
Caudal vertebra	2
Thoracic rib	14
Gastralia	5
Tibia	1
Fibula	5**
Metatarsal	2
Pedal phalanx	4

* Data from Anonymous 1997a,b; Currie 1997; Hanna 2000; Harris 1997;
Lambe 1917; Larson 2001; Madsen 1976; Molnar 2001; Poling 1996;
Psihoyos 1994; Rothschild 1997; Rothschild 1999; Stromer 1915; Tanke 1996;
Tanke and Currie 1998.
**Exclusive of Tanke's 1996 data.

indeed been found. Direct fracture-related pathologies, stress fractures, and bite marks evidence an "enthusiastic animal" pursuing life with gusto.

Fractures. Unfortunately, the literature essentially contains isolated reports of fractures, without providing a denominator (minimum number of individuals; Table 18.1) (Tanke and Rothschild 2002). Thus the pertinent question of frequency usually cannot be addressed. Tanke (1996) and Tanke and Currie (1998) contributed the only known frequency information, describing fibular fractures in 10–15 percent of Albertan tyrannosaurids. Tanke (1996) suggested that most fibular fractures were in sub-adults. Although fractures of thoracic ribs and fibulae were most commonly reported (Fig. 18.1), there is no statistically significant difference between the frequency of breaks in these bones and the frequency in other affected bones. Although larger theropods account for most of the reports (Table 18.2), the role of bias (in preservation, collection, and analysis) must be considered. Allosaurs and tyrannosaurs are prominent in the reports (Anonymous 1997a,b; Claessens 1996; Currie 1997; Hanna 2000; Lambe 1917; Larson 2001; Madsen 1976; Stromer 1915; Tanke 1996; Tanke and Currie 1998; Rothschild 1999), perhaps reflecting both specimen frequency and popular interest?

Reported fracture complications in humans include pseudoarthrosis, wherein the fracture components do not fuse but rather form a false

TABLE 18.2

Taxonomic Distribution of Injuries*

Taxon	Fractures	Bite marks	Infection	Stress fractures	Abnormal teeth
Acrocanthosaurus	+	+			
Albertosaurus	+	+		+	+
Alectrosaurus					+
Allosaurus	+		+	+	
Carcharodontosaurus		+			
Ceratosaurus				+	
Chirostenotes				+	
Daspletosaurus	+	+	+		+
Deinonychus	+			+	
Dilophosaurus			+		
Dromaeosaurus					+
Gorgosaurus	+	+			
Herrerasaurus		+			
Megalosaurus	+				
Monolophosaurus	+	+			
Neovenator	+				
Ornithomimidae	+			+	
Oviraptoridae	+				
Poekilopleuron	+				
Saurornitholestes		+			
Sinraptor	+	+			
Spinosaurus	+				
Syntarsus	+				
Tarbosaurus		+		+	
Troodon		+	+		
Tyrannosaurus	+	+	+	+	+
Velociraptor		+			

* Data from Anonymous 1997a,b; Currie 1985; 1997; Currie and Zhao 1993; Erickson 1995; Fiorillo and Gangloff 2000; Hanna 2000; Harris 1997; Jacobson 2001; Lambe 1917; Lang 2000; Larson 2001; Madsen 1976; Molnar 2001; Poling 1996; Psihoyos 1994; Rothschild 1997; Rothschild 1999; Rothschild et al. 2001; Sereno and Novas 1993; Stromer 1915; Tanke 1996; Tanke and Currie 1995, 1998; Webster 1999; Williamson and Carr 1999.

joint (Fig. 18.1) (Tanke and Rothschild 2002). Rib and neural spine pseudoarthroses are most common, as exemplified by *Allosaurus* DMNH 2149 (Fig. 18.1). Healing may also result in limb-element shortening with malpositioning of elements, as exemplified by *Alberto-saurus* MOR 379 (Fig. 18.2). Manal phalanx injuries in *Acrocantho-*

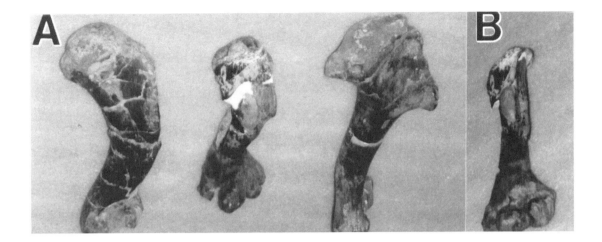

saurus (NCSM 14345) provide a unique window on variable bone response to injury in a sub-adult. The distal portion of one healed phalanx was diminished in size, whereas the distal portion of the other was enlarged.

Other evidence of injury includes exostoses (Fig. 18.3, top), wherein a portion of the bone is spalled free at one end. Growth from the retained base produces an external bony overgrowth, the exostosis (Rothschild and Martin 1993). Neural spine and scapular exostoses have been noted in *Acrocanthosaurus* and *Allosaurus* (Fig. 18.3), respectively (Moodie 1917; Rothschild 1997; Stovall and Langston 1950), a humeral exostosis in *Tyrannosaurus* FMNH PR2081, and a metatarsal IV exostosis in *Albertosaurus* ROM 807 (Molnar 2001). Laws's (1997) diagnosis of infection in the left metatarsal III of *Allosaurus* MOR 693 ("Big Al") appears to actually to represent an exostosis. If such damage involves an articular area, a cartilage cap may grow on the exostosis (Resnick 2002; Rothschild and Martin 1993). Referred to as an osteochondroma, this pathology was found in *Gorgosaurus* pedal phalanx TMP 91.36.500. Fibular shaft twisting in a *Procompsognathus* (Sereno and Wild 1992) may represent abnormal position healing of a fracture.

Although muscle attachments to bone are relatively strong, forces of injury occasionally overcome them. The result is an avulsion, wherein the muscle actually pulls out of the bone (Fig. 18.3, bottom) at its insertion (Martin and Rothschild 1993). An avulsion was noted in the humerus of the *Tyrannosaurus* Sue (FMNH PR2081; Carpenter and Smith 2001). It left a hole and a spicule of bone (exostosis) (Fig. 18.4). Such a humeral tendon avulsion evidences strenuous use of the forearms in predation activities and provides further evidence that *Tyrannosaurus* was not simply a scavenger.

Stress Fractures. Stress or fatigue fractures are caused by strenuous repetitive activities (Resnick 2002; Rothschild and Martin 1993), in contrast to the above discussed fractures, which were the result of acute trauma. Stress fractures have a highly characteristic appearance (Resnick 2002), presenting as an osseous "bump." They are present (Fig.

Figure 18.2. Albertosaurus *humeri. (A) Comparison of MOR 379 (central image in oblique view) with normal humeri in oblique (left) and lateral (right) views. (B) Anterior view of MOR 379.*

Figure 18.3. Allosaurus USNM
4734 (top) lateral view, showing
that the normal rectangular
scapula has a more triangular
shape related to a large exostosis.
Allosaurus scapula UUVP 6023
(bottom) lateral view of tendon
avulsion located one-eighth of
distance from proximal end.
Close-up view reveals an area of
bone grown up as a collar around
a hole created when a muscle
tendon was pulled free.

Figure 18.4. Lateral view of
humerus of Tyrannosaurus
FMNH PR2081. Spicule of bone
(arrow) identifies location of
tendon avulsion.

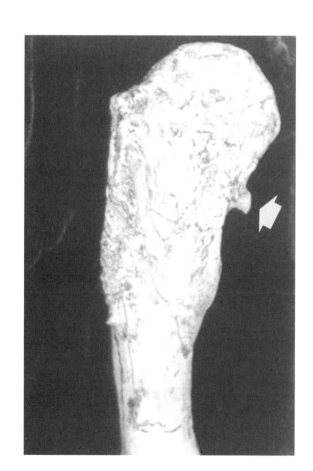

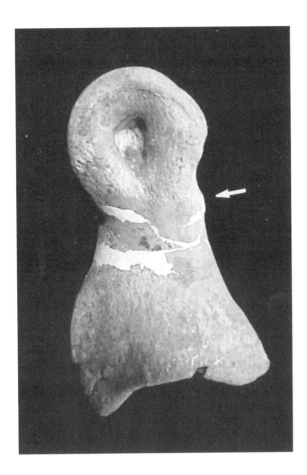

Figure 18.5. Lateral view of Tyrannosaurus LACM 23844 proximal pedal phalanx. Bump (arrow) indicates presence of stress fracture.

18.5) across the spectrum of theropod size (Rothschild et al. 2001) and are easily distinguished from osteomyelitis (bone infection) by the lack of bone destruction (Resnick 2002; Rothschild and Martin 1993). The frequency in the *Allosaurus* and dromaeosaurid manus and pes was significantly greater than that noted in other theropods, tyrannosaurids excepted. Stress fractures affecting the manus were common, providing further evidence that tyrannosaurs were not simply scavengers. Active resistance of prey is required to overstress the manus.

Bite Marks. Intra- and interspecific interactions are also identified in the form of bite marks. These are recognized as short circular punctures and gouges or drag marks, often with healing noted (Tanke and Currie 1995). Again, predominantly larger theropods are represented (Table 18.2) (Anonymous 1996; Currie 1985; Currie and Zhao 1993; Sereno and Novas 1993; Tanke and Currie 1995, 1998; Webster 1999; Williamson and Carr 1999).

Face and skull biting are prominent (Currie 1985; Currie and Zhao 1993; Hanna 2000; Jacobson 2001; Sereno and Novas 1993; Tanke and Currie 1995, 1998; Webster 1999). Especially curious is unilateral loss of the occipital crest in the *Tyrannosaurus* Stan (BHI 3033), attributed to a "love bite," similar to that rarely observed in *Panthera leo.* Tooth marks have also been noted on bone present in theropod copro-

Theropod Paleopathology • 357

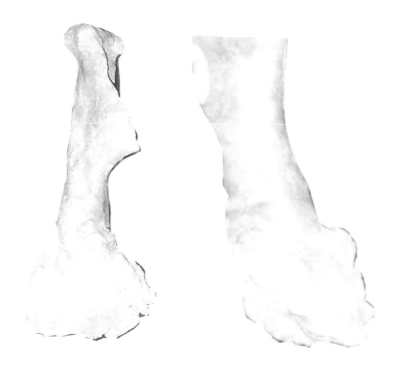

Figure 18.6. Lateral (left) and
anterior (right) views of
Allosaurus BYUVP 5099. Lateral
view reveals missing bone
segment. Overgrowth of bone
around edges is sharply defined at
its juncture with residual normal
bone. Reactive component of
bone in this overgrowth is
indicative of an infected fracture.
It lacks the continuity of cancer.

lites (Chin et al. 1998). Interspecific biting (without healing) of *Saurornitholestes* by a tyrannosaurid was documented by Jacobson (2001). Recognition of intraspecific biting injuries in tyrannosaurids shows that they did bite living animals that were armed with claws and teeth, making it very likely that they also attacked living prey, such as unarmored hadrosaurs. Intraspecific biting in tyrannosaurs is compelling evidence for active predation, supplemented by scavenging when opportunities arose.

Tooth damage was noted in the form of transverse parallel striations in Campanian tyrannosaurids from Alberta (Tanke and Currie 1995) and broken, polished teeth with wear facets in *Saurornitholestes*, *Sinoraptor*, *Allosaurus*, *Ceratosaurus*, and *Tyrannosaurus* (Carpenter 1979; Currie and Zhao 1993; Farlow and Brinkman 1994; Madsen 1976; Molnar 2001; Mongelli et al. 1999). Quantitative information is available only for tyrannosaurids (Mongelli et al. 1999), in which 47 percent of broken premaxillary teeth had wear facets, compared to 8.4 percent of lateral teeth. Most teeth were probably damaged by biting on prey bone during feeding.

Infection. Infections are predominantly reported as isolated phenomena at bite sites, including skull, vertebrae, scapula, ilium, ischium, humerus pedal, and manus phalanges (Hanna 2000; Rothschild 1997; Tanke and Rothschild 2002; Taylor 1992; Welles 1984; Williamson and Carr 1999; Rothschild 1999). Involvement of a humerus was so severe in one *Allosaurus* (BYUVP 5099; Fig. 18.6) that cauliflower-like growth was mistaken for cancer (Taylor 1992). The growth had disorganized internal structure characteristic of infection. Infection in the

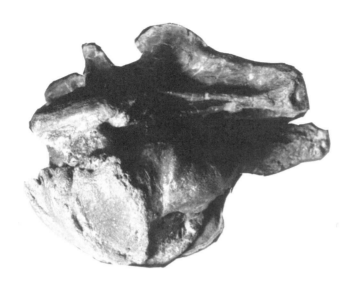

Tyrannosaurus Sue (FMNH PR2081) was more generalized (pers. obs.), as it was in the *Allosaurus* Rip Van Al (WDIS) (Bob Bakker, pers. comm.). Fusion of two thoracic vertebrae occurred with reactive new bone formation on the anterior surface (Fig. 18.7). This is just one area of infection in this individual. Examination of the total picture reveals an intriguing story, which awaits publication (Bakker, pers. comm.). Alleged fungal infection of the mandible in the *Tyrannosaurus* Sue (Webster 1999) was actually characteristic of a healing bite injury.

Figure 18.7. Anterior oblique view of Allosaurus *Rip Van Al vertebrae (WDIS). Fusion of vertebrae with draining sinuses and reactive new bone on the anterior surface are prominent.*

Congenital

Abnormal Teeth. Split or extra carinae (serration rows) predominate, as reported in *Allosaurus, Albertosaurus, Daspletosaurus,* and *Tyrannosaurus* (Erickson 1995; Fiorillo and Gangloff 2000; Molnar 2001; Tanke and Currie 1995). In a critical epidemiologic study, Erickson (1993) reported that 10 percent of tyrannosaurids had carinal splitting (in some geological formations splitting exceeded 25 percent), and 0.4 percent had extra serration rows. Erickson (1995) reported split carinae as more common in *Albertosaurus* than in *Tyrannosaurus rex.* He also reported rates of 10 percent in Campanian albertosaurs (in both highland and lowland environments), contrasted with 27 percent in the Maastrichtian. One wonders about the effect on this congenital pathology of environmental stresses or population bottlenecks, which are often responsible for such observations in recent animals (Rothschild and Martin 1993). Unfortunately, such information is not yet available for other groups.

Vertebral Fusion. Fusion of thoracic vertebral centra has been noted in *Tyrannosaurus* AMNH 5027 (Fig. 18.8) (Anonymous 1997b; Dingus 1996; Rothschild 1997), of mid-caudal and posterior caudal centra in *Allosaurus* (Madsen 1976; Rothschild 1997), and of a chevron to a proximal caudal vertebra in *Poekilopleuron* (Eudes-Deslong-

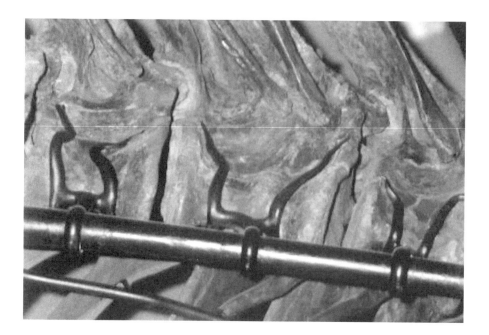

champs 1838). Such fusion is distinguishable from diffuse idiopathic skeletal hyperostosis because of the absence of ligamentous fusion and from spondyloarthropathy because of the absence of segmentation (Rothschild and Martin 1993). Fusion appears to represent a congenital abnormality.

Arthritis

"Osteoarthritis." Osteoarthritis had been erroneously diagnosed in dinosaurs because of semantic confusion. Osteoarthritis is correctly diagnosed on the basis of recognition of osteophytes at articular surfaces. Confusion arises because the term "osteophyte" is also used to describe overgrowth of vertebral centra, a condition called spondylosis deformans, which is unrelated to osteoarthritis (Resnick 2002; Rothschild and Martin 1993). Spondylosis deformans is occasionally noted in large theropod dinosaurs, but osteoarthritis has not been found (Rothschild 1990); and diffuse idiopathic skeletal hyperostosis (DISH; Rothschild and Berman 1991) has yet to be reported.

Gout. Gout is a metabolic disorder in which uric acid crystals accumulate as masses, producing bony erosion (Resnick 2002; Rothschild and Martin 1993) characterized by overgrowth at its margins. Such features (Fig. 18.9) have been found in FMNH PR2081 metacarpals I and II and in an unspeciated tyrannosaurid pedal phalanx from TMP 92.36.328 (Rothschild et al. 1997).

Tumors

Apparent bone overgrowth in an *Allosaurus* humerus (BYUVP 5099) was originally thought to be caused by cancer (Taylor 1992). Re-

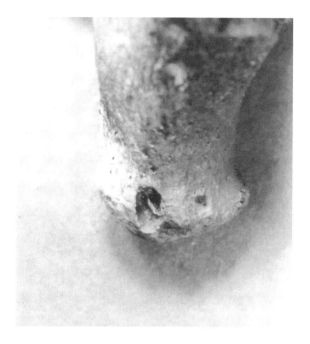

examination of the specimen revealed an infected fracture (McWhinney, in prep.). Filigree surface reaction is characteristic of an infectious process, and the disorganized underlying architecture is classic for infection. There was no evidence of the space-occupying mass one expects with cancer. A recent X-ray survey of North American theropods revealed no evidence of actual tumors or cancer (Helbling et al. 2001).

Figure 18.9. Tyrannosaurid (TMP 92.36.328) proximal phalanx. Hole at distal portion with reactive bone margin identifies gout.

Conclusions

The long literature association of paleopathology and theropods has continued to the present day. While the spectrum of observed pathologies has been limited to those related to injury, infection, congenital acquisition, and arthritis, only injury and tooth deformities lend themselves to potential epidemiologic assessment (statistical significance) and therefore to hypothesis testing. Localization of bites may identify intraspecific interactions (hierarchal, territorial, or mating related). Examination of patterns of tooth wear may identify feeding or prey-acquisition techniques. Patterns of congenital tooth deformities may allow identification of environmental stresses or genetic bottlenecks. Dental osteopathy in Late Cretaceous tyrannosaurids supports the view that they were very active and capable of bringing down live prey.

Recording the number of affected specimens and quantifying the denominator will provide information amenable to statistical analysis, allow theropod paleopathology to advance as a science, and contribute to our understanding of the lives and habitats of these formidable ancient predators.

Acknowledgments. Appreciation is expressed to Robert Bakker, Dave Berman, Kenneth Carpenter, Sankar Chatterjee, Richard Cifelli, Nick Czaplewski, Kyle Davies, Mary Dawson, Ray DiVasto, Margaret Feuerstack, Richard Harrington, Elizabeth Hill, Jack Horner, Juan Langston, Peter and Neal Larson, Kyle S. McQuilkin, Lorrie McWhinney, Mark Norrell, Pamela Owen, Robert Purdy, Scott Sampson, Kevin Seymour, Kieran Shepherd, Bill Simpson, Allison A. Smith, Ken Stadtman, J.D. Stewart, Hans-Dieter Sues, and Mary Ann Turner for assistance in accessing the collections they curate at the American Museum of Natural History, New York (AMNH); Black Hills Institute, Black Hills, S.D. (BHI); Brigham Young University, Provo, Utah (BYU); Canadian Museum of Nature, Ottawa, Ontario; Carnegie Museum of Natural History, Pittsburgh, Penn.; Denver Museum of Science and Nature, Denver, Colo.; Field Museum of Natural History, Chicago (FMNH); Los Angeles Museum of Natural History, Los Angeles (LACM); Museum of the Rockies, Bozeman, Mont. (MOR); National Museum of Natural History, Washington, D.C. (USNM); North Carolina State Museum of Natural Sciences, Raleigh (NCSM); Oklahoma Museum of Natural History, Norman, Okla.; Royal Ontario Museum, Toronto, Ontario, Canada (ROM); Royal Tyrrell Museum, Drumheller, Alberta (TMP); Texas Tech Museum, Lubbock; University of Kansas Museum of Natural History, Lawrence; University of Texas Museum, Austin (TMM); University of Utah, Salt Lake City; Yale Peabody Museum, New Haven, Conn.; and the Wyoming Dinosaur International Society (WDIS).

References Cited

Anonymous. 1996. Two new African theropods. *Dinosaur Discoveries* 1: 3.

Anonymous. 1997a. Tyrannosaurus rex: *A Highly Important and Virtually Complete Fossil Skeleton.* Sale 7045 auction catalogue. New York: Sotheby's Auction House.

Anonymous. 1997b. The case of the hole in the head. *Bones: Big Horn Basin Foundation,* February 1997, p. 5. Thermopolis, Wyo.: The Wyoming Dinosaur Center.

Carpenter, K. 1979. Vertebrate fauna of the Laramie Formation (Maestrichtian), Weld County, Colorado. *Contributions to Geology, University of Wyoming* 17: 37–49.

Carpenter, K., and M. Smith. 2001. Forelimb osteology and biomechanics of *Tyrannosaurus rex*. In D. H. Tanke and K. Carpenter (eds.), *Mesozoic Vertebrate Life,* pp. 90–116. Bloomington: Indiana University Press.

Chin, K., T. T. Tokaryk, G. M. Erickson, and L. C. Calk. 1998. A king-sized theropod coprolite. *Nature* 393: 680–682.

Claessens, L. 1996. Dinosaur gastralia: Morphology and function. Master's thesis, Utrecht University.

Currie, P. J. 1985. Cranial anatomy of *Stenonychosaurus inequalis* (Saurischia, Theropoda) and its bearing on the origin of birds. *Canadian Journal of Earth Sciences* 22: 1643–1658.

Currie, P. J. 1997. *Gorgosaurus?* Hip and tail. In *Field Experience— Summer 1996,* p. 3. Royal Tyrrell Museum of Palaeontology Field Experience 96 update.

Currie, P. J., and Zhao X.-J. 1993. A new carnosaur (Dinosauria, Theropoda) from the Jurassic of Xinjiang, People's Republic of China. *Canadian Journal of Earth Sciences* 30: 2037–2081.

Dantas, P., V. F. dos Santos, M. G. Lockley, and C. A. Meyer. 1995. Footprint evidence for limping dinosaurs from the Upper Jurassic of Portugal. In M. G. Lockley, V. F. dos Santos, C. A. Meyer, and A. Hunt (eds.), *Gaia: Aspects of Sauropod Paleobiology*, vol. 10, pp. 43–48. Lisbon: Museu Nacional de História Natural.

Dingus, L. 1996. *Next of Kin: Great Fossils at the American Museum of Natural History.* New York: Rizzoli.

Erickson, G. M. 1993. The mystery of the "split" toothed tyrannosaurs. California Paleontology Conference Abstracts. *Paleobios* 14 (4 suppl.): 5.

Erickson, G. M. 1995. Split carinae on tyrannosaurid teeth and implications of their development. *Journal of Vertebrate Paleontology* 15: 268–274.

Eudes-Deslongchamps, M. 1838. Mémoire sur le *Poekilopleuron bucklandii*, grand saurien fossile, intermédiaire entre les crocodiles et les lézards. *Mémoires de Société Linnéenne de Normandie* 6: 37–146.

Farlow, J. O., and D. L. Brinkman. 1994. Wear surfaces on the teeth of tyrannosaurs. In G. D. Rosenberg and D. L. Wolberg (eds.), *Dino Fest: Proceedings of a Conference for the General Public, March 24, 1994*, pp. 165–175. Paleontological Society Special Publication, no. 7. Indianapolis: Geology Department, Indiana University–Purdue University, and Knoxville: Paleontological Society.

Fiorillo, A. R, and R. A. Gangloff. 2000. Theropod teeth from the Prince Creek Formation (Cretaceous) of Northern Alaska, with speculations on Arctic dinosaur paleoecology. *Journal of Vertebrate Paleontology* 20: 675–682.

Hanna, R. R. 2000. Dinosaurs got hurt too. In G. S. Paul (ed.), *The Scientific American Book of Dinosaurs*, pp. 119–126. New York: St. Martin's Press.

Harris, J. D. 1997. A reanalysis of *Acrocanthosaurus*, its phylogenetic status, and paleobiogeographic implications, based on a new specimen from Texas. Master's thesis, Southern Methodist University, Dallas.

Helbling, M., II, B. M. Rothschild, and D. Tanke. 2001. Tertiary neoplasia: A family affair. *Journal of Vertebrate Paleontology* 21: 60A.

Ishigaki, S. 1986a. Dinosaur footprints of the Atlas Mountains. *Nature Study* (Japan) 32 (1): 6–9.

Ishigaki S. 1986b. *Morokko no kyōryū.* Tokyo: Tsukiji shokan.

Jacobson, A. R. 2001. Tooth-marked small theropod: An extremely rare trace. In D. H. Tanke and K. Carpenter (eds.), *Mesozoic Vertebrate Life*, pp. 58–63. Bloomington: Indiana University Press.

Lambe, L. 1917. The Cretaceous theropodous dinosaur *Gorgosaurus. Geological Survey of Canada Memoir* 100: 1–84.

Lang, G. H. 2000. Dinosaur stress. *Natural History* 3: 8.

Larson, P. L. 2001. Paleopathologies in *Tyrannosaurus rex*: Snapshots of a killer's life. *Journal of Vertebrate Paleontology* 21: 71A–72A.

Laws, R. R. 1997. Allosaur trauma and infection: Paleopathological analysis as a tool for lifestyle reconstruction. *Journal of Vertebrate Paleontology* 17: 59A–60A.

Madsen, J. H., Jr. 1976. Allosaurus fragilis: *A Revised Osteology.* Utah Geological and Mineral Survey Bulletin, no. 109. Salt Lake City: Utah

Geological and Mineral Survey, Utah Department of Natural Resources.

Molnar, R. E. 2001. Theropod paleopathology: A literature search. In D. H. Tanke and K. Carpenter (eds.), *Mesozoic Vertebrate Life*, pp. 337–363. Bloomington: Indiana University Press.

Mongelli, A., Jr., D. J. Varricchio, and J. J. Borkowski. 1999. Wear surfaces and breakage patterns of tyrannosaurid (Theropoda: Coelurosauria) teeth. *Journal of Vertebrate Paleontology* 19: 64A.

Moodie, R. 1917. Studies in paleopathology: I. General consideration of the evidences of pathological conditions found among fossil animals. *Annals of Medical History* 1: 374–393.

Poling, J. 1996. Dinosauria On-line, Dinosaur Picture Gallery. http://www.dinosauria.com/gallery/darren/fibula.jpg (accessed June 6, 2004).

Psihoyos, L. 1994. *Hunting Dinosaurs*. New York: Random House.

Resnick, D. 2002. *Diagnosis of Bone and Joint Disorders*. Philadelphia: Saunders.

Rothschild, B. M. 1990. Radiologic assessment of osteoarthritis in dinosaurs. *Annals of the Carnegie Museum* 59: 295–301.

Rothschild, B. M. 1997. Dinosaurian paleopathology. In J. O. Farlow and M. K. Brett-Surman (eds.), *The Complete Dinosaur*, pp. 426–448. Bloomington: University of Indiana Press.

Rothschild, B. M. 1999. Do we know anything about the kinds of diseases that affected dinosaurs? Ask the Experts: Medicine, posted October 21, 1999. http://www.sciam.com/askexpert_directory.cfm (accessed May 7, 2004).

Rothschild, B. M., and D. S. Berman. 1991. Fusion of caudal vertebrae in Late Jurassic sauropods. *Journal of Vertebrate Paleontology* 11: 29–36.

Rothschild, B. M., and L. D. Martin. 1993. *Paleopathology: Disease in the Fossil Record*. London: CRC Press.

Rothschild, B. M., D. Tanke, and K. Carpenter. 1997. Tyrannosaurs suffered from gout. *Nature* 387: 357.

Rothschild, B. M., D. H. Tanke, and T. Ford. 2001. Theropod stress fractures and tendon avulsions as a clue to activity. In D. H. Tanke and K. Carpenter (eds.), *Mesozoic Vertebrate Life*, pp. 331–336. Bloomington: Indiana University Press.

Russell, D. A. 1970. *Tyrannosaurs from the Late Cretaceous of Western Canada*. National Museums of Canada, National Museum of Natural Sciences, Publications in Palaeontology, no 1. Ottawa.

Sereno, P. C., and F. E. Novas. 1993. The skull and neck of the basal theropod *Herrerasaurus ischigualastensis*. *Journal of Vertebrate Paleontology* 13: 451–476.

Sereno, P. C., and R. Wild. 1992. *Procompsognathus*: Theropod, "theocodont" or both? *Journal of Vertebrate Paleontology* 12: 435–438.

Stovall, J. W., and W. Langston Jr. 1950. *Acrocanthosaurus atokensis*, a new genus and species of Lower Cretaceous Theropoda from Oklahoma. *American Midland Naturalist* 43: 696–728.

Stromer, E. 915. Ergebnisse der Forschungsreisen Prof. E. Stromers in den Wüsten Ägyptens. II. Wirbeltier-reste der Baharije-Stufe (unterstes Cenoman). 3. Das Original des Theropoden *Spinosaurus aegyptiacus* nov. gen. nov. spec. *Abhandlungen Konig bayer Akademische Wissenschraft Mathematische-Physiche Klasse* 28: 1–32.

Tanke, D. 1996. Leg injuries in large theropods. Message posted to the

Dinosaur Mailing List, March 9. http://www.cmnh.org/dinoarch/
1996Mar/msg00150.html (accessed May 7, 2004).

Tanke, D. H., and P. J. Currie. 1995. Intraspecific fighting behavior inferred
from toothmark trauma on skulls and teeth of large carnosaurs (Dinosauria). *Journal of Vertebrate Paleontology* 15 (3 suppl.): 55A.

Tanke, D. H., and P. Currie. 1998. Head-biting behavior in theropod
dinosaurs: Paleopathogical evidence. In B. P. Pérez-Moreno, T. Holtz
Jr., J. L. Sanz, and J. Moratalla (eds.), *Gaia: Aspects of Theropod
Paleobiology,* vol. 15, pp. pp. 167–184. Lisbon: Museu Nacional de
História Natural.

Tanke, D. H., and B. M. Rothschild. 2002. *An Annotated Bibliography
of Dinosaur Paleopathology and Related Topics, 1838–1999.* New
Mexico Museum of Natural History and Science Bulletin, no. 20.
Albuquerque: New Mexico Museum of Natural History and Science.

Taylor, P. 1992. Doctors try to diagnose dinosaur cancer. *Toronto Globe
and Mail,* January 5: a1–a2.

Webster, D. 1999. A dinosaur named Sue. *National Geographic* 1995 (6):
46–59.

Welles, S. P. 1984. *Dilophosaurus wetherilli* (Dinosauria, Theropoda):
Osteology and Comparisons. *Palaeontographica Abt. A* 185: 85–180.

Williamson, T. E., and T. D. Carr. 1999. A new tyrannosaurid (Dinosauria: Theropoda) partial skeleton from the Upper Cretaceous Kirt-
land Formation, San Juan Basin, New Mexico. *New Mexico Geology*
21: 42–43.

KENNETH CARPENTER

is the dinosaur paleontologist for the Denver Museum of Natural History and author of *Eggs, Nests, and Baby Dinosaurs*, editor of *The Armored Dinosaurs*, and co-editor of *Mesozoic Vertebrate Life*, all published by Indiana University Press. He is also co-editor of *Dinosaur Systematics*, *Dinosaur Eggs and Babies*, and *The Upper Jurassic Morrison Formation*.